Applications of Artificial Intelligence for Decision-Making

Patrick J. Talbot
Dennis R. Ellis

Novel Ideas

- Each chapter showcases an idea. Some are profound, others are novel, and all are practical.

- Focus is on decision-making under (12 types of) uncertainty

- A Decision-Centered methodology enhances applications.

- Decisions are State Transitions from a current to a desired state.

- A Hierarchical, Characteristics Based Semantic Network unifies multi-strategy reasoning.

- Knowledge is represented as executable stories.

- The knowledge base is self-aware: it knows what it knows.

- A Wavelet Transform fosters multi-level Data Triage granularity.

- Common representation encompasses analogical, genetic, evidential, abductive, inductive, and simulation-based reasoning.

- Multi-strategy reasoning combines algorithms graphs and simulations.

- Deep Learning Belief Networks are trained using backpropagation

- Automated Discovery of Unknown Unknowns uncovers "surprises".

- Emergence of a market characterizes the behavior of an internet forum.

- 15 applications, ranging from sensor fusion to space control cyber warfare, a retail problem solver and longevity prediction, are presented.

- Starship Cybernetics describes the use of artificial intelligence on an interstellar voyage.

Practical Applications

- Storyboarding gets the visualization right for human decision-making.

- Long Processing Threads provide decision loops to monitor, assess, plan, and execute a mission during routine and crisis operations.

- Information Extraction technology is extended to include uncertainty by parsing hedge words such as maybe, probably, and not.

- Belief Networks for evidential reasoning provide Bayesian, Dempster-Shafer, and six other data fusion algorithms.

- A Taxonomy of social network analysis algorithms is provided with detailed examples of use.

- A Simulated Commander is prototyped using fuzzy rules.

- Lack of information is an impediment to decision-making and shown explicitly.

- Offense and Defense are combined synergistically, producing a force multiplier.

- A detailed discussion of pattern-preserving extrapolation of complex data, using the Space Catalog as an example, is addressed.

- The impact of Fog-of-War on decision-making is quantified.

- Monte Carlo techniques are presented for numerical integration, threat discovery, benchmarking in the cloud, and perturbation analysis.

- A Deep Belief Network is combined with Case-based Reasoning for Longevity Prediction.

Copyright © 2015 by Patrick J. Talbot

First Edition

All rights reserved

ISBN 10: 1502907593

ISBN 13: 978-1502907592

We dedicate this book to the memory of Terry Trout, our dear friend and coworker

Acknowledgments: We thank Kevin Dillon for the graphic embedded in the cover, Danny Hillis and the talented folks at Applied Minds for the idea of structuring knowledge as stories, for their contributions to the genetic optimizer visualization, and for the Attention Cuer icon mockup, Bert Lavenhar for being the best Section Head ever, Tim Trowbridge for paving the way as the first researcher in our organization, Joe Santa for introducing us to algorithms, Max Williams for research on Dempster-Shafer theory for fusion of sparse evidence, Andy Trimble for his original work in wavelet data triage and threat discovery, Marie Anderson for her perturbation analysis of Bayesian Networks, Jeff McCartney for his information extraction contributions, Martin Hyatt for the semantic classification algorithm, Alan Siegel for the replan trigger algorithms and Bayesian Network research, Scott Sage for facilitating our social network analysis work, Dorothea Morrow for Dempster-Shafer Belief Network prototype, Terry Trout for adding back-propagation and many other features to our General Purpose Evidence Fusion Engine, Erni McMullen for his software expertise, Sunjir Bir and Tim Ferguson for their technical leadership, Ryan Sanders for his insights into automated discovery of unknown unknowns, Mary Cheek for defining fuzzy rules for the simulated commander, Mark Nixon for his contributions to abductive reasoning and best next observation ideas, Keith Mathias for partnering with us on the uncertainty management project, Scott Van Dyke for valuable team interactions, Joe Colangelo and Ken Freeman for the innovative Lambert Guidance algorithm, Mike Papay for featuring our team in the Command and Control of the Future research project, Cindy Trowbridge for the Monte Carlo integration idea and software prototype, Dale Gardner for his sponsorship of the Offense/Defense Integration research project, Ken Henry for his sponsorship and encouragement in the computer network defense domain, Larry Stahler and Frank Fischer for their marketing pizzazz, and the many other talented professionals we've had the opportunity to work with.

Table of Contents

Part I. Fundamental Processes
Chapter I-1: Requirements Definition
Chapter I-2: Designing Automated Processes
Chapter I-3: Decisions as State Transitions
Chapter I-4: Visualization for Understanding
Chapter I-5: Decision Algorithms
Chapter I-6: Pattern-Preserving Operations on Data Sources

Part II. Representation: Framing the Problem Space
Chapter II-1: Knowledge Capture
Chapter II-2: Semantic Networks for Multi-Strategy Reasoning and Learning
Chapter II-3: Knowledge Management via Stories
Chapter II-4: Executable and Self-Aware Knowledge Bases

Part III. Algorithms for Computer Reasoning and Learning
Chapter III-1: Data Triage
Chapter III-2: Information Extraction
Chapter III-3: Hypothesis Association
Chapter III-4: Threat Discovery
Chapter III-5: Social Network Analysis
Chapter III-6: Analogical and Terrain Reasoning
Chapter III-7: Genetic Reasoning and Plan Optimization
Chapter III-8: Evidential Reasoning
Chapter III-9: Inductive Reasoning: Generalizing from Data
Chapter III-10: Deductive Reasoning: Rules and Logic
Chapter III-11: Abductive Reasoning to the Best Explanation
Chapter III-12: Reasoning with Physics: A Few Astrodynamic Algorithms

Part IV. Multi-Strategy Reasoning and Learning
Chapter IV-1: Uncertainty Management
Chapter IV-2: Combining Algorithms, Graphs and Simulation
Chapter IV-3: Processing Synergy and Feedback Loops
Chapter IV-4: Automated Discovery of Unknown Unknowns
Chapter IV-5: Network Metrics Define Centers-of-Gravity
Chapter IV-6: Emergent Behavior

Part V. Applications
Chapter V-0 Sensor Fusion
Chapter V-1: Simulated Commander for Missile Defense War Games
Chapter V-2: Strategic Offense/Defense Integration
Chapter V-3: Fog-of-War
Chapter V-4: Systems of Systems Analysis
Chapter V-5: Defensive Space Control
Chapter V-6: Space Situation Awareness
Chapter V-7: Computer Network Defense

Chapter V-8: Uncertainty Management – Army Scenario
Chapter V-9: Pattern Preserving Extrapolation of the Space Catalog
Chapter V-10 Starship Cybernetics
Chapter V-11 Longevity Prediction
Chapter V-12 Fusing Data to Estimate Total Uncertainty
Chapter V-13 Monte Carlo Sampling for Benchmarking in the Cloud
Chapter V-14 Incorporating Uncertainty in a First Responder Simulation
Chapter V-15 Retail Problem Solver

List of Figures

I-2.1: Computer and User Strengths
I-2.2: Design Components
I-3.1: Decisions are State Transitions
I-4.1: Main Display – The Dashboard
I-4.2: High Interest Event for situation Awareness
I-4.3: Belief Network Editor for Combat Assessment
I-5.1: Sample Decision Algorithms
I-6.1 Current Process for Data Set Extrapolation
I-6.2: Approach to Pattern Preserving Operations
I-6.3: Example – Extrapolation of the Space Object Catalog

II-1.1: Different Algorithms, Different Representations
II-2.1: Lots of Algorithms, Lots of Interactions
II-2.2: Substantive Integration Is Required
II-2.3: Data Interface for Substantive Integration
II-2.4: Knowledge Base Design Using Protégé
II-2.5: Reporter's Questions Template
II-2.6: Algorithm Interfaces
II-3.1: Knowledge Structure for Multi-Strategy Reasoning
II-3.2: Union and Intersection of Stories
II-3.3: Automatic Generation of Stories from Text
II-4.1: Self-Aware Knowledge Base Ontology

III.1-1: Document Size Scalability
III.2.1: Information Extraction Flow Diagram
III-2.2: Information Extraction with Underlined Tags
III-3.1: Hypothesis Association Overview
III-3.2: Semantic Mean Example
III-4.1: Confidence in Having a Complete Set
III-4.2: Cooperation Produces a Force Multiplier
III-4.3: Threat Distribution Perturbation Analysis
III-4.4: Overall Perturbation Analysis
III-5.1: Metrics Taxonomy
III-5.2: Kite Network: Webmaster Has Highest Betweenness
III-5.3: Example Calculation of In-Degree and Effective In-Degree
III-5.4: Example Calculation of Clustering Coefficients (Undirected Graph)
III-5.5 Importance Under Uncertainty
III-5.6: Kamada-Kawai Visualization
III-5.7: Sample Eigenvalue Calculation
III-5.8 Fractal Dimension for Two Time Series
III-6.1: Case-Based Solution Strategy
III-6.2: Optimum Locator Input Display
III-6.3: Optimum Locator Geographic Display
III-6.4: Optimum Locator Output Display

III-6.5: Plan Update Strategy Selection Screen
III-7.1: Approach to Genetic Optimization
III-7.2: Genetic Optimizer Output Screen
III-7.3: Course-of-Action (COA)Flow Diagram
III-8.1: Data Fusion Hierarchy
III-8.2 Dempster-Shafer Combination Rule
III-8.3: Data Fusion Flow Diagram
III-8.4: Terrorist Threat Assessment
III-9.1: Rule Tree
III-9.2: Weka Clustering
III-10.1: Decision Making Process
III-11.1: SUBDUE Hierarchical Clustering – Example
III-12.1: Conjunction Analysis Geometry
III-12.2: Conjunction Probability Versus CPU Time
III-12.3: Space-to-Space Intercept

IV-1.1: Taxonomy of Types of Uncertainty
IV-1.2: Uncertainty Management Applications Architecture
IV-1.3: Kiviat Diagram Identifies Problematic Uncertainties
IV-2.1: Combining Graphs and Simulations - Processing Flow
IV-2.2: Belief Network Interface with Simulation Tools
IV-2.3: Augmented Time Critical Targeting Graph
IV-3.1: Double Monitor-Assess-Plan-Execute (MAPE)Loop – Example
IV-4.1: Uncertainty Pyramid
IV-4.2: Approach to Automated Discovery of Unknown Unknowns (ADUU)
IV-4.3: Implementation of Automated Discovery of Unknown Unknowns
IV-5.1: Info-Structure Network, 1 Forum, 6 Threads
IV-6.1: Agent Interactions in Internet Forums
IV-6.2: Sensitivity Analysis

V-1.1 Object-Oriented Design
V-1.2: Belief Network for Data Fusion
V-2.1: Approach to Deriving Performance Metrics
V-2.2: Ratio of Damage Expectancies
V-2.3: Ratio of Damage Expectancies versus Leakage
V-2.4: Target Definitions
V-2.5: Example of Offense Defense Integration (ODI) as a Force Multiplier
V-2.6: Hierarchy of Performance Metrics
V-3.1 Fog-of-War Approach
V-3.2: Belief Network Perturbations
V-3.3: Automated Decision Support System
V-3.4: Fog-of-War Example Results
V-4.1: Systems Perspective of the Operational Environment
V-4.2: Dimension Minimization
V-5.1: Anti-Satellite Scenario
V-5.2: Belief network for Iroma Attack
V-6.1: Sensor Fusion Processing Flow
V-6.2: Semantic Network for Data Fusion

V-6.3: Data Fusion Summary
V-7.1: Rule Induction Provides Attack Patterns
V-7.2: Case Based Planner
V-7.3: Computer Network Defense Belief Network
V-8.1: Dashboard for Army Scenario
V-8.2: Army Scenario Belief Network
V-9.1: Satellite Growth by Type
V-9.2: Earth Orbiting Objects
V-9.3: Superposition of Results for Three Catalogs
V-10.1: Rule Induction Applied to Space Weather
V-10.2: Fisheye Nomogran for Interstellar Time - Distance Calculations
V-11.1: Deep Belief Network Sketch
V-11.2: Sample Data For Worked Scenario
V-13.1: Competitive Filtering Strategy Minimizes Monte Carlo Runs
V-14.1: Belief Network for Space Scenario
V-14.2   Data Mining Produces Patterns in Maneuver Data
V-14.3   Nomogram to Determine Maneuver Time versus Lifetime Prediction
V-15.1: Automated Problem Solver Flow Diagram
V-15.2: Automated Problem Solver Questions

Preface

This book is an outgrowth of research and development projects that we conducted over 40-year aerospace engineering careers. Practical techniques for developing smart applications using artificial intelligence, with emphasis on coping with uncertainty, started with classical decision support approaches and evolved to a comprehensive methodology for automated decision-making in a wide range of domains.

The motivation for tackling this topic is finding ways to automate human decision-making. The techniques do not replace the human, but rather seek to combine what people do best (for example, common sense) with what computers do best (for example, computation). The mantra, reinforced throughout this book, is that uncertainties complicate decision-making. We never have all the data, are seldom certain that the information is accurate, labor under conflicting opinions, and rarely get timely updates. Yet, we must decide.

A significant shift has occurred over the past 40 years. Back in the 1970's, computers worked on highly structured numerical data, and most uncertainty was confined to variations in a random variable and confidence intervals associated with data set size. Now, over 95% of the actionable data we receive is in the form of unstructured content. This introduces new sources of uncertainty (vagueness, unknown unknowns, conflicts, opinions) that require new strategies, such as natural language processing and belief networks.

We worked for TRW, the company that invented systems engineering. Our company evolved this process over many decades, driven by TRW's role as the Minuteman Intercontinental Ballistic Missile integration contractor. Systems engineering is the fundamental process from which the need for decision support tools is derived. Requirements are the starting point for the systems engineering process. Ideally, we receive these requirements from our customer – the Department of Defense or the Intelligence Community. Practically, we often work with the customer to translate needs to requirements and to identify solutions to problems. The goal of all engineering, we are taught, is to reduce risk. Here, risk is defined as a function of the probability of an issue occurring and the impact of the issue. To do good systems engineering, an understanding of probability and mission impact is essential.

Rather than use the word "cybernetics" or the phrase "artificial intelligence", the engineering phrase "decision support tools" is preferred and will be used, with few exceptions, throughout the book.

Part I
Fundamental Processes

Chapter I-1: Requirements Definition
Chapter I-2: Designing Automated Processes
Chapter I-3: Decisions as State Transitions
Chapter I-4: Visualization for Understanding
Chapter I-5: Decision Algorithms
Chapter I-6: Pattern-Preserving Operations on Data Sources

Part I
Introduction

This part of the book focuses on fundamentals. Requirements definition is the key engineering process. We tailor this process – to a decision-centered methodology - to derive requirements for decision support. Decisions are defined as a state transition from a current state to a desired state according to a plan, with assessment as a feedback process to determine whether the desired state is achieved. A chapter on visualization stresses the need for visual displays to help users understand the mission and to make decisions – especially in the face of uncertainty and risk. A preview of decision algorithms is provided, with emphasis on a mission thread that spans the various decision types. Finally, operations on data sources are characterized.

Chapter I-1
Requirements Definition

Systems engineering is based on the definition, allocation, and satisfaction of requirements. Functional requirements specify what the system must do. Performance requirements define how well the system must perform. Interface requirements identify the other systems that the system of interest must interact with. Rather than detail the systems engineering process as it is formally defined and instituted, for our purposes we will describe a tailored version that is well-suited to an engineering formulation of decision support tools.

Mission

A decision-centered approach is useful for deriving requirements for decision support tools. The process begins with mission requirements. These are either received from or coordinated with the customer. More recently, mission requirements are articulated with use cases and a scenario. For many space and cyber missions, we use the formal Department of Defense process, referred to as the Joint Operations Planning and Execution System.

Operational Processing

Mission requirements are then decomposed to identify operational processing requirements, including the constraints on the legacy system, the degree of automation required based on system throughput, input ingest rate, and product timeliness. Operational processing requirements sort the tasks that the system performs into allocations for hardware, software, and users. At his level of requirements decomposition, we are specifying who does what, without regard to how the task is accomplished.

Decisions

Operational processing requirements result is a set of tasks, or decisions, allocated to the user. These decisions are central to the process of determining what decision support tools are required, their degree of automation, and the functionality that they provide. Fortunately, across many mission domains, users are required to make the same set of decisions. At a high level, these decisions are: what's going on? what to do?, do it?, and how well did we do?

Displays

The user interface with system hardware and software is characterized as a display. It could be a computer screen, an iPhone, or the knobs on an oscilloscope (an ancient device). The sole purpose of the content of displays is to help the user make decisions. Not much "real estate" is devoted to unadorned maps, "orbitology" (those senseless three-dimensional maps with pictures of satellites and sensor cones), icons, and other eyewash. If it doesn't support a user decision, it doesn't belong on the display.

Algorithms

This is the good stuff – what makes decision support tools useful. These applications provide what computers do best; for example, math, memory, and search, and users don't. Having said this, the use of algorithms is highly constrained: their only role is to provide information to displays – that help users make decisions. This realization is essential – it separates engineers from the scientists and mathematicians – and helps minimize rather than add risk.

Knowledge

Although available incoming information and display outputs generally bound what a system does, it is useful for decision support purposes to enhance information to produce knowledge as an input to algorithms. Knowledge drives algorithms that populate displays that help users decide.

Chapter I-2
Designing Automated Processes

Why do we need automated decision support tools? Why are computer-automated applications necessary? Answers to these questions are important to the design process, which requires a synergistic relationship (Figure I-2.1) between what users do best and what computers do best.

- **Easy tasks for users are difficult for computers and vice versa!**

	Cognitive Strategy	Computer Algorithm	Example
Easy for Users	Common Sense	Circumscription Event Calculus	Everyday Physics
	Applying Wisdom	Case-Based Reasoning	Brainstorming
	Reflex/Instinct	Artificial Neural Network	Hitting a fastball
	Creativity	Framing, Case-based	General Relativity
	Remembering	Case-Based Reasoning	Revising a plan
	Rules-of-Thumb	Expert System/Heuristics	IF… Then…Rules
	Search	Genetic Algorithm	Finding "best" solution
	Hypothesize	Abductive reasoning	Providing explanation
	Story Telling	Truth Maintenance	Combining Evidence
	Generalize	Rule Induction tree	Finding a pattern
Difficult for Users	Uncertain Reasoning	Bayesian Inference	Making a bet
	Mathematics	Arithmetic, Trigonometry	Large multiplications

Figure I-2.1 Computer and User Strengths

As a recap, decision support tools are designed to complement user's strengths while forming long mission threads. "Top down" mission requirements are satisfied by a set of operational processes that help users make decisions. Data drives algorithms that populate displays that, from the "bottom up", help users make decisions. An example mission thread showing the flow of data => algorithms => displays provides a glimpse of common design components (Figure I-2.2). Content, in the form of free text, semi-structured reports, and database files is posted to the knowledge base and made available for user interaction via visualization tools. Text processing includes ingest of the content, segmentation, information extraction, and classification (or 'tagging" by concept) of the input. This is registered in the knowledge base for follow-on automated inference which may include data mining to discover patterns, data fusion to reduce uncertainty, and plan update to revise ongoing courses-of-action.

Translated to the fundamental decisions mentioned earlier, visualization of text and reports helps the user decide what's going on. Data mining and plan update help the user determine what to do and to do it. Evidence that is processed using data fusion algorithms helps the user decide how well the mission is going.

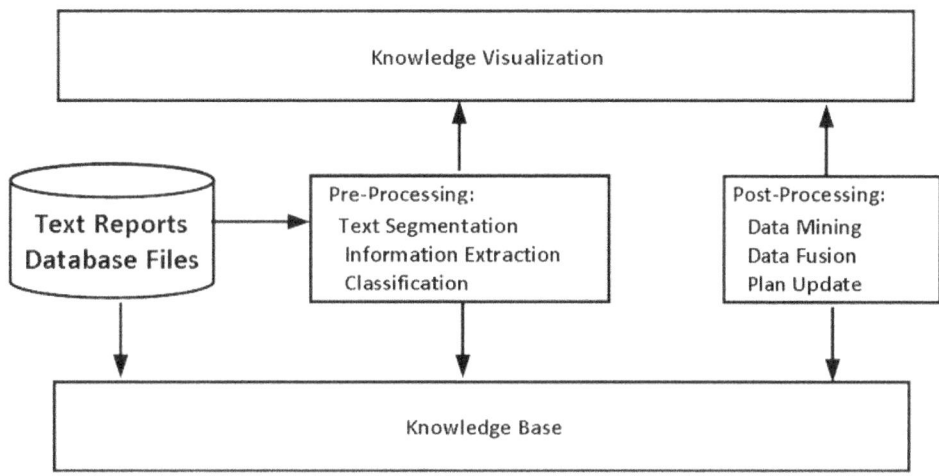

Figure I-2.2 Design Components

An orthogonal view of users and computers that further refines the question of roles is based on the Reporter's Questions. This approach is based on getting the answers: who, what, where, when, why, how, and how much.

Questions	Sample Problems	Human Deficiencies	Decision Type	Decision Tools
Who	Who's in charge? Who's the Threat?	None, proficient Uncertain Reasoning	Organization Chart Assess Threat	None required Belief Networks
What	Goal, concept, strategy	Data overload, limited memory	Diagnose Situation Determine Goal	Planning Tool Situation Map
When	Event Timing Task Duration	Few cognitive skills, Uneven time flow, Time pressure	When to act, When to update, Monitor timeline	Timeline Enforcer, Planning Horizon, Replan Trigger
Where	Asset Placement, Target Location	Few cognitive skills, Memory overload	Determine location Detailed planning	Planning Map Plan Optimizer
Why	Enemy Intent Cause & Effect	Handling uncertainty Conflicting evidence	Assess situation, Determne goals	Data Fusion, Case-based Plan
How	Detailed Plan Know-how	Lack of experience, Unanticipated effects	Detailed Planning Optimize Plan	Knowledge Base Schedule Optimizer
How much	How good is the plan, # resources	Quantify Success, Handle uncertainty	Detailed Planning Monitor Execution	Figure-of-Merit Schedule repai

Users have few cognitive skills in deciding when a situation is ready for diagnosis, when to change goals, and when to plan in detail. Consequently, as shown, many decision tools have been provided. When getting answers to what, why, and how much, users have difficulty coping with uncertainty. On the other hand, users gain proficiency in knowing how to accomplish a task and quickly learn who is in charge. However, users do have difficulty discovering who the adversary is and who enemies are partnered with.

Here's a distilled list of impediments to getting answers to the Reporter's Questions – used as a simple device to assure a completeness checklist:

- **Time Pressure:** Unsound decisions are often the result of deadlines.
- **Uncertainty:** Users across many domains fail to cope with sparse evidence, conflicting data, group-think, stale data, and unconscious biases.
- **Memory Limitations**: Users, especially under pressure, fail to remember and retrieve potentially useful preplanned options and previously successful plans.
- **Information Overload**: We have the data – and that is the problem. Too much data, poorly distilled information, sparse knowledge, and incomplete understanding.
- **Math Skill:** Users, even mathematicians, don't have computer-like ability.

This set of five impediments does, in fact, mirror and extend the previous view of innate user versus computer strengths and serves as a reasonable set of issues to drive the development of automated decision support tools. We revisit the very useful concept of Reporter's Questions as a device for characterizing hypotheses in the knowledge base, information extraction, and for categorizing displays.

In this chapter, we motivated the need for automated decision support tools, and provided a sample mission thread that orchestrates the various decision types. We saw (Figure I-2.1) that computers are good at solving problems that are difficult for users, and vice versa. To simplify, users excel at tasks that require commonsense and wisdom, while computers are best at memory and math. The goal of decision support tools is to blend user and computer processing to perform missions faster, better, and cheaper (pick two).

Chapter I-3
Decisions as State Transitions

The design of a mission thread is critically dependent on the nature of the decisions required of the user. Because of this, it is worthwhile to take a closer look at what constitutes a decision. What does it mean to decide? What is the tangible or practical result of a decision? Can a decision and its impact be measured? The answer to these questions is closely connected to the concept (Figure I-3.1) of a decision as a transition from a current state to a desired state, according to a plan. We learned this in the early 1990's and continue to use it to explain what a decision is. A more recent reference[1] links state transitions to uncertainty. What we add is a feedback loop from the desired state to a new current state based on assessment to determine if the desired state is obtained.

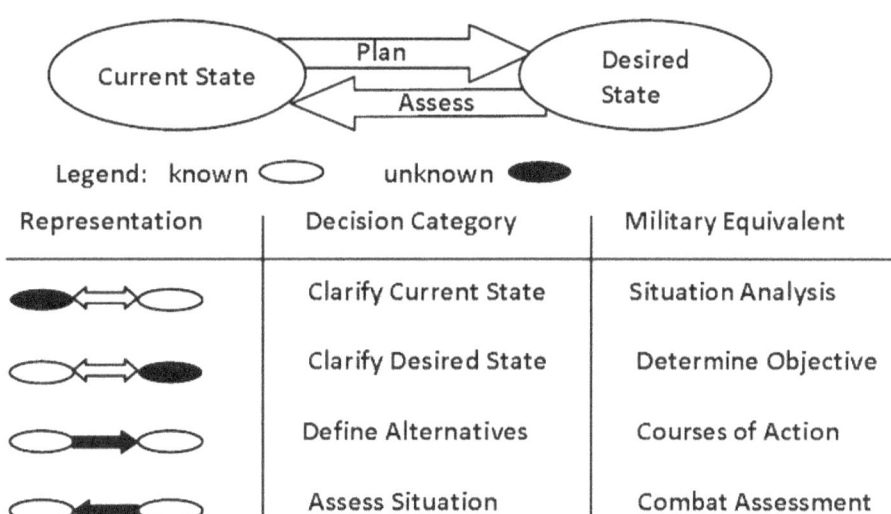

Figure I-3.1 Decisions are State Transitions

Just as we earlier saw that the need for system engineers to minimize risk leads to probabilistic reasoning and the need to deal with uncertainty, so too does the notion of decisions as state transitions! Why? Because the four components of the state transition - current state, plan, desired state, and assessment - may, or may not, be known. In virtually all cases, they are only partially known. This corresponds to our intuition. Decisions are easy if we completely understand our current state (what's going on?), have the perfect plan (what to do?), act to reach a perfectly conceived desired state (do it?), and are able to fully assess whether the desired state is achieved (how well we do?). Unfortunately, these aspects of the decision process are never understood with certainty.

---

[1] http://groups.csail.mit.edu/belief-dynamics/pdfs/BrianS_draft.pdf, accessed 03/25/2015.

We once drew a Venn diagram that showed the overlap in functionality for algorithms we were using for a space operations project. It was before PowerPoint, so we don't have it anymore, but it showed massive overlaps - more thought is required to sort algorithms by function. However, the "decisions as state transitions' paradigm not only sorts decisions into categories, but also sorts algorithms and displays into these same categories.

The first representation shown (Figure I-3.1) has the current state unknown, so the task is to "Clarify Current State", which corresponds to the military equivalent "Situation Analysis". Here, the algorithms are text segmentation, information extraction, classification, and data base registration. The displays are maps with icons, tables that organize data files, and event timelines.

The 2nd decision category- "Clarify Desired State"- corresponds to an unknown desired state and the military equivalent is "Determine Objective". Example algorithms are template-filling planning tools, and case-based reasoning applications for remembering appropriate plans or plan updates. Example displays are hierarchical plans and performance curves.

The 3rd decision category -"Define Alternatives"- corresponds to an unknown plan and the military equivalent is "Develop Courses-of-Action". Example algorithms are template- search optimizers, schedulers, and expert systems. Example displays are Gantt charts, and block diagrams.

The 4th decision category, our favorite - "Execute/assess Plan" – corresponds to an unknown assessment and the military equivalent is combat assessment. Example algorithms are data mining and data fusion tools. Example displays are induced rule trees, cluster diagrams, and belief networks. All of these tools compute and display uncertainty explicitly.

In this chapter, we provided a formal definition of a decision as a state transition and provided examples of how this paradigm neatly sorts decisions, displays, and algorithms.

Chapter I-4
Visualization for Understanding

We see that computer algorithms help human operators make decisions. The complexity and sophistication of these algorithms provide sufficient computational support for successful outcomes. The real challenge is designing a visual delivery system that clearly presents this information and thereby enhances the decision-making process. The primary focus of legacy interface displays is on showing the processing performed by the underlying algorithms and not on presenting the information in a manner that facilitates human understanding. Our goal is to redesign the Human-Computer Interface to make it much easier for the user to access, understand, and use the information provided by the algorithms. Using the story metaphor, we identify and link three basic visualization layers for delivery of the information. These are the geographic, temporal, and logical views.

The resulting design provides high-level geographic views linked to logical views. Both geographic and logical views incorporate temporal information through animation and timelines. A dashboard summary and textual explanations complement the orchestration of views. The underlying technology is visualization techniques that allow humans to better understand the output of automated reasoning algorithms in support of military and intelligence community decision making. Metrics are defined to quantify the effectiveness of this technology for knowledge visualization in decision making. The design is demonstrated with a New York City Terrorism scenario. The knowledge visualization is found to be intuitive and easy to navigate. We also quantify[2], based on experimentation and workflow instrumentation, the extent to which decisions are made faster, with more accuracy, and with smaller crews.

The challenge is to design an interface that satisfies requirements to achieve clear articulation of the knowledge found within the collective algorithms of the network. Four concepts are derived from the challenge. Graphic User Interface (GUI) design must be consistent throughout to seamlessly connect algorithm results. To be effective the knowledge visualization must be accurately linked both temporally, logically, and geographically. It must maintain only that level of complexity necessary to convey the story and this complexity must be layered down to the algorithmic source.

Our solution is to enhance user understanding of computer-generated knowledge by presenting it as a story. Simply stated, we design a user interface that tells a story through linked geographic, logical, and temporal views. These views, capturing the "who, what, when, where, why, and how" of a successful story, are summarized in a mission summary and supported by evidence. Stories are dynamic in nature, allowing for the input and update of information. Output comes in the form of changes in the story as the response to new input. Confidence in the knowledge presented by the storyline is also displayed and updated as the underlying algorithms assimilate new information in the form of evidence. To validate our solution, we conceive of metrics to quantitatively and qualitatively assess to utility of the story metaphor.

---

[2] http://www.idemployee.id.tue.nl/g.w.m.rauterberg//amme/dillon-et-al-2005.pdf, accessed 04/07/2015.

Three types of decisions are supported by the new knowledge visualization design. The decisions are: what's going on, understood using the Attention Cuer; what to do, understood using the interactive Plan Analyzer; and how well are we doing, understood using the Belief Network Editor. These are discussed in significant detail.

Visualization Concept Development – A Storyboard Approach

To improve knowledge visualization, our technical approach consists of six steps. We begin with a storyboard that is driven by military and anti-terrorism scenarios. The legacy screens help us identify shortfalls and define a mitigation strategy. We derive requirements to produce objectives and define tasks. With goals in sight, we form a team composed of multiple organizations to produce prototype visualizations in the form of mockups – typically in Photoshop, Macromedia Flash, or Java. For visualizations of significant interest to our military and intelligence community customers, we integrate existing algorithms with the displays to provide working demonstrations. To quantify the effectiveness of the knowledge visualization design, we define metrics and determine effectiveness in terms of quantified Measures of Effectiveness. These steps are now discussed in more detail.

Step 1: Constructing the Storyboard.

In combination with one other, the legacy displays show computer-generated content. We see this process as telling a dynamically changing story, delivered in structured and textual forms. In the film industry, the use of *storyboards* is universal for designing the flow of the story. The paradigm is used here to aid in our interface design. It gives us a visual flow of the displays. This method is used from the early stages of the project and helps in ongoing refinement of the look and usability of the interface, as well as in the presentation of knowledge with the goal of telling a story.

Step 2: Identifying Visualization Shortfalls

We then define shortfalls of the storyboard views and identify mitigation strategies. Typically, each of the legacy displays was developed in applications that generated disparate interfaces for accessing, viewing, and editing content. The only way to extract understanding from the system was to manually search out each display, extract the pertinent information, exit that application, enter another, extract another data set, exit that application, etc.

Compounding problems are the confusing nature of the inconsistent displays and the lack of flexibility inherent in a single input/output layer. As a final issue, we look at the text-based visualization of the knowledge: the system relies heavily on delivering knowledge textually, even system navigation depends on textual input. We abandon this tedious and error-prone interface for a metaphorically based system that lends itself to a more intuitive operator interaction. By mitigating these shortfalls, we have ideas for a robust system for visualizing a given story.

Step 3: Designing to Objectives

The primary goal is to provide an interface to deliver knowledge; that is, to tell a story, so that decisions are made quickly and with the highest confidence. To satisfy this requirement, the interface has to be consistent throughout. From screen to screen, dialogue box to dialogue box, and algorithm-to-algorithm, the "look and feel" has to be consistent The interface has to be intuitive, but not so simple that it fails to deliver understanding. Along with this ease of use, we require the ability to delve deeper into the details. This allows the user to "drill down" to an explanation of results computed by automated reasoning tools.

The final major requirement is to preserve the integrity of the evidential information as it propagates through the processing thread. We implement the story metaphor using maps, graphs, and timelines linked together to allow more robust user understanding of the underlying questions that a story answers: who, what, where, when, why, and how of the computer-generated data. In our interface design we include domain-specific geographic (what, where), logical (what, who, how, why) and temporal visualization (what, when).

The concept of "the smallest effective difference"[3] is also part of our design philosophy. This "makes all visual distinctions as subtle as possible, but still clear and effective". Within the network of algorithms lies the knowledge needed to make accurate decisions, the key is being able to discern the knowledge and not be distracted by the graphics delivering that knowledge.

Step 4: Prototyping Visualizations

Collaboration with visualization specialists leads us to a design synthesis that includes visual consistency for a system that incorporates all of our available knowledge-based algorithms with a clear, intuitive interface from which the information is accessed and edited by the operator. It also produces three compelling visualizations and associated algorithms: the Attention Cuer, Belief Network Editor, and Plan Analyzer.

Step 5: Integrating Algorithms

We identify applications that form a processing thread. The processing is automated by integrating information extraction and knowledge base technology to automatically populate the displays. These are integrated into three sequenced applications: the Attention Cuer for situation awareness, the Plan Analyzer for strategic planning, and the Belief Network Editor for mission assessment.

Step 6: Assessing Effectiveness.

To date, the primary means of assessing display effectiveness is the "Cool Factor". We show our visualization technology, hoping to extract a collective 'That's Cool" from the audience, and declare success. We seek engineering discipline (in addition to the Cool Factor) by defining

---

[3] http://www.edwardtufte.com/tufte/books_visex, accessed 02/15/2015.

quantitative metrics to assess effectiveness. Of particular interest is determining how much more accurate, faster, and cheaper the decisions based on knowledge visualizations are, compared to legacy displays and ground truth; for example, statistical results or empirical wisdom. This question is more easily asked than answered. Our approach to determining how much faster our automated knowledge visualization products are than alternatives is to design stopwatch experiments that time human decision-makers in performing relevant tasks using legacy versus knowledge-based visualization. To determine how much more accurate the knowledge visualization is than our legacy displays, we postulate trials to validate statistical or empirical correctness. To assess how much cheaper the knowledge visualization approach is, compared to the legacy displays, we identify workflow analysis to compute differences in man-hours to perform pertinent tasks.

Applications: Mission-Oriented Perspectives

Knowledge Visualization describes storyboards for military and intelligence community scenarios, prototype visualizations, and algorithm integration. The technology emphasis is on the uniting of automated reasoning algorithms and knowledge visualization to produce user understanding based on the story metaphor.

Knowledge Visualization:

We use the story metaphor to graphically represent knowledge. The story begins with situation monitoring. In this vignette, the crew chief is briefed on the status of a terrorist attack and the crew monitors message traffic that is posted to a map display. As tensions increases, the operations center receives tasking directives from higher authority. In this vignette, the crew shifts focus to mission planning and plan optimization. Authority to execute the plan is received, the battle begins, and combat assessment evidence is received at the operations center. In the final vignette, the crew fuses this evidence to perform mission assessment.

We produce a storyboard, consisting of sketches and legacy displays, to describe this story. Mission summaries, evidence, maps, graphs, and timelines are linked to make the story coherent. We begin with a high level display as the point of entry to our knowledge visualization software. The main display, or dashboard, (Figure I-4.1) consists of a mission summary, linked geographic, logical, and temporal views, and an evidence log. Depending on the mission parameters, these primary indicators direct an operator's attention in a specific direction. The goal is to enhance the judgment of an operator, while minimizing keystrokes, as he accesses the interface. The system moves the operator through the knowledge set as the operator sees fit at whatever time he needs it to do so, as explained next. For demonstrations, scenario drivers consisting of situation reports, planning directives, and mission assessment evidence, are posted and acknowledged by clicking boxes in the display.

At this juncture the operator has choices: view the information on each screen or combine two or all three screens to consolidate the information depending on need or preference. The three primary screens are the geographic view provided by the Attention Cuer, the logical view provided by the Belief Network Editor, and the mission metrics window driven by the Plan Analyzer. These are discussed in the upcoming sections.

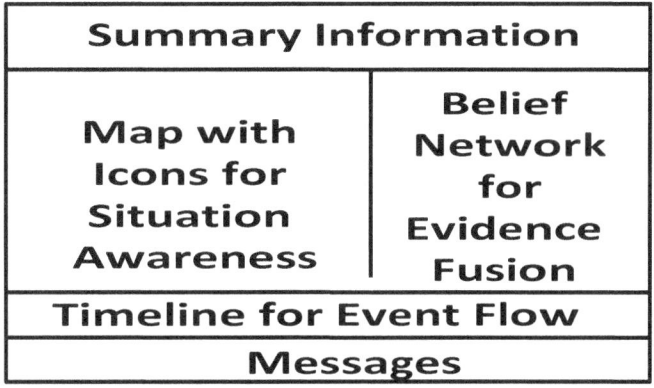

Figure I-4.1: Main Display – The Dashboard

High Interest Event for Situation Awareness

Many present day situations have related, concurrent real time events requiring attention that stress available resources. Thus, we prioritize and organize these many events to efficiently make timely decisions based on relevant information. The Attention Cuer (Figure I-4.2) is an application that organizes and presents 15 dimensions, or characteristics, of an event. This includes three dimensions of position, three dimensions of velocity, time, time duration, event identifier (icon label), event description (text field), importance (icon size), urgency (icon glow), event category (icon color), effort required (icon shape), and event uncertainty (link to belief network) for a large set of concurrent events. These features are displayed spatially (on a map) and temporally (on a timeline) and are updated in real time, providing a "dashboard" summary of the overall situation. Potential situation domains include battlefield decision-making, resource allocation, urban disaster emergency support (shown in Figure I-4.2), National Park Service fire hazard planning, and personal "To Do" task planning.

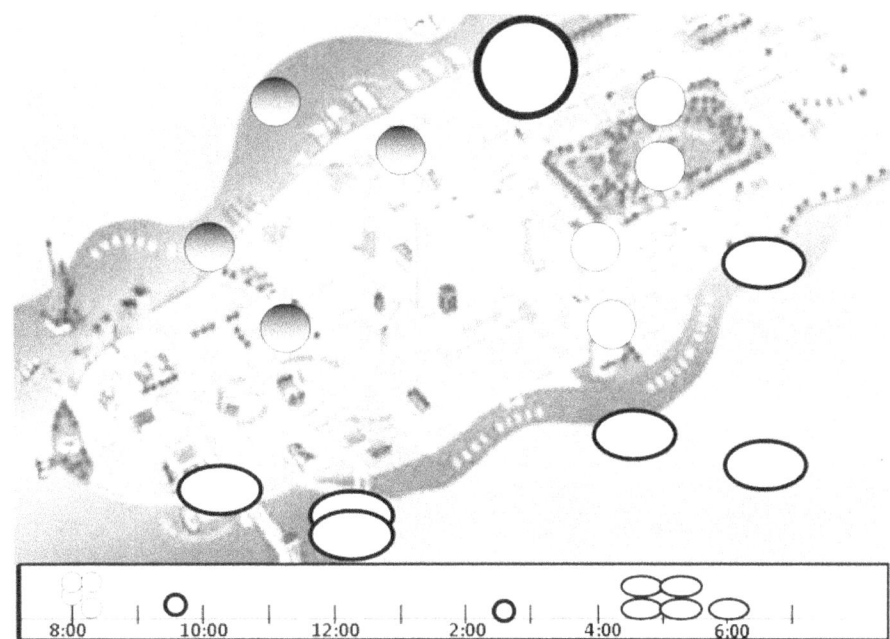

Figure I-4.2: High Interest Event for Situation Awareness

This application demonstrates visual shorthand for organizing and prioritizing events. It allows users to represent events as nodes with visual characteristics corresponding to meta-data of interest to decision makers. The Attention Cuer is designed as a new way of organizing and prioritizing events that gives the user more depth of information about what is going on, and also, by combining information graphically, provides a clearer, simpler picture of what is happening in the situation as a whole.

Features include the ability to create nodes with various characteristics at scenario specified times. Nodes can be created on a user specified background and positioned anywhere on the application canvas. Changes to nodes can be set at specific times on the timeline. The user-created visualization can be animated; for example, for mission rehearsal, to show addition of new nodes and changes to existing nodes. We interface the Attention Cuer to our information extraction application to automatically extract node attributes from unstructured text. This relieves the user of the tedious and time-consuming process of reading a message, extracting the information, and filling a template. Instead, the user edits the default metadata supplied by the information extractor

Belief Network Editor

Analysts, researchers, and investigators often must take imprecise and unstructured information and form hypotheses and structured arguments. However, existing tools did not let these researchers work with sparse evidence, imprecise estimates, confidence in hypotheses, and links between hypotheses in an intuitive way. The Belief Network Editor (Figure I-4.3) allows users to structure arguments and evidence to produce quantitative values in a more intuitive way.

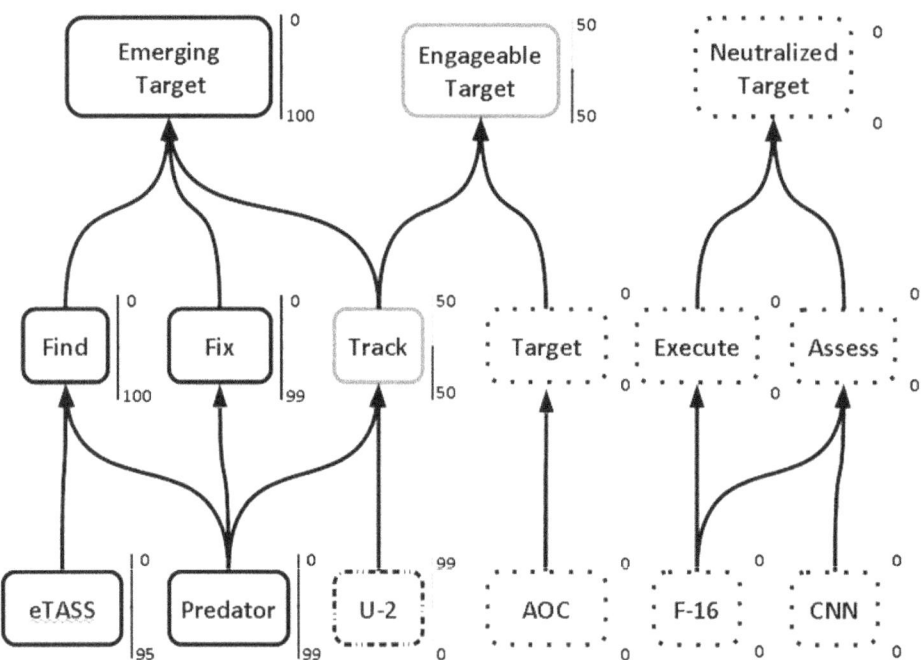

Figure I-4.3: Belief Network Editor for Combat Assessment

This Belief Network Editor is found useful for visualizing and interacting with belief networks. The novelty is that it allows exploration of a complex mathematical model using an easy-to-use interface; for example, prior probability distributions are not required – in the absence of evidence, "ignorance" is the default value for the hypothesis. Belief networks consist of nodes and links, with each node and line having visual representations (size, color, value) corresponding to their numeric weights. The user can perform sensitivity analysis to see the influence that uncertainties in one hypothesis have on another hypothesis by adjusting sliders.

The Belief Network Editor (BNE) is a tool for quantitative exploration of a space of lattice-shaped (multiple influences on a hypothesis) uncertainties. Once invoked, the user can add or edit nodes and links, propagate beliefs, arrange nodes in the network, collapse sub-trees, and scroll through a large network.

Decision algorithms are combined with knowledge visualization to construct decision-support systems. These systems exploit the unique information-crunching capabilities of the algorithms, while providing understanding of computer-generated content to the user. We rated the knowledge visualization provided against our objectives and found our requirements to be satisfied – these requirements and other metrics are our criteria for success.

Lessons Learned

- The classic Windows environment is not well suited to risky, real time decision-making: in mission-critical settings, no time is available to drill down through multiple menus. Uncertainty must be explicitly presented.

- Simplicity versus features is often a good trade-off when the application domain is complex and/or hard to understand: we relegated many "bells and whistles" to expert menus.

- What-if analysis is important to planners: real-time sensitivity analysis in the BNE and Plan Analyzer is a very useful tool for detecting important influences and eliminating unimportant factors.

- Scalable graphic user interfaces are a very important in making complex decision processes more tractable and accessible: we focused on scalable belief networks and attention cueing.

Chapter I-5
Decision Algorithms

An algorithm is an effective method for solving a problem. Most practical decision-making problems are not amenable to "brute force" approaches because the underlying problem is known to be "non-polynomial hard" (NP-Hard). This means that the problem cannot be solved in polynomial (quadratic, cubic, quartic, etc.) time of the parameters of the problem. This fact – that most practical decision-making problems are NP-Hard – makes algorithms the critical component of decision support systems.

This chapter introduces algorithms as a fundamental process: algorithms accept data and drive the displays that facilitate user decision-making. In this role, algorithms pre-process and ingest structured and unstructured content. They figure out what the content is about and "tag" content with a concept. Algorithms register this content in a knowledge base. In fact, algorithms operate on the knowledge base to make it executable. Once data is registered in the knowledge base, multi-strategy reasoning algorithms perform automated manipulation, distillation, and discovery on the data and post the results to displays.

A sample set of decision algorithms (Figure I-5.1) consists of background and foreground tasks. Background tasks do the computations that "set the stage" for foreground tasks that are more highly user interactive. In this example, knowledge management provides content in context, dat triage filters large texts, information extraction parses content, and fog-of-war perturbs already uncertain data. Foreground tasks (Figure I-5.1) have been arranged by decision type and connected to form a mission thread.

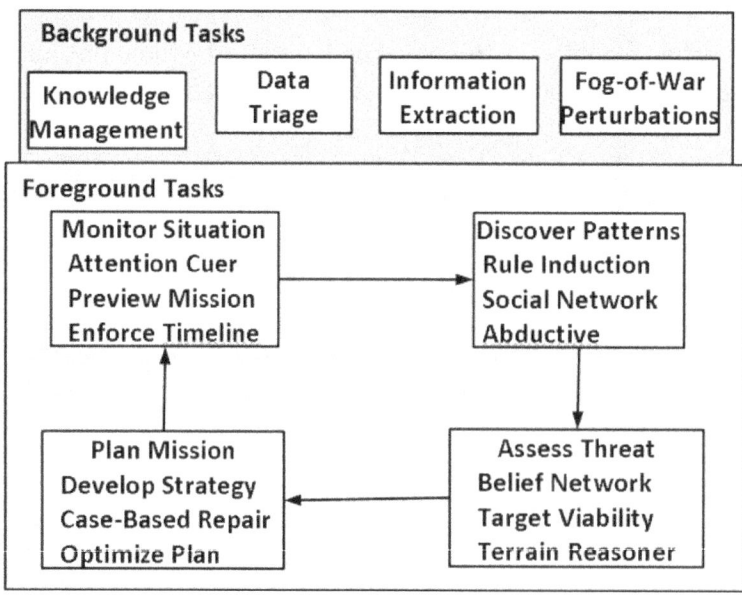

Figure I-5.1: Sample Decision Algorithms

Monitor Situation consists of an Attention Cuer that puts icons on maps, Preview Mission that simulates the expected mission events, and Enforce Timeline that shows event deadlines.

Discover Patterns consists of a Rule Induction algorithm that finds IF...AND...AND...THEN rules, Social Network analysis tools that find adversary vulnerabilities, and Abductive reasoning algorithms that reason to the best explanation.

Assess Threat consists of a Belief Network to fuse uncertain evidence to form indicators and reason about outcomes, Target Viability to retrieve target characteristics dynamically, and a Terrain Reasoner to find best routes.

Plan Mission consists of a template-filling tool, the Strategy Development Tool, to develop a strategy, objectives, and tasks, A Case-Based Repair algorithm that updates plans, and an Optimize Plan application that uses genetic algorithms to optimize actions against targets, subject to constraints.

Chapter I-6
Pattern-Preserving Operations on Data Sources

Three types of data formats, geared to their relationship with decision support tools, are discussed. Recall that in the decision centered approach, data is what actuates algorithms that populate displays that help the user decide. In this sense, data includes data sources, parameters, and interfaces. Data actuates algorithms in different ways, with unstructured text requiring natural language processing and structured text requiring formatted reads and writes. Semi-structured text requires both natural language process and formatting software, along with smart code to determine which portions are structured and which are not. Data types, such as voice, picture, video, gesture, thought (Brain-Computer Interfaces), and other non-character based modalities, are not discussed.

Introduction

Natural language processing, to include data triage, text retrieval, and information extraction are, on the other hand, fascinating, and will be discussed in Part III. Details of particular structured data sets are domain specific, but a critical need is for operating on the data to make it useful for analysis. This requires that patterns within the data be preserved.

Need

An electronic procedure is needed for extrapolating or performing other operations on a data set that has multi-dimensional, multi-resolution patterns. The goal is an arbitrarily large data set that maintains or modifies those patterns. We sometimes need to extrapolate a small data set to a large one, while maintaining distinguishing characteristics or adding additional anticipated characteristics. We need large data sets with provably correct properties for analysis, system sizing, and processing metrics collection.

Objective

Produce a suitably large data set for system sizing studies, topology analysis, and metrics collection.

Technical Description

The patterns in a wide variety of data sets are discoverable as multi-dimensional (based, for example, on the multiple parameters that comprise a data entry), and multi-resolution (bands and sub-bands of data entries). These are analyzed and extrapolated based on temporal trends, the attributes of the system using the data set, and domain expertise to capture the patterns.

Currently, extended data sets are produced manually (Figure I-6.1). Semi-automated techniques consist of choosing and duplicating existing data entries. In some cases (for example, social network analysis), single characteristics of nodes and links are automatically extended using Monte Carlo sampling from a single distribution: for example, a scale-free distribution.

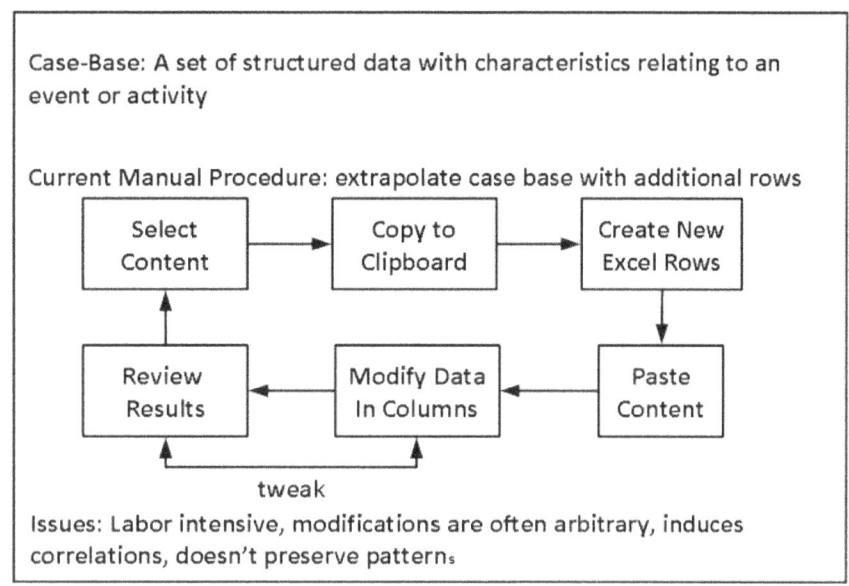

Figure I-6.1: Current Process for Data Set Extrapolation

Discovery of underlying spatial, temporal, and logical patterns in an existing data set is key. These patterns (Figure I-6.2) may encompass one or more dimensions (attributes of the data set) and are expressed as a rule-tree, a graph with quantifiable node and link characteristics, or a set of differing properties at different levels of resolution; for example, locally the members of a network communicate peer-to-peer, but globally, the internet forum has a scale-free distribution. Wavelet multi-resolution segmentation is (optionally) used to discover these patterns. The second inventive concept is to extrapolate the discovered patterns to a larger data set that provably has the identified patterns. What is new is the process and electronic procedure for discovering and extrapolating patterns to produce a large data set with "guaranteed" properties.

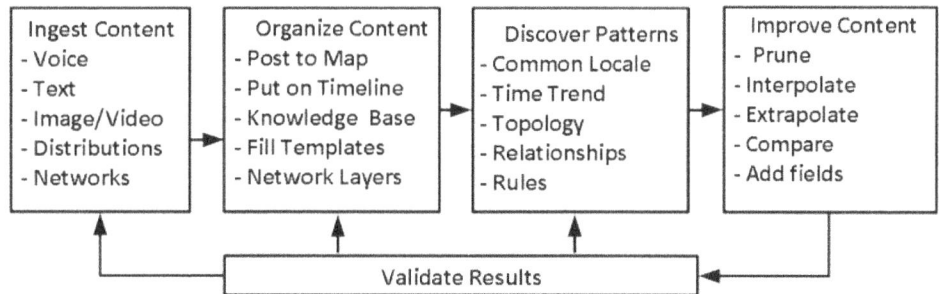

Figure I-6.2: Approach to Pattern-Preserving Operations

Results

We test this idea on an Internet forum data set. Nodes consist of groups of individuals of about eight different types and importance. Links consist of the strength of influence that one group has on another. A small data set is extrapolated to a larger one with specified connectivity rules.

In a second case, system anomaly data, with each data entry consisting of about 12 attributes, is extrapolated to a larger data set, based on the numerical and logical characteristics of each of the attributes. The resulting larger data set is processed using a rule induction algorithm to test the resulting rule tree for compliance with extrapolation goals.

In a third case, the Space Catalog, consisting of ~13,000 objects is extrapolated to a 100,000 object catalog with appropriately represented orbit types, altitude bands, sub-bands (multi-resolution), and object sizes. A new debris pattern is added that represents objects between 5 and 10 cm in size that are detectable by a new sensor. The process and procedure for this catalog extrapolation work (Figure I-6.3) is formulated and implemented on a laptop computer. Note that the extrapolated catalog, shown for high earth orbit (HEO) satellites, mirrors the present day catalog. A more detailed discussion is provided in Chapter V-9.

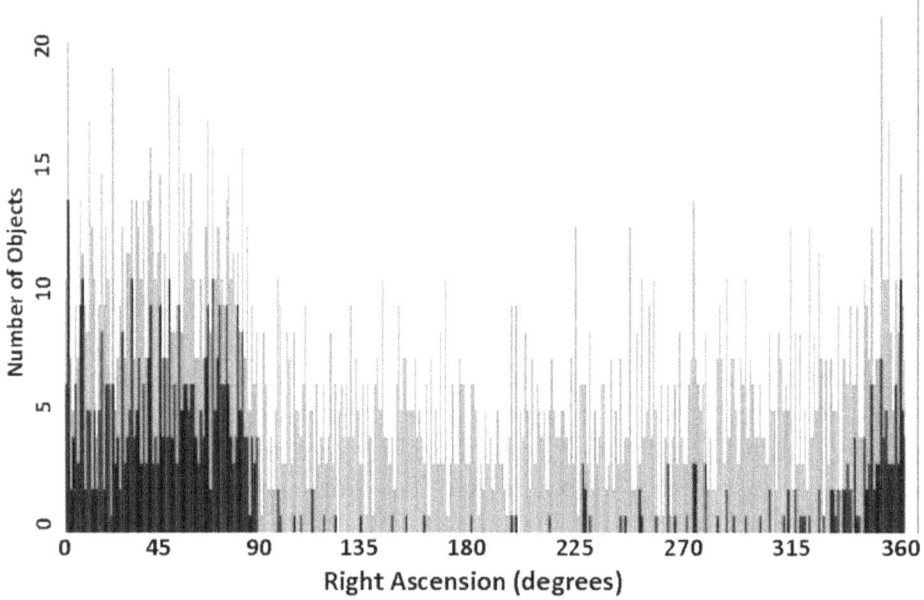

Figure I-6.3: Example – Extrapolation of the Space Object Catalog

Part II
Representation: Framing the Problem Space

Chapter II-1: Knowledge Capture
Chapter II-2: Semantic Networks for Multi-Strategy Reasoning and Learning
Chapter II-3: Knowledge Management via Stories
Chapter II-4: Executable and Self-Aware Knowledge Bases

Part II
Introduction

This part of the book focuses on establishing a reference frame for decision support tools. We discuss, in more detail, the roles of humans and computers in decision support. A common framework – a knowledge base structured as a semantic network – is defined. This knowledge base provides a sufficiently rich representation to accommodate all types of algorithms of interest. The ontology allows substantive integration for multi-strategy reasoning. Knowledge management is addressed. The knowledge base is shown to consist of a collection of stories which are updated with evidence extracted from text. These stories are updated dynamically as new evidence arrives and is fused and propagated. A self-aware knowledge base provides both static and dynamically-computed metadata.

In engineering terms, *representation* means the formulation of the solution to a problem. Not surprisingly, the ability to solve a problem is often critically dependent on the formulation. To optimally frame a problem, we take advantage of rules-of-thumb, symmetries, shortcuts, and analogies to solved problems. An example that will be discussed in excruciating detail – because it is very important to producing a decision support toolset – is the representation of data and information as knowledge. This representation must be sufficiently rich to serve as the basis for multi-strategy reasoning. That's because practical problems typically require substantive integration of multiple algorithms. We want a single content representation with a single interface to all algorithms so that new algorithms can easily be plugged in without requiring content restructuring.

Chapter II-1
Knowledge Capture

Definitions of the terms that place "knowledge" in context are the first topic of discussion:

- Data is raw evidence with minimal refinement.
- Information is data that is applicable to support a processing goal.
- Knowledge in the "chunking" of processed information - a concept with context.
- Understanding is the human ability to assimilate data, information, and knowledge.

Input to a decision-making system is typically data. With pre-processing refinements, which may segment, prioritize and filter the data, it is fair to call this refined data information. The goal is to automate both pre-processing refinements and the conversion of information to knowledge. The resulting knowledge representation, given that it is sufficiently rich, forms the basis for multi-strategy algorithmic reasoning. Understanding is a user ability that cannot (currently) be automated; however, displays that present knowledge to a user can facilitate user understanding.

Many organizational schemes, called ontologies, are in use to structure knowledge. The resulting knowledge bases may or may not support automated decision support tools. Here are a few:

- **Declarative:** Knowledge is stored as a set of statements about the world. These statements are static but can be added to, deleted or modified.
- **Procedural:** Knowledge is stored as a set of procedures which can themselves determine when they should be executed. Their execution is the intelligent behavior that is expected in the situation.
- **Symbolic:** Storage of the knowledge uses symbols to represent objects of the outside world or sets of perceptions about the outside world.
- **Sub-Symbolic:** knowledge is stored without the use of symbols. This typically means the architecture uses direct mapping from the inputs to outputs.
- **Uniform Representation:** one method is chosen for representing the knowledge (e.g. frames, semantic nets etc) and used exclusively.
- **Non-Uniform Representation:** Many different representation methods are used.

A brief history of knowledge bases indicates a shift from their use as the basis for an expert system to the current hierarchical, frame-based Protégé[4] which supports many types of automated inference. As will be later discussed - at length - we use a uniform representation: the knowledge base is a semantic network, structured as a hierarchical, characteristics-based (Reporter's Questions) ontology.

Imagine working with three dozen algorithms. Suppose that some are found, some are developed, and others are tailored to specified goals. We find that computer algorithms that support human decision-making arise from different fields of research and consequently vary

---

[4] http://protege.stanford.edu/, accessed 02/24/2015.

significantly (Figure II-3.1) in the types of reasoning performed. Fields of research include: artificial intelligence, machine learning, data mining, knowledge management, probability theory, and computational linguistics. Reasoning types include: abductive, analogical, deductive, inductive, and probabilistic. We implement algorithms to automate these reasoning schemes using algorithm types such as: ontologies, rules, neural networks, evolutionary (including genetic) algorithms, belief networks, and case bases. Each algorithm type has a unique representation; for example, genetic algorithms use the biological chromosome metaphor, while case based reasoning uses tables to format options, selection criteria, weights, and scores. This diversity of formulations is a problem.

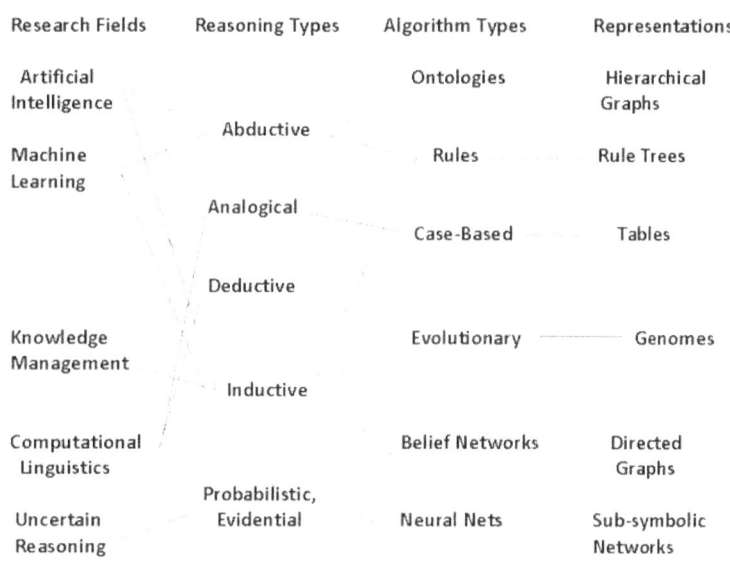

Figure II-1.1: Different Algorithms, Different Representations

The difficulty with this wide variety of research fields, reasoning types and algorithm representations is that the algorithms don't work well together. The goal of substantive integration at the algorithm level – a combination of algorithms is usually required in "real world" applications - is elusive. The idea of semantic networks to provide a unifying framework first surfaced in the machine learning community[5], based on a largely unsupported proposition that five different reasoning strategies (rules, belief networks, genetic algorithms, case-based systems, and neural networks) have algorithmic representations that are expressible as semantic networks. Other approaches to a unified framework for multi-strategy reasoning include rule-based methods[6], introspection[7], and simulation[8].

In previous work, our algorithms were organized by function and accessed from a menu: the human analyst remembered the result of a decision algorithm and decided what to do next. Tighter integration across multiple algorithms is required to ease the "knowledge acquisition bottleneck. This directly translates to offloading drudge-work from the operator – perhaps the most worthwhile of roles for computers.

---

[5] http://citeseerx.ist.psu.edu/showciting?cid=161321, accessed 02/24/2015.
[6] http://lalab.gmu.edu/publications/1994/TecuciG_Multistrategy_Learning%20_ML.pdf, accessed 04/08/2015.
[7] http://www.cc.gatech.edu/faculty/ashwin/papers/git-cc-92-19.pdf, accessed 04/08/2015.
[8] http://www.modelbenders.com/papers/RSmith_OneSAF_KIDA.pdf, accessed 04/08/2015.

Chapter II-2
Semantic Networks for Multi-Strategy Reasoning and Learning

A semantic network is a graph that consists of links and nodes. These graphs provide a powerful representation for processing by computer algorithms. We have found or developed about three dozen algorithms that support human decision-making, and because they arise from diverse fields of research, the algorithms are historically executed as standalone applications. Making these algorithms work together is difficult because they employ widely varying reasoning paradigms and structural representations – belief networks don't appear to have much in common with evolutionary algorithms! However, based on the theoretical premise that a semantic network is lurking within each of these algorithm types, we structure a knowledge base and an algorithm interface to achieve a unified framework. A practical advantage is that each algorithm interfaces only with the knowledge base via a single import and export interface. This allows algorithm work to proceed in parallel because it removes dependencies among algorithms. Our results indicate that a semantic network provides the desired unified framework for multi-strategy reasoning algorithms.

A semantic network is, literally, a network with meaning. It is a graph with nodes connected by links. Nodes represent concepts and links associate them. These graphs provide a general, and quite powerful, representation of problems for solution by computer algorithms.

The result of our decision to use semantic networks as a unifying framework for multi-strategy reasoning is to define a knowledge base schema. The purpose of the knowledge base is to provide a single interface for all algorithms and a central repository for all information used by the algorithms. Data specific to an algorithm is not included in the knowledge base. The structure of this repository is a hierarchical, frame-based set of nodes and links; specifically, nodes are concepts, links are impacts of one concept on another, frames contain characteristics, and concepts are arranged hierarchically, from general to specific, to facilitate inheritance.

Algorithms are recast to leverage the semantic network structure that is postulated to reside within them. This theoretical structure is obvious for belief networks, rule trees, and abductive reasoning algorithms, - the knowledge base interface is straightforward. For other algorithms, association of concepts in algorithms with concepts in the knowledge base provides way to connect algorithms with the knowledge base.

The semantic network is found to be viable as a unifying framework. The knowledge base abstracts this structure to provide a central knowledge repository. All algorithms interface with one another through the knowledge base, thus providing substantive integration. A practical advantage is that algorithms are developed in parallel, based on a single interface, rather than each algorithm having a custom interfaces for import and export with every other one.

Approach to Multi-Strategy Reasoning

We have a decision algorithm test bed with about 36 algorithms (Figure II-2.1) with significant interactions among them. Each of these is organized under a functionally oriented menu. Any algorithm is executable at any time by merely double-clicking on an algorithm sub-menu.

Results are computed based on stored input and defaults. Each algorithm is standalone. As the test bed evolves, message-passing interfaces are constructed between algorithms. This works well for a while, but becomes unmanageable with increases in the number of algorithms, the number of interfaces between algorithms, the number of developers, and the dynamic change in algorithm interfaces.

We assess the way in which content is passed between algorithms: it is highly customized. Worse, it makes the execution sequential with no feedback. Each interface requires substantial rework as new data items are added. No central repository to track the collective content in the test bed existed. The number of interfaces scales as the square of twice the number of algorithms - every algorithm importing and exporting information to every other.

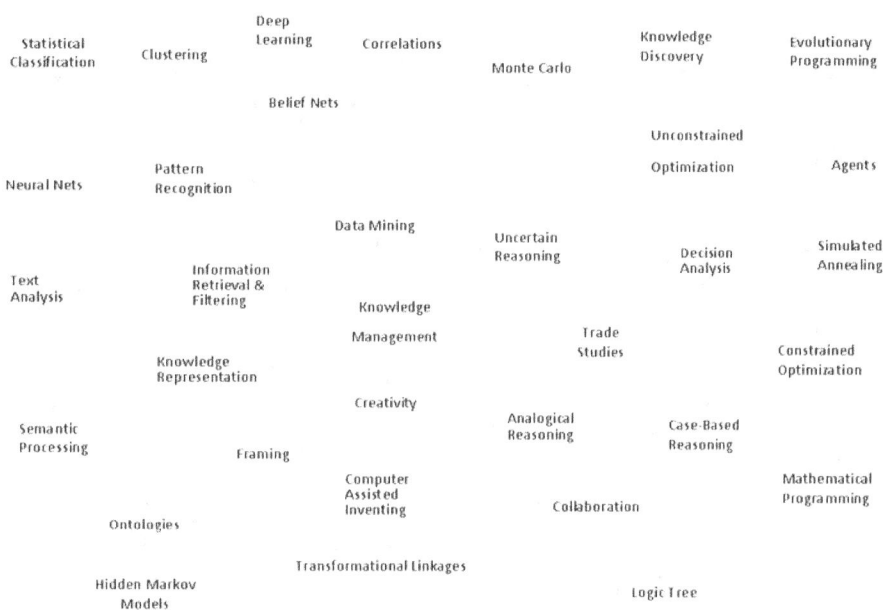

Figure II-2.1: Lots of Algorithm, Lots of Interactions

We first levy a few requirements on the inevitable upgrades. In order of importance, they are: substantively integrate algorithms so that they work together, minimize interfaces, register all knowledge in the knowledge base, and retain the ability to execute algorithms in standalone fashion. Although the knowledge base is an algorithm (it relates concepts to one another), it is identified as the hub of the processing architecture, so we design its semantic network structure first. We then minimally populate the knowledge base to understand what it needs to store, what inputs are required from other algorithms, and what outputs it provides to other algorithms. From this analysis, and an understanding of the underlying knowledge base application code (Protégé[9]), we define a single input and output format. Finally, we modify each of the decision algorithms to meet the knowledge base interface.

Substantive Integration

The objective of the processing loop (Figure II-2.2) is to reduce the data overload on the

---
[9] http://protege.stanford.edu/, acccessed 04/08/2015.

operator by providing decision support for two questions: what's going on, and what to do?

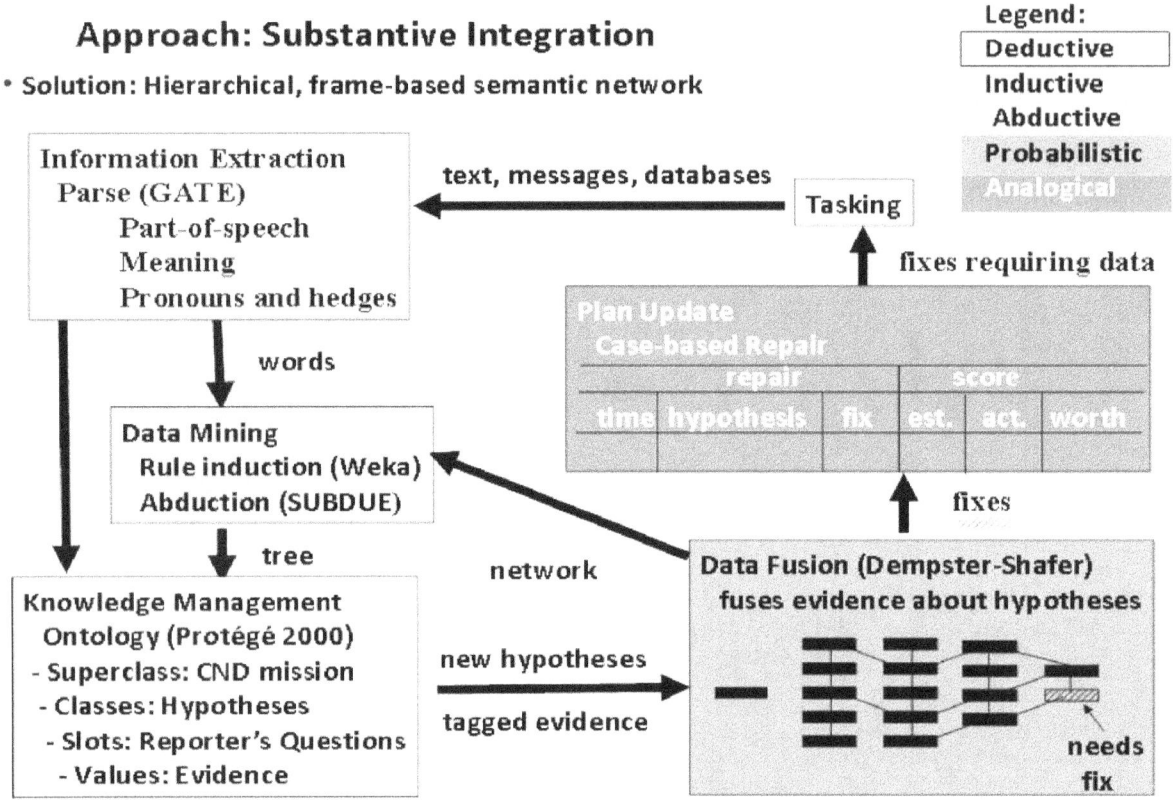

Figure II-2.2: Substantive Integration Is Required

All processing occurs interactively with an analyst. We use information extraction to collect information from text, reports, and disparate databases and put it into evidence templates, which are then "tagged" with the concept that the evidence supports. The format of the evidence template defines the common interface. Data mining algorithms then discover interesting patterns and new concepts in the data. Belief networks reason probabilistically based on confidence in evidence, and trigger plan updates, which produce tasking for more information, thus completing the processing loop.

The resultant information flow between algorithms and the knowledge base (Figure II-2.3) consists of evidence and hypotheses. Both the evidence and the hypothesis it supports are contained in a single record. As indicated earlier, the format of this record defines the single interface between the knowledge base and the algorithms. As the central repository, the knowledge base makes the most current content available to the algorithms: all of them need this in order to construct an explanation – a longstanding requirement for all algorithms in the test bed.

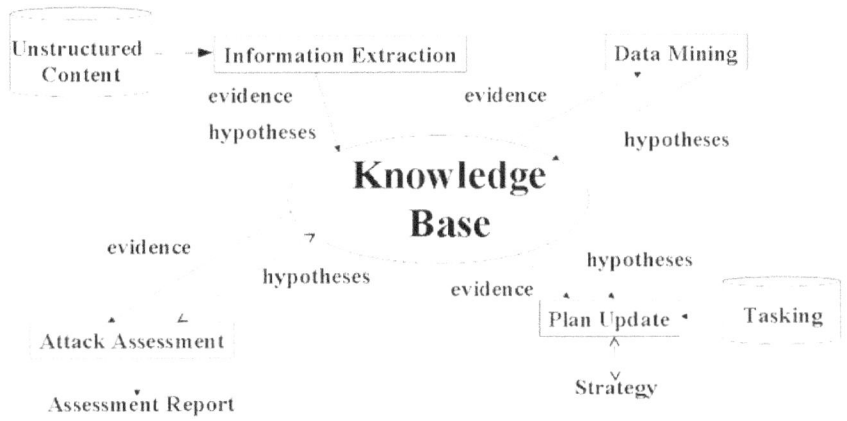

Figure II-2.3: Data Interface for Substantive Integration

Knowledge base design (Figure II-2.4) is based on the notion of a semantic network. For organizing knowledge we get hints, which result in a hierarchical structure, from our belief networks and abductive data mining algorithms. We derive an understanding of the characteristics associated with concepts from the idea of a reporter's questions template to answer the detailed who, what, when, why, how, how much, how certain, etc.

The design is for a domain (superclass) with hierarchically nested classes (concepts or hypotheses) that inherit slots (Reporter's Questions), which contain sub-slots (Reporter's Questions template), and are populated with instances of evidence. The structuring of the knowledge base (sometimes referred to as an ontology) is greatly facilitated by the graphic user interface contained in Protége.

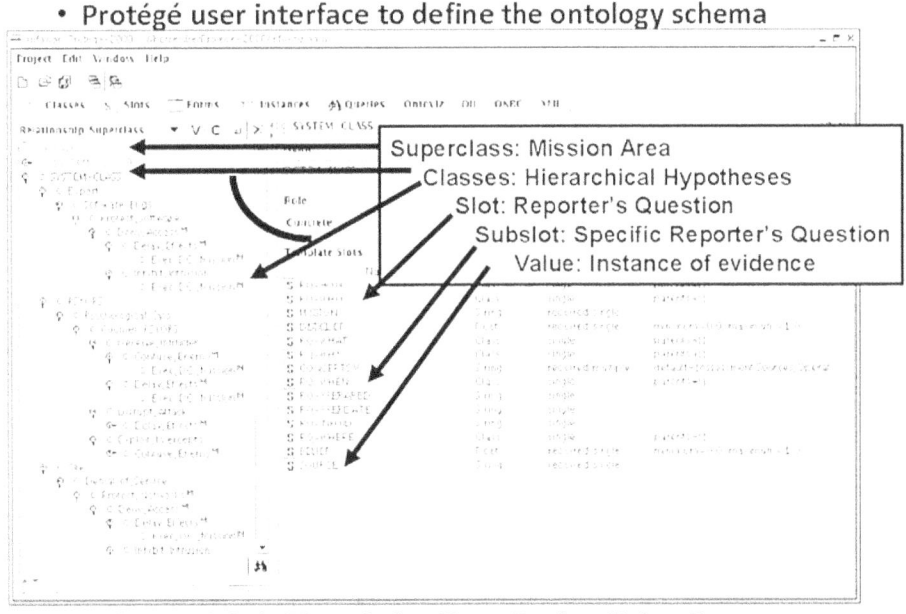

Figure II-2.4: Knowledge Base Design Using Protege

The interface to the knowledge base is an electronic equivalent of the content in the Reporter's Questions template (Figure II-2.5). For compact storage, it is published as a flat file with multiple records. Each record has fields that are space delimited and labeled with the detailed reporter's questions. An evidence record can be tagged by rule; for example, if evidence comes from a log file, it is tagged as "LOGFILE" evidence. Evidence can also be tagged by an analyst or by an evidence classifier algorithm. The "Prepared by" field (upper left) states how it is tagged. The "tagged" concept is an entry in the Reporter's Questions Template – typically the "what" or "why" fields. Another characteristic of the information extraction algorithm is confidence (belief, disbelief) in a tagged concept. These characteristics are also derived by rule, analyst input/override, or by an algorithm that co-references and translates "hedge" words to confidences.

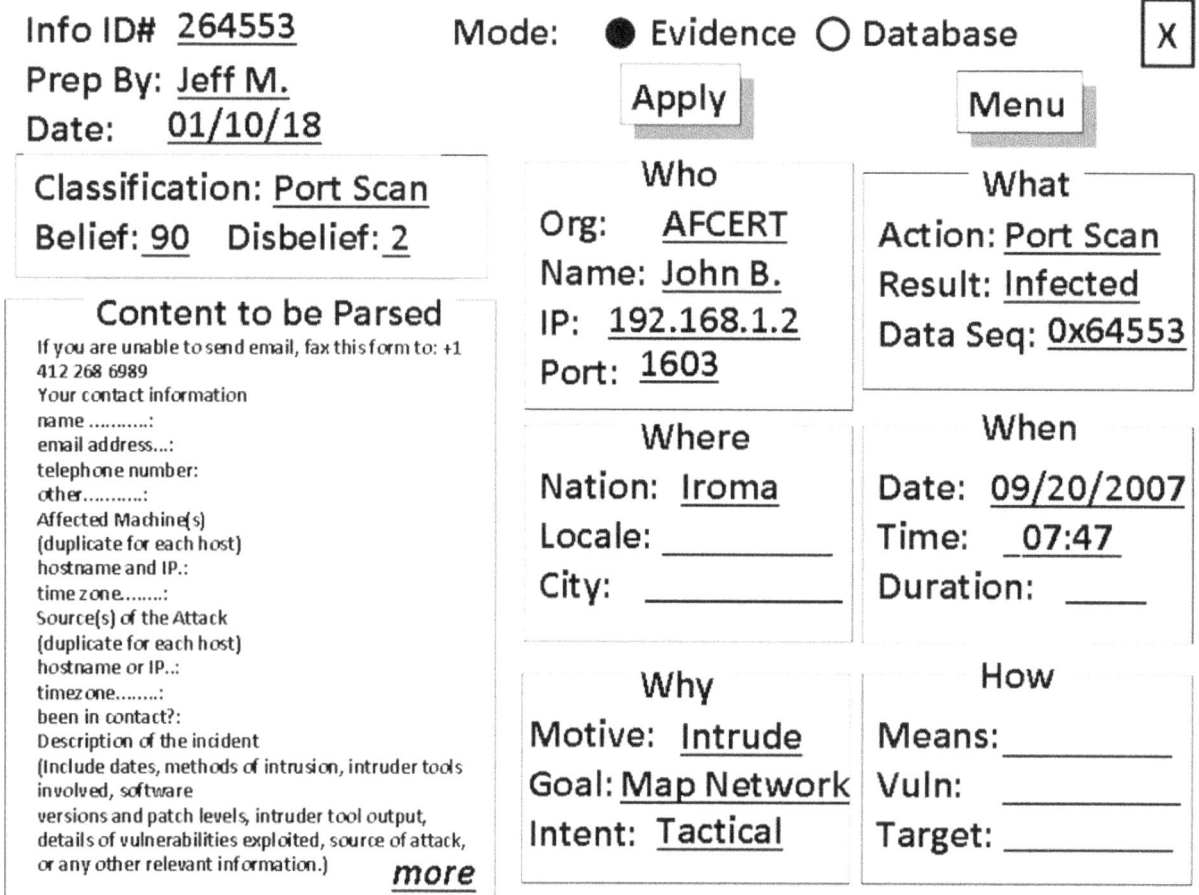

Figure II-2.5: Reporter's Questions Template

An internal representation of the evidence is also required. Extensible Markup Language (XML) is chosen because it allows the content to "know about itself". It also allows clean interfaces with the information extraction and data mining tools that already use XML for import and export.

Results with algorithm interfaces (Figure II-2.6) are now reported. Relying on the theoretical notion that any processing algorithm can be represented as a semantic network, the challenge is to convert the diverse representations of algorithms from various research "fiefdoms" to a semantic network formulation. A practical result is that meeting the knowledge base interface tends to suggest the underlying semantic network: concept nouns in the knowledge base are associated with hypotheses in the algorithms. Evidence in the knowledge base is associated with descriptive parameters required by the algorithms.

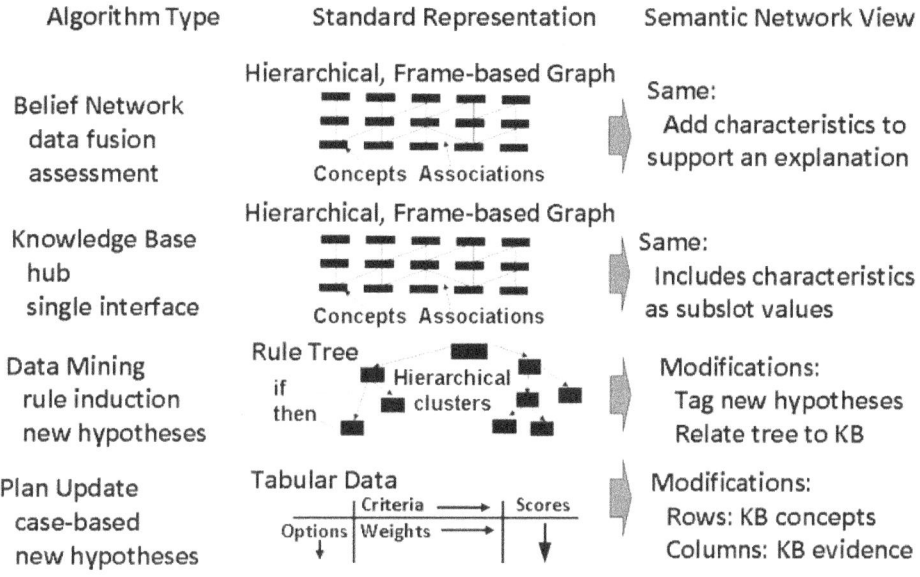

Figure II-2.6: Algorithm Interfaces

The Belief Networks[6] are directed graphs with hierarchical nodes that are connected with links. Nodes are hypotheses with characteristics. Links are directed from one node to another and signify the impact that one node has on another. Comparison of this structure with the description of a semantic network given earlier indicates that a belief network is a semantic network – so this is easy! We build a plug-in that imports needed information from the knowledge base and exports processing results to the knowledge base. More characteristics than are needed for the belief networks are extracted from the knowledge base – the additional characteristics allow the belief network algorithm to provide an explanation for the results and such explanations require a bit of context.

Information extraction algorithms feed the knowledge base. This information is automatically put into the Reporter's Questions template, the evidence format that the knowledge base uses. This interface is straightforward. This thread is discussed in more detail later.

Data mining algorithms (we use Weka, Subdue, and UCINet) vary widely in their import and export requirements. Weka[7] is a collection of data mining algorithms supported by a graphic user interface. We use a clustering algorithm for unsupervised learning, and a rule induction algorithm for supervised learning. Fortunately, all Weka algorithms use a common (.arff) format, which consists of a header record followed by data records that contain the fields, or

---

[6] https://www.e-education.psu.edu/drupal6/files/sgam/Evidence_Fusion_02SS_Talbot.pdf, accessed 02/20/2015.
[7] http://www.cs.waikato.ac.nz/~ml/weka/ , accessed 04/08/2015.

dimensions, to be mined. These dimensions are mapped to the Reporter's Question Template for insertion into the knowledge base. Subdue is a graph-based abductive reasoning algorithm – it reasons to the best explanation. It either accepts a graph or constructs one. Given a graph that represents our understanding of how the data is hierarchically organized, Subdue finds new hypotheses and organized them hierarchically, based on new evidence.

A significant challenge with the data mining applications (Weka, Subdue, and UCINet) is to export results to the knowledge base. The results consist of hypotheses supported by evidence and theoretically fit directly into the knowledge base. The issue is that all three tools only output plots, so we need to deconstruct the applications to get the content required to update the knowledge base.

Our Plan Update algorithm relies on case-based reasoning to "remember" a worthwhile "fix" to a plan that requires repair. Case-base reasoning is akin to an engineering trade study: Cases are options, indices are selection criteria, weights are used to rank the importance on the indices, and options are scored. The typical representation of content is tabular with rows of options and columns of selection criteria. The last column provides scores for each of the options. We convert this tabular representation to a semantic network by associating the options with hypotheses and the selection criteria with characteristics of evidence. This retains the linkage between the belief networks that perform mission assessment and the plan update algorithm that identifies repairs based on the assessment.

Summary

Overall, the semantic network concept provides a unified framework for interfacing all algorithms with one another through the knowledge base. To recap, the knowledge base is structured as a frame-based hierarchical ontology.

Chapter II-3
Knowledge Management

The previous chapter describes our use of semantic networks to structure knowledge. Here, we discuss how to manage it. As might be expected, since the knowledge structure has hierarchical stories (Figure II-3.1), the management of multiple knowledge bases is also hierarchical. We have multiple knowledge bases, each consisting of multiple stories. A story is implemented as a single semantic network consisting of nodes connected by links. Tools like Protégé load one story at a time.

A story fragment is a subset of nodes and links, perhaps representing an adversary's vulnerability (a.k.a. Center-of-gravity). Stories have evidence records that are stored as flat files and more recently as data services. Evidence records are linked to the text or structured data file from which content is extracted. This provides a drill-back from high level mission impact inferences to the raw text evidence. Stories have nodes that state hypotheses and these have degrees of [belief, unknown, disbelief]. Hypothesis characteristics are linked to nodes by the node name. Stories have links that define the impact one node has on another. Links have node values, which may be either supporting or detracting, for both belief and disbelief.

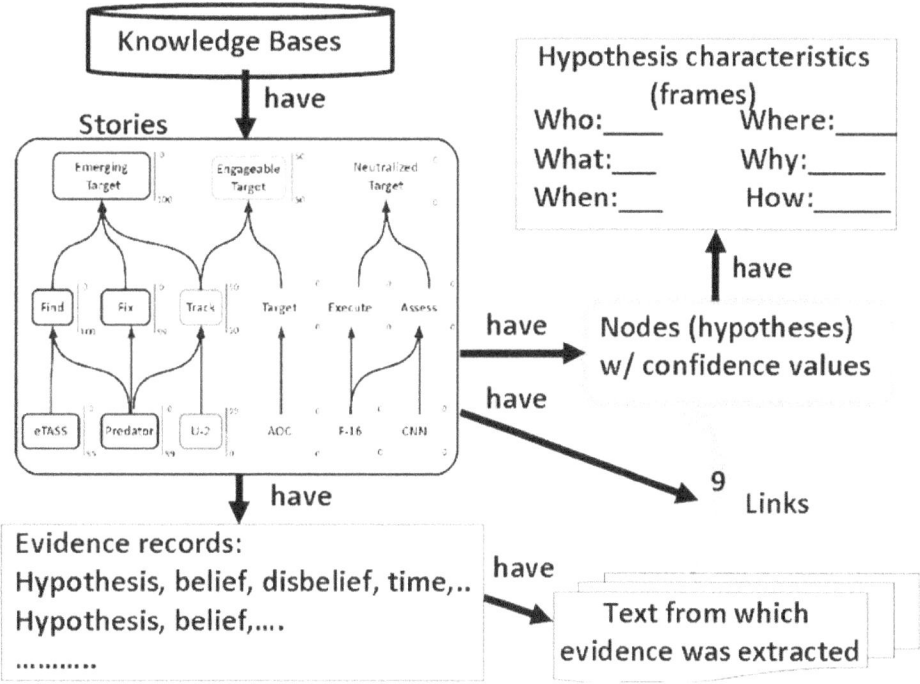

Figure II-3.1: Knowledge Structure for Multi-Strategy Reasoning

We have built many knowledge bases. Each usually serves a mission area within a domain. For example, we have a knowledge base for time critical targeting as shown above. A sample story is structured to help the user reason about whether a target is emerging, engaged, and neutralized. For the belief network shown above, we conclude that we are able to find and fix the emerging target with certainty, but currently have only a 50/50 chance of engaging it.

Associated with the outcomes is the need to determine whether we can find, fix, target, track, engage, and assess a target. The indicators are based on evidence from various sources at the bottom of the belief network.

Story Generation

A user working in a mission domain, say terrorist activity in Los Angeles, will have occasion to investigate many potential plots, will track many indicators, and will have many sources of evidence. It is therefore useful for this user to have multiple stories in play simultaneously – even multiple versions of a single plot, with some differing and some common evidence, indicators, and possible outcomes. In this chapter, we discuss how stories are constructed, combined, and sorted.

Dynamic Representation of Stories from Text

Stories change over time. New evidence is added, old evidence is de-rated. New indicators are identified, and unintended effects become evident. We treat stories as sets that can be operated upon by set operations such as taking the union or intersection (Figure II-3.2) to form a new story that may identify concepts that the originals had in common (intersection) or determine the combination of concepts from the originals (union).

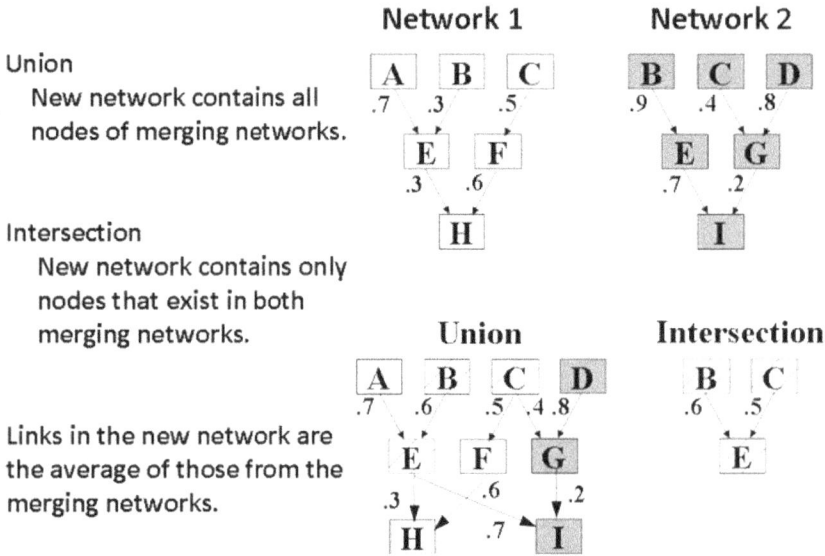

Figure II-3.2: Union and Intersection of Stories

Another useful operation on stories is to overlay an adversary story on a story associated with friendly forces. This produces a hierarchical exchange ratio, discussed later, for concepts common to both. Treating the belief network as a function allows useful mathematical operations. For example, we can differentiate the belief network; that is, we can perturb each node to determine the sensitivity on each node to changes in other nodes. Analysts find it difficult to efficiently share findings. Our concept (Figure II-3.3) allows textual information to be passed as an executable model representing a "chunk" of knowledge.

This graphic shows that a document is parsed using information extraction from a tool such as the General Architecture for Text Engineering (GATE) and put in an evidence template for classification using a semantic distance, AutoClass (Bayesian Classifier), or rules to tag it with a hypothesis and hierarchical level. The evidence, hypothesis and level are inserted into the knowledge base such as Protege and passed to an abductive data mining algorithm such as Subdue in parallel. Multiple algorithms, including Subdue and a Case-Based Reasoner add links and nodes to the belief network supplemented by analyst interaction and restructuring of levels as necessary.

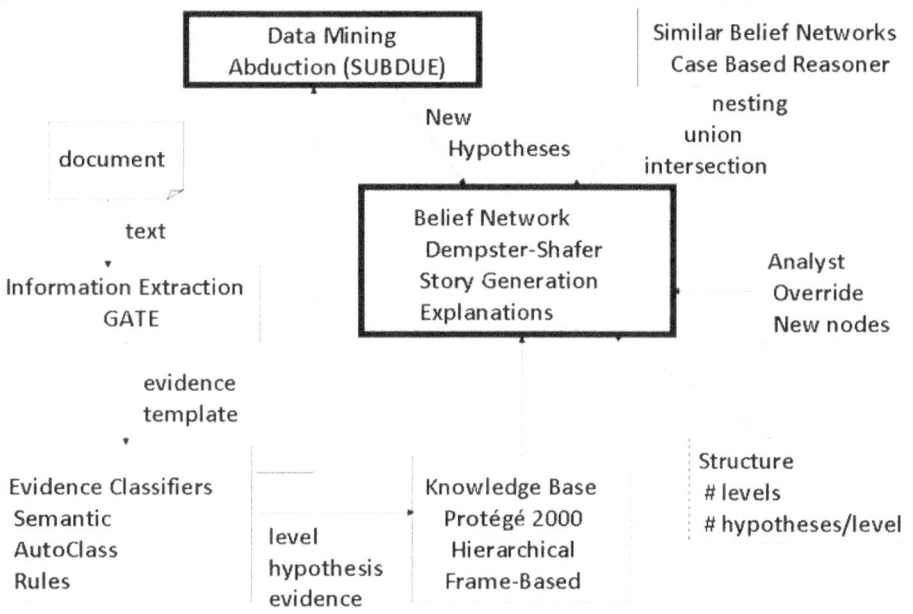

Figure II-3.3 Automatic Generation of Stories from Text

Chapter II-4
Executable and Self-Aware Knowledge Bases

Knowledge Management via stories is accomplished using an hierarchical, characteristics-based ontology. This has the same structure as our general purpose evidence fusion engine. Consequently, we combine the two technologies (Figure II-4.1) to make an executable knowledge base with a self-aware domain. The self-aware domain of the knowledge base consists of a self-diagnosis module, self-aware stories, and static content. Self-aware stories are structured as belief networks with hypotheses and links that are driven by evidence. Data mining tools and analyst interaction modify hypotheses and links. A reconciliation module deletes redundant links. Text, rules, and wordlists produce evidence from text.

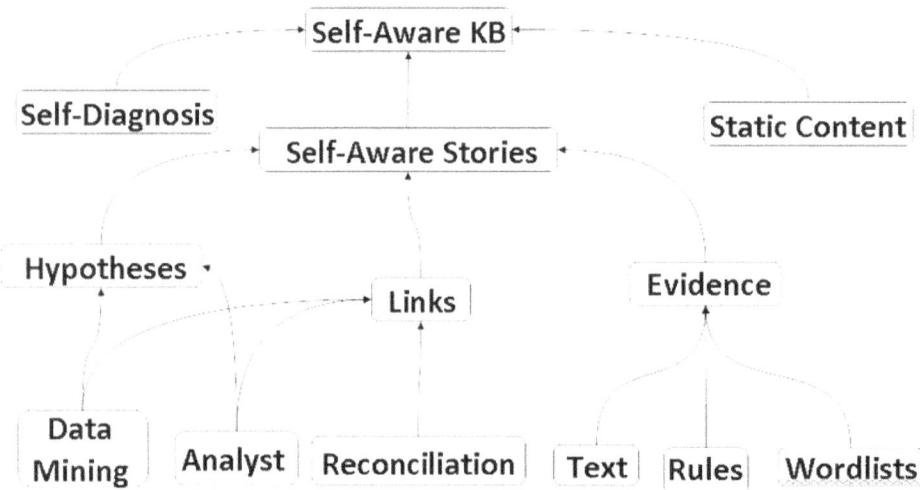

Figure II-4.1: Self-Aware Knowledge Base Ontology

A first step is to build a "self" domain. It organizes and reasons about concepts such as: who am I, who built me, who do I know and work with, what am I, what can I do, what can't I do, what is my matrix of states, what are the values of those states, what additional information do I need, what is my task list, what are my priorities, what are my goals, what should I do next, where do I reside, where am I connected to, when I built, when was I last updated, why am I built, why do I interact in various domains, why did I answer a query in a particular way, why can't I answer a query, how do I do my job, how well am I doing, how accurate is my content, how old is my content, how connected is my content, how am I maintained, how busy am I, how do I know what I know....The self domain will interact with mission domains to answer these questions using static information and dynamically computed content.

Many of these tasks are reasonably straightforward because a computer application, as compared to a human who has sub-symbolic state information, popularly characterized as residing in neuron pathways, has an explicit representation of state; that is, it can answer the questions above based on static data and access to dynamic data. Other, more reflective tasks are more difficult. Tasks for the self-aware knowledge base are prioritized based on expected payoff for a "smarter" content repository; for example, one with self - diagnostic ability.

The technical approach again relies on the Reporter's Questions, in this case for the self domain. This is a meta-domain: it has knowledge about knowledge. A sample of the Reporter's questions with either static or computed answers (bold) follows:

<u>Who</u>

- Who are you? **A self-aware domain of domains called Meta**
- Name? **Meta**
- Who built you? **Dennis Ellis**
- Who do you work with? **ARIES Team and users**
- Who are your current users? Technologists, **RAIDRS, Clockworks, UM, C2OTF, and other IR&Ds**
- Who are you related to? **Other knowledge bases such as DOTS, TUCANA (free-ware version is Kowari), and databases extended to knowledge-based ontologies (MySQL) are related.**

<u>What</u>

- What are you? **Self-aware knowledge base that knows what it knows**
- What are potential uses? **Meta-domain inference, Creative computer reasoning, reasoning by analogy, concept fusion, identifying emergent behavior, automated discovery of unknown unknowns, simulation-based explanation, creative web browsing, guess and check problem-solving, creative engineering solutions, and constraint relaxation.**
- What is the underlying application? **Protégé**
- What can you do? **Answer questions about knowledge base domains, answer questions about plug-ins to the knowledge base, launch applications based on questions, update answers based on the dynamic state of the knowledge base, and perform self-diagnostics.**
- What can't you do? **Replace an analyst's common sense reasoning**
- What are your domains? **Space, Cyber, Climate, Command and Control, Intelligence, Retail,..** (dynamic list)
- What interaction occurs among and within domains? **Domains are like books on a shelf talking to one another. Data mining and data fusion tools operate across domains. Within a domain, constituent stories are simultaneously updated with evidence. Stories can also be combined through union and intersection, nested, and overlaid.**
- What do you know? **Information about the stories in the knowledge base, and plug-ins to the knowledge base, including information about underlying algorithms, displays, ontologies, and architectures.**
- What are stories? **Augmented belief network, constructed of links and nodes. Links and nodes have characteristics that answer detailed Reporter's Questions, such as these, and are driven by evidence.**
- What state information? **State information about the application, the knowledge domains within it, the stories in each knowledge domain, and the hypotheses, links,**

evidence, and explanations that form a story.
- What application information? **Application information about the size of the underlying Protégé code, runtime performance (CPU time, storage, and memory, and a visual view of the screens. (dynamically updated)**
- What information about the knowledge domains? **Domain information about the number of domains, and summary information about the number of stories in each domain** (dynamic list).
- What information about stories? **Names, display showing "best" stories, hierarchical lists of hypotheses, matrices of links, views of evidence, links to explanations.**
- What information about hypotheses? **Names, Stories containing them, domains containing them, Degrees of belief, disbelief, and ignorance; time histories, associated evidence, and associated explanations.**
- What information about links? **Name of hypothesis that influence another hypothesis, name of hypothesis being influence, weights of influence for belief and disbelief.**
- What information about evidence? **label, hypotheses supported, stories supported, domains supported, link to source, source, parsed words – including hedge words**
- What information about explanations? **Label, link to source, hypotheses supported, stories supported, domains supported.**
- What information is static? **Explanatory information that doesn't change as domains, stories, hypotheses, links, evidence, and explanations change**
- What information is dynamic? **Explanatory information that does change as domains, stories, hypotheses, links, evidence, and explanations change**.
- What are your goals? **Provide an intuitive user interface to the content of the knowledge base, answer FAQs, dynamically update information about the content of the knowledge base.**
- What is your Version Number? **3.4** (dynamically computed)

When

- Last User Update? **Date/time group** (dynamically computed)
- Last Developer Update? **Date/time group** (dynamically computed)
- Now? **While responding to this question, the following processes are executing ___,___ (dynamic list). The following users are interacting with the knowledge base: _____,____ (dynamic list). The following domains, stories, hypotheses, links, evidence, and explanation are or have recently been modified (dynamic, hierarchical list with timestamps).**
- Recently? the following processes were recently executed ___,___ (dynamic list). The following users recently interacted with the knowledge base: _____,____ (dynamic list). The following domains, stories, hypotheses, links, evidence, and explanation have recently been modified (dynamic, hierarchical list with timestamps).
- Past? **Historical summaries of the knowledge base, domains, stories, hypotheses, links, evidence, and explanations**
- Next? **Task list pending processing is empty** (default – replace with dynamic list)
- Priorities? **Processing is first-in, first-out, with no priority processing**

Where

- Where are you? **On a Windows laptop, serial number 11xxxxxx. Also accessible from the website, IP address: 158.114.xx.xxx.**
- Where are you connected: **IP address: 158.114.xx.xxx.**
- Where have you been deployed? **Air Force Information Warfare contract, Effects-Based Information Operations Proposal, Technology Demo, Space Defense Contract – Company Solution, Information Operations Research & Development, Offense Defense Integration test bed, Command and Control of the Future demonstration, Uncertainty Management Reseach Task, Advanced Concepts Demonstration Laboratory**

Why

- Why was this Self-Aware Knowledge Base built? **To answer questions about knowledge base domains, to answer questions about plug-ins to the knowledge base, to launch applications based on questions, to update answers based on the dynamic state of the knowledge base, and to perform self-diagnostics.**
- Why does it interact with various domains? **To find patterns, common story fragments, and larger contexts.**
- Why are queries answered in a particular way? **Static queries are answered based on "remembering" underlying developer input, and dynamic queries are answered based on calculations of parameters using the current state of the knowledge base**

How

- How are answers assembled? **For static information, by matching questions to similar questions with answers (Case-Based Reasoning). For dynamic information, by matching questions to similar questions that call a calculation method.**
- How aware is the knowledge base? **Metrics are collected for Answer accuracy, answer relevance, activity understanding, currency of static answers, currency of dynamic calculations, accuracy of self-diagnosis, and usability of the meta domain.**
- How connected is knowledge base content, domain content, story content, hypotheses, evidence, and explanations? **Click on a hyperlink to get the link analysis of belief network answer.**
- How busy is the knowledge base? **Click to see history of: CPU usage, Memory usage, storage usage**
- How well maintained is the Knowledge Base? **Click here to get build history** .
- How do you know what you know? **By remembering and computing.**
- How certain about knowledge base overall, domains, stories, hypotheses, evidence, explanations? **Click on hyperlink to find out.**

No Match

- Wow, that's a hard question, I don't know. Please call Dennis Ellis at (___) - ___-

____ for the answer.
- Please rephrase, using simpler words.

Partial Match

- Can you please rephrase the question?
- What's available is,_____
- Not sure, but _____

Part III
Algorithms for Computer Reasoning and Learning

Chapter III-1: Data Triage
Chapter III-2: Information Extraction
Chapter III-3: Hypothesis Association
Chapter III-4: Threat Discovery
Chapter III-5: Social Network Analysis
Chapter III-6: Analogical and Terrain Reasoning
Chapter III-7: Genetic Reasoning and Plan Optimization
Chapter III-8: Evidential Reasoning
Chapter III-9: Inductive Reasoning: Generalizing from Data
Chapter III-10: Deductive Reasoning: Rules and Logic
Chapter III-11: Abductive Reasoning to the Best Explanation
Chapter III-12: Reasoning with Physics: A Few Astrodynamic Algorithms

---

Part III
Introduction

The next collection of Chapters deals with the many types of reasoning. Emphasis is on when these different inferences are appropriate, how users and computers interact, and how performance is measured. The sequence of Chapters follows a typical mission thread:

- Data Triage (Chapter III-1) segments, prioritizes and filters unstructured data
- Information Extraction (Chapter III-2) uses natural language processing to parse data
- Hypotheses are associated (Chapter III-3) with parsed information
- Threat Discovery (Chapter III-4) determines the number and types of unseen threats.
- Social Network Analysis (Chapter III-5) defines threat network connectivity.
- Reasoning by analogy (Chapter III-6) remembers preplanned options and optimizes routes
- Genetic reasoning and plan optimization (Chapter III-7) find best pre-planned options.
- Evidential reasoning (Chapter III-8) interleaves planning and combat assessment.
- Inductive reasoning (Chapter III-9) discovers general rules in structured data
- Deductive reasoning (Chapter III-10) provides rules of engagement for mission execution
- Abductive reasoning (Chapter III-11) uses machine learning to cite the best explanation
- Physics algorithms (Chapter III-12) are used for predictive awareness.

These automated reasoning techniques provide a foundation for multi-strategy reasoning (Part IV). The goal is to help the user to make better decisions in complex mission environments.

Chapter III-1
Data Triage

In this chapter we present a wavelet-based analysis technique to assist in coping with information overload. Specifically, this method is applied to segmenting, classifying, and filtering unstructured text. Segmentation allows unstructured text to be divided into smaller, more manageable sections. Classification extracts the subject keywords from each segment, and filtering assigns a priority value to each segment. These three processes are performed using the wavelet transform of signals formed using the information content of text and provide an automated means of reducing large amounts of unstructured text into segments prior to information extraction or other data processing.

Coping with a large amount of unstructured text is an increasing difficult problem in the field of data processing. While we regularly perform information extraction, when large amounts of text are present, this processing yields results which cannot be easily understood. In addition, unstructured text poses a unique problem, because it lacks identifiers or keywords. This adds an additional layer of work for the data extraction tool, since it must identify key information without any user input. In this paper, we present a new method to segment, classify, and filter unstructured text.

Segmentation is a process in which large amounts of text are decomposed into smaller, more manageable segments. A book is logically divided into sections called chapters. These divisions are made by an author, because of some difference between the various sections of text. What if we need to further divide the text into areas defined by content? This is the heart of segmentation.

We develop a process to segment text. Data Triage uses an original algorithm, which was initially developed in Matlab and Perl, and then ported to Java while adding additional functionality. The Java platform offers the advantage of platform independence, easier usability, and significant speed increases. Segmentation is divided into three parts: generating the signal, smoothing the signal, and segmenting the signal.

<u>Generating the Signal</u>

Prior to segmentation, we transform text into a numerical signal. This is accomplished using WordNet, a lexical database provided by the Princeton University Cognitive Science Laboratory[10]. Before we use WordNet's database, every word in the unstructured text must be tagged with its part of speech. In our current implementation we use the MontyTagger API[11], a "commonsense-informed part-of-speech tagger," developed at the MIT Media Lab[12], to tag words in unstructured text. While this is not the only available speech tagger, it is one of the best for interfacing with a Java program.

Once the text is tagged, we submit words and their corresponding parts of speech to WordNet.

---

[10] http://wordnet.princeton.edu/, accessed 01/21/2015.
[11] http://web.media.mit.edu/~hugo/montylingua/ , accessed 01/21/2015.
[12] http://www.media.mit.edu/, accessed 02/20/2015.

It returns either the hyponym (a word whose meaning is included in the meaning of another, more general, word)count or the synonym count, depending on the submitted word's part of speech. These counts helps in determining where a topic break occurs, because words with the least meaning have the most hyponyms (hypo) or synonyms (syns).

The sequence of information content values for each word in a sequence of unstructured text forms the text signal, which is used for all further processing.

Smoothing the Signal (De-Noising)

The text signal obtained in the signal creation process described above is a very irregular signal, where every maximum represents a lack of information. Because content tends to clump in language, the lack of information points represents content breaks. However, we are only interested in some of the content breaks. By limiting the number of content breaks we limit the "granularity" of the representation. Put simply, high granularity gives more topic breaks and a finer view of the text, while low granularity gives less topic breaks and a summarized view of the text. We change the "granularity" of the analysis using the discrete wavelet transform (DWT). This transform is a set of filters which splits the original signal into two new signals. The first signal contains the details (high frequency information), and the second contains the averages (low frequency information). The averages represent a smoothed version of our original text signal. We can continually decrease the granularity of the analysis by taking more DWTs of the signal. The level of granularity in processing can be controlled by our application.

Segmenting the Signal

With the signal smoothed to the desired granularity, the text is segmented at its topic breaks. However, we do not just simply split the text at the exact content-break location. We dictate that a content break cannot occur in the middle of a sentence, so all content breaks are moved either to the nearest beginning or end of a sentence. While this is not exactly accurate to the true content breaks, it makes sense grammatically. Rarely, if ever do topics change within a sentence.

In our Data Triage application, the segmented text is displayed on the screen for the user. The algorithm breaks previously clumped text into multiple segments; these segments are delineated by paragraph breaks.

Classification

With the text decomposed into topic segments, we then add a "classifier" or "identifier" to each segment. This is a word or words that best describe the subject, or main topic, of each segment. A number of methods to determine this classifying word are available (e.g. a Bayesian Classifier or latent semantic analysis). However, we choose a fairly simple, yet effective, method, which uses the smoothed text signal.

As previously described, WordNet assigns a number to each word based on how much information content it contains. Therefore the word with the largest corresponding value would, with respect to content, be the most important. However, not every part of speech will prove helpful when trying to identify the subject. For example, the verb "is" could theoretically have the highest WordNet value, but the word does not shed light on the subject. In light of this, we choose to restrict the classifying words to nouns.

An underlined word or phrase preceding the text segment is the output from the classification stage. As mentioned before, the key words, in italics and underscored, are the most important words in each segment; they have the most information content as defined by their number of synonym and hyponyms and our information content calculation.

Since it has very little computational overhead, the classification step is fast in comparison to the rest of the program, because we have already gathered the WordNet signal values and calculated the information content of each word in the program's segmentation phase. Therefore, we just have to identify the largest signal values of each segment and find the corresponding word or words. These words classify that segment.

Filtering

Currently we have only one filtering scheme, and like the classification method, it filters based on the calculated information content value. This filtering scheme weights the segments by how important they are to the document, or how many context bearing words a segment contains.

In order to eliminate a larger segment's dominance over a smaller segment, we take the average of all the WordNet values. The segment with the highest average value is then deemed the most important, or more correctly, the segment with the most content-bearing information. The example below shows a small sample of the filtering output. Note how a number has been added to each segment. This number corresponds to the segment's overall topic information.

Shuttle Text:

20 *Launch*: Despite ongoing troubleshooting throughout the weekend and into Monday, NASA engineers still have not reached any firm conclusions as to what's wrong with a faulty hydrogen fuel sensor that prompted the space agency to scrub the shuttle Discovery last Wednesday.
14 *safety*: But, as the first space shuttle mission since the Columbia disaster nearly 3 years ago comes down to the wire, NASA officials reiterated their determination not to let "launch fever" compromise safety.
25 *sensors*: Discovery's crew was on-board and strapped in on July 13 when one of the four redundant engine cutoff or "ECO" sensors malfunctioned. The faulty sensor indicated the shuttle's external fule tank, which had been filled just hours before, was running dangerously low on liquid hydrogen propellant.
24 *liftoff:* Though the three others sensors were functioning normally, and only two are needed to achieve orbit, NASA scrubbed the launch because procedures require all ECO sensors to be operational for liftoff.
11 *conditions*: NASA sources familiar with the ongoing engineering discussions told CNN that

shuttle managers are considering several options on how to proceed. Engineers could decide to do another "tanking test" in which super-cold liquid oxygen and hydrogen fuel would be pumped into the external tank to see how the tank responds to launch – day conditions.

21 *tests*: NASA conducted two tanking tests earlier this spring, with the hydrogen fuel sensor problem cropping in the first but not the second.

Scalability Testing

The key to any algorithm is speed. While any task can be accomplished with an unlimited amount of time, data processing algorithms need to be performed quickly and preferably in real-time. With Data Triage, the number of words or tokens (including punctuation) drives the speed of our computations. This first analysis involved over 100 documents run in series to determine the relationship between the number of tokens and the time to run the computation.

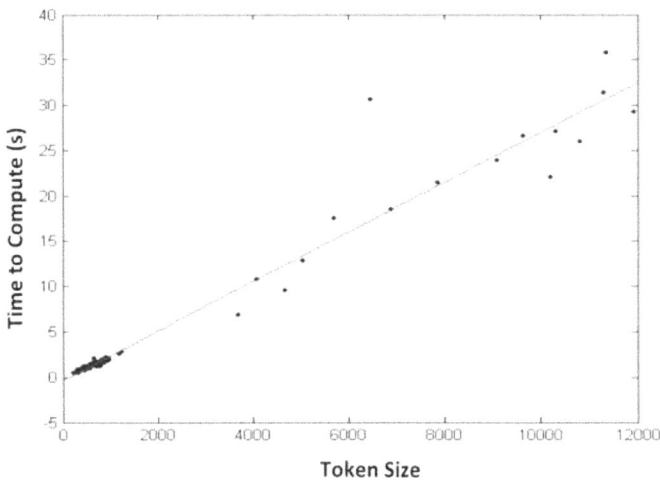

Figure III-1.1: Document Size Scalability

In Figure III-1.1 excluding a few outlying data points, the relationship has a linear characteristic, which is very good for scalability. The Data Triage application's speed will always have a linear relationship to the size of the document being processed. In other words, the Data Triage application can easily and efficiently handle large unstructured text documents.

While doing this initial analysis, we also assess the individual program processes to see the percentage of time each process takes. This analysis indicates which parts of the program are taking the majority of the processing time. Segmentation dominates classification and filtering. In light of this, we further analyze the individual segmentation processes. The part-of-speech tagging process takes the majority (74%) of the segmentation processing time. Since this task is performed by the MontyTagger API, for which we don't have source code, it isn't possible for us to optimize this process short of creating our own part-of-speech tagger.

We see that processing time increases with both level granularity and number of classification words. Although, this time increase in negligible when compared to the time increase associated with the number of tokens processed. With each addition level, the transform only processes half of the previous level's signal values (dyadic down-sampling); therefore, each additional level takes half the time of the previous level. This provides for a converging effect for the computation time as the number of levels increases.

Level of Computation	Level Computation Time	Total Time
1	T_0	T_0
2	$T_0/2$	$T_0 + T_0/2$
3	$T_0/4$	$T_0 + T_0/2 + T_0/4$
4	$T_0/8$	$T_0 + T_0/2 + T_0/4 + T_0/8$

In short, even if an infinite number of levels are calculated, that total calculation time is never greater than twice the calculation time of the first level.

Accuracy. To determine whether the Data Triage application correctly segments, classifies, and filters text, we have an expert segment, classify, and filter twenty text files by hand. When this is complete, another expert compares the hand segmented text with the Data Triage application's output. The results of this experiment follow.

Segmentation. We first compare the segmentation of the expertly segmented text to the application segmented text of twenty files. We count the number of times a segment appears in exactly the same place in both segmented documents, when an application generated segment appears within one sentence of the expert's segment, and when an application generated segment appears within two sentences of the expert's segment. This experiment performed with the application segmentation level set to both three (sentence level segmentation) and four (paragraph level segmentation). Segmentation accuracy is shown in the Table.

	Segmentation Level 3	Segmentation Level 4
Exact	43.251 %	45.413 %
Within 1 Sentence	82.129 %	83.707 %
Within 2 Sentences	90.690 %	91.182 %

Future Exploration

There are a number of avenues to expand this application. Most notably is in the classification and filtering sections. To improve the classification, a learning algorithm would supplement WordNet. Further, our classification algorithm sometimes returns words or numbers that do not help classify the document. With a learning filter, we could easily recognize these words, remove them, and search for more helpful classification words.

Currently the filtering section is quite limited; however, we could implement a wavelet based search or sorting algorithm using the idea of a query's energy to assist us[13]. Put simply, we could transform a query using WordNet and wavelets to find the queries bearing content words.

---

[13] http://users.cis.fiu.edu/~taoli/pub/home.html, accessed 02/24/2015.

We could then either search the documents, or search the keywords for these content bearing words.

In addition to modification to the automated data triage itself, we would also like to interface this application with our information extraction tools. Done as a preprocessing step to information extraction, we believe data triage could significantly increase the performance and results of our information extraction tools.

Conclusion

Given that the need to process unstructured text is an evolving challenge in data mining, any method to give context and structure to text is of great benefit. Data Triage gives a logical structure to the text by splitting the text at its topic breaks. Furthermore, Data Triage provides a means of classification and filtering these newly formed segments.

Chapter III-2
Information Extraction

Introduction

We set out to discover a method for taking information from text and converting it into an evidence form called the Reporter's Questions Template. In the process of looking for a way to accomplish this task, we come across a technology called Information Extraction. Information Extraction is exactly what we are looking for. This area is well defined and has been around since the early 1950's, but only recently has the topic turned into an applicable product for use within broad domains. Once we discover the underlying theory, we start researching the actual computer programs available to the public. The number of applications available leads to a trade study of these products. The best is chosen and studied more closely. This chosen product is then refined to interact with an evidence form. To have it interact with the form, we create new modules that go beyond the standard packaging of the product. We also add a module that uses hedge words to capture the confidence associated with the information being extracted. Finally we discuss how the product interacts with the form and the modifications needed for current use of this tool.

Information Extraction

Today we have thousands of text messages stored in databases, scattered everywhere. Information Extraction is designed to help convert these texts into a more usable form for processing by the computer. Information Extraction (IE) is the process of taking words from the body of text and putting them into a template. Basically it is a way to have the computer answer questions using information from a text. The steps (Figure III-2.1) to do IE are: tokenizing, sentence splitting, part of speech tagging, named entity tagging, co-reference, and template filling.

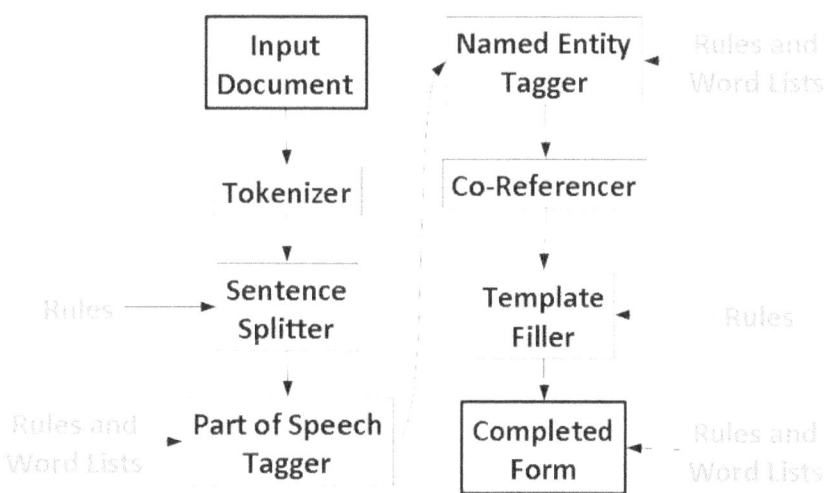

Figure III-2.1: Information Extraction Flow Diagram

Tokenizing is the process of taking a text and making a distinction between words and spaces. After that has completed, sentence splitting tags the different sentences. Sentence splitting doesn't just find the punctuation and declare the words between them a sentence. It also has to differentiate between abbreviations and title punctuation. Next is part-of-speech tagging. This process defines the part-of-speech each word belongs to; that is, noun, verb, or pronoun. This process takes into account the surrounding words to determine which part of speech the word is.

Name entity tagging identifies words belonging to predefined categories such as persons, quantities, organizations, time, and others. The name entity process uses word lists and rules to define what words go in which groups. Then, the co-reference process attempts to match pronouns within the text to nouns to which they refer. Finally, template filling is the process of extracting answers to the Reporter's Questions from the text and entering them in a template.

To measure the accuracy of Information Extraction, the IE community has defined metrics. They take into account Precision, the fraction of extracted words that are relevant, and Recall, the fraction of relevant instances that are achieved. The F-measure is often used in conjunction with Precision and Recall, as a weighted average of the two. This gives a percentage of accuracy. Researchers test and measure three processes using this system: part of speech tagging, name entity tagging, and template filling. The first two, part of speech and name entity, generally score around 80% - 90%. To compare that to a human, people generally score around 87% on all processes. Template filling, on the other hand, generally scores around 60%. Currently[14] the focus in Information Extraction is to raise the template filling number to be closer to 90%. The accuracy of these capabilities can be increased if the domain is very narrow. How narrow, depends on the user of the application. If we allow a larger number of errors to occur then our domain allows more sources for processing. But, if we want accuracy over the ability to process a larger domain, then we narrow the domain down to one source. Another way to increase accuracy is through the number of rules and the size of the word lists that are used. But using this technique has its own problems. These problems include the speed at which the program can process all of this information and the time involved in creating all the rules and finding all of the words that pertain to a certain domain.

Based on the results of a detailed trade study, we chose to work with the General Architecture for Text Engineering[15] (GATE). It is a development environment for creating modules that parse text. Researchers at Sheffield distribute GATE with the basic modules: tokenizer, sentence splitter, part of speech tagger, name entity tagger, co-reference, and a lot of other modules. However, it does not come with a module for template filling. GATE executes by loading the included modules or by incorporating the user-created modules. When GATE is executed, the modules must be in a certain order for it to properly parse the text. The text is imported in almost any format, but internally the text is handled as an .xml document. The tags are contained within the .xml when GATE is done processing. The user either displays the text with the tags highlighted or saves the document in .xml format. GATE also has a tool that allows the user to see how accurately it tagged a text. The tool takes an untagged text and the same text with tags from another source as the baseline and then tags the untagged text and compares it to

---

[14] https://web.stanford.edu/class/cs124/lec/Information_Extraction_and_Named_Entity_Recognition.pptx+&cd=1&hl=en&ct=clnk&gl=us, accessed 02/24/2015.

[15] https://gate.ac.uk/ie/, accessed 02/24/2015.

the tagged text. Then GATE displays the difference. Finally, the user can modify the modules within GATE, or the user-created modules, to increase parsing accuracy.

Hedge Words and Template Filling

GATE's ability to incorporate user-created modules is one of its prominent features. We have created a few modules. The first one is based on a concept of hedge words. Hedge words are the words that people use within a text to emphasize their degree of belief (or disbelief) in a particular fact. The idea is to make a word list of these hedge words and assign a value of belief (or disbelief) to that word. Within a given text, hedge words give the program some idea of how the person felt when they wrote the document. The program assigns a value of belief (or disbelief) to the passage. To see if this is a viable way to determine the belief in a text, we take the mean value of the hedge words contained within that text. In testing this module we conclude that one cannot judge the belief of a text solely on the words within the text. When people believe or disbelieve a text, they not only take into account the words but also the source, their own past experience with the topic of the text, and also their belief in the source. In other words, people put their knowledge of the world behind the belief in a given text. Here are examples of beliefs associated with hedge words.

Absolutely	1	Actively	0.78	Actually	0.76	Admittedly	0.89	Almost	0.83
Apparently	0.93	Certainly	1	Characteristically	0.88	Already	0.97	Conceivably	0.78
Conceptually	0.75	Hardly	0.23	Hopefully	0.43	Clearly	1	Conceptually	0.75
Eventually	0.89	Inevitably	0.67	Might	0.52	Ideally	0.5	Impossible	0
Likely	0.72	Maybe	0.54	Perhaps	0.4	Normally	0.67	Occasionally	0.51
Possibly	0.6	Predictably	0.7	Presumably	0.45	Probably	0.55	Rarely	0.23
Supposedly	0.64	Theoretically	0.48	Unlikely	0.03	Strongly	0.98	May	0.50

Note that the current interest[16] in fine-grained sentiment analysis is readily accommodated in GATE using the same technique described above for hedge words.

The other module we created does template filling. This module is simplistic compared to the conceptualized version of template filling. The module works by use of a combination of word lists and rules. It also uses the existing name entity tags that GATE provides. The module runs after a set of the GATE modules. When the module runs it first tries to get a more detailed description of the tagged words from the GATE modules (detailed answers to the Reporter's Questions; for example, who is the threat, who is the evidence provider, who is the analyst) through the use of rules. Then the module uses its own word list and rules to find other words that may fit a particular field. Once it is done tagging the words, the module goes through the text and pulls out the words in the order defined by the template and formats the output for the evidence form.

The evidence form (Figure II-2.2) is a standardized input to a knowledge base. It has slots that accept values that answer a tailored set of Reporter's Questions (who, what where, when, why, how and how certain). The evidence helps the knowledge base to explain conclusions drawn

---

[16] http://aclweb.org/anthology/I/I11/I11-1038.pdf, accessed 9/24/2014.

about the topic of interest..

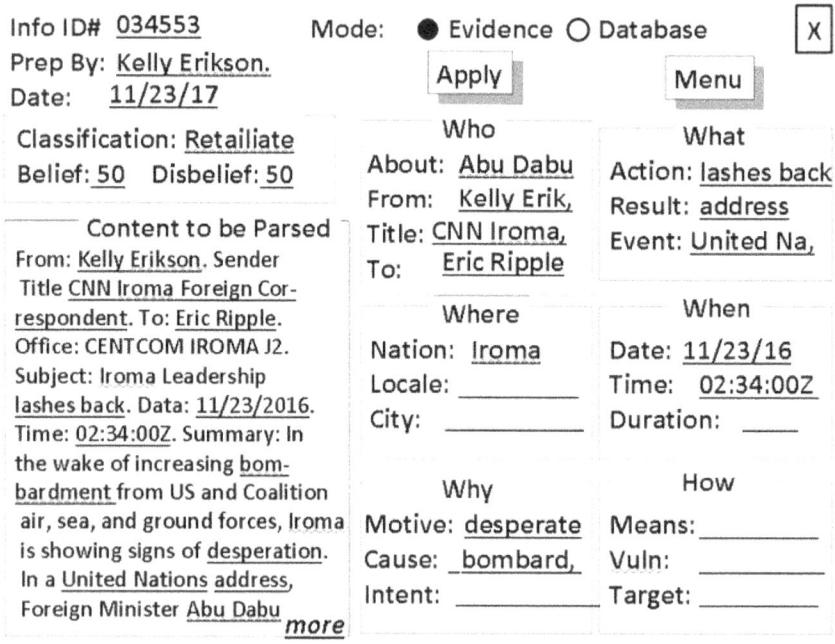

Figure III-2.2: Information Extraction with Underlined Tags

Our domain is intrusion detection analysis, so the evidence form is tailored to the characteristics of packets sent over the Internet.

In testing GATE with the modules that we create and the interaction with the evidence form, we decide that GATE needs to have supervision from an analyst. This front end to GATE hides the application from the user to make getting a tagged document a lot easier. Also it only suggests to the user what might best fit into a particular field, but the user still has to make the decision of whether the suggested information should be there or should have more words to describe the field or something completely different. This front end also shows the text with the tags highlighted for the user to quickly scan the text and make quick decisions. This tool speeds up the analyst process of taking a text and converting it to the evidence form. After the analyst has edited the fields, he sends the information to the evidence form via message passing over TCP/IP, or has the option to save it as a file. We envision the ability for GATE to learn correct tags based on operator interaction.

Conclusion

The work described in this chapter shows that GATE is a powerful tool for Information Extraction. GATE and its associated library of modules are capable of demonstrating and teaching the concepts of IE while giving the user the ability to create their own modules, as we describe. The modules that we create extract the needed information and put it into the evidence form. These modules, being the Hedger and Template Filling modules, use a combination of rules and word lists to accomplish this task. Accuracy is not high enough to automatically fill a template. This is the reason for creating an Assisted Evidence form for analysts. This form displays the possible words that match a given field and puts the decision making in the hands of a human while allowing the computer to learn from this interaction.

The integrated package illustrates how everything works together. Text is input into GATE, which tags the content associated with the evidence form. The tagged words are automatically input into the Reporter's Questions Template and provided to the analyst for review. The resulting content is entered automatically into the knowledge base.

Chapter III-3
Hypothesis Association

We are faced with a proliferation of data. This information must be grouped and organized before any type of analysis can be performed. Doing this manually would be a daunting task. We are interested in an automated procedure that will allow for the expedited classification of textual documents. The methods presented here provide a simple, innovative way to perform hypothesis association via text classification. Using a rule based classifier, a Bayesian classifier, and a semantic classifier; we are able to automatically tag a document with its underlying concept. Each of the three techniques requires a specific set of tools. The rule-based classifier relies on a set of simple inductive rules for the tagging process, the Bayesian classifier relies on a corpus of training data, and the semantic distance classifier relies on two semantic distance algorithms and a knowledge base. The two semantic algorithms used are semantic mean and semantic mode. Through the use of all three classification schemes, we are able to provide an accurate model of the document's conceptual content while avoiding the pitfalls of each individual technique.

Introduction

When dealing with data analysis, it is often imperative that incoming information is grouped in a meaningful way. Organizing all the data by hand is both time and cost prohibitive. We need an automatic system for information classification. In order to group like data, an understanding of the content is necessary. For humans, this is a trivial task. We do this when reading the newspaper, talking to a friend, or even thinking to ourselves. Unfortunately, machines are very bad at this sort of thing. They cannot simply collect and synthesize information in the fashion that we can. This makes the automation of the process a bit tricky. We are looking for an effective way to force the computer to simulate content recognition and classification. Specifically, we are looking for a simple and efficient way to classify incoming text. Often, we are given a deluge of information with no real idea of what is contained in the collection. Instead of sorting through by hand, we want the computer to automatically tag each piece with the appropriate concept.

We implement an innovative way (Figure III-3.1) of coupling a piece of text with its underlying concept. Each document is parsed into a set of comprehensive fields. The templates are then analyzed using a combination of rule based (heuristic) classification, Bayesian classification, and semantic distance classification. If a rule-based classifier is used, the template values are individually evaluated against a set of user-defined rules to find any directives specifying a value-concept pair. If rule-based classification fails, a Bayesian clustering scheme is employed to find a probability of correctness of the document's concept. If the Bayesian classifier fails to provide a sufficiently high probability of correct classification, a semantic classifier is used, the templates are mapped into a hierarchical concept space and two semantic algorithms are applied. Once tagged with the appropriate concept, the templates are inserted into a domain specific knowledge base.

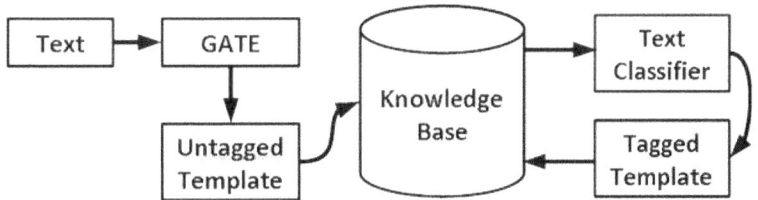

Figure III-3.1: Hypothesis Association Overview

We use a tool called AutoClass[17] for the Bayesian classification scheme. This tool takes a document and uses a naïve Bayes classifier to probabilistically determine the class to which the document belongs. Although effective, this method is time intensive and it requires a set of statistical training data. For the semantic distance classifier, we employ two algorithms, semantic mean and semantic mode. Analogous to their statistical counterparts, the algorithms provide a useful way for evaluating the concepts present in a document. They do this without the difficulty of collecting a corpus of statistical data.

Standard Template

The first step in the process (Figure III-3.1) is to parse the document into a standard form. We input text to be classified and extract the pertinent information through the use of a parsing tool called GATE. The software processes the content and identifies the pieces of evidence. The information it extracted is used to fill in multiple standard document templates. The template encapsulates and summarizes the information contained in the document. It contains information such as who, what, where, when, how and how certain. In this format, the conceptual content is easily pinpointed and mapped into our knowledge base. Once the documents are parsed into the template, they are inserted into a knowledge base as untagged evidence. From here, the user can use any combination of classification methods on either a single piece of evidence or a batch of evidence.

Knowledge Base

Our knowledge base (KB) is a graph with links and nodes that hierarchically encapsulates relationships between concepts. Each node in the graph represents a concept related to an event or outcome. Links in the graph represent an association between different concepts. Since the links give us an idea of the relationships between ideas and how they interrelate, it is thus possible to understand an idea based solely on its position in the graph. From this conceptual structure, we are able to take the various ideas presented in a paper and determine the governing concept. We label this the hypothesis.

The KB's that we deal with take a slightly different form from what is normally considered when posing this problem. Usually, an ontology is used. In our implementation, the KB's consist of hypotheses arranged in a hierarchical fashion. Ontologies represent an 'is-a'

---

[17] http://ti.arc.nasa.gov/tech/rse/synthesis-projects-applications/autoclass/, accessed 02/24/2015.

relationship between words. Our KB represents correlations between ideas. The representation of concepts instead of simple words allows us to represent more information than a simple ontology allows. In turn, this gives us an element of predictive analysis. Most ontologies are trees. That is, each node in the graph has at most one parent. When discussing concepts, however, it is necessary to allow for multiple parents. It is incorrect to assume that an event or concept has but one predecessor. There might be many factors contributing to a particular outcome. It might be the case that every possibility that is allowed by the domain is required to produce a given outcome. For this reason, we abstract the idea of an ontology into a general knowledge base.

For the semantic distance metrics, we require a method to map a template onto the KB. Each hypothesis in the KB has an associated set of attributes used in matching keyword hits. We consider every field in the template to be a keyword. A hit occurs if there is an attribute associating a contained keyword with a hypothesis. In this way, a document's concept space is transposed onto the KB. Once the hits are tallied, the concept is tagged.

Three Classifiers

We employ three classifier processes for the basic classification functionality. These are rule-based tagging, Bayesian classification, and semantic distance classification. The three can be used in any combination to provide a more robust solution. Once the documents are parsed into the system, the user has the option of deciding which, if any, to employ. If no classifiers are selected, the user has the option to classify manually.

The first process is a rule based tagging module. Our implementation defines an interface to create user-defined directives. These rules consist of simple if-then statements. If a document template contains a certain word or phrase, then it is tagged with the hypothesis that the rule indicates. Directives are thus defined to associate hypotheses with specific key words. For example, a user can create a rule that dictates that the document pertains to a network scan if the key word "scan log" is found in the source field of the template. It is often the case that the documents source dictates which class it belongs to. Using this tagging scheme, the end analyst can automatically filter out and classify documents that have an obvious relationship to a class of data.

The second classification technique involves a tool, developed by NASA, called AutoClass. This software suite defines a comprehensive Bayesian classification scheme. Using historical data, it determines which class the document will probably fit into.

The third technique involves two semantic distance algorithms. The ideas are unique to our group, thanks to Martin Hyatt, and not found in the literature. The semantic mean algorithm provides a way to evaluate the average concept in a document and the semantic mode provides a way to evaluate the most common concept in the document. We can think of a semantic mean in terms of the statistical mean. Recall that one way of interpreting the statistical mean is the point with minimum sum of the distance to all points. Similarly, the semantic mean is the concept that is the minimum sum of the distance to all concept hits in the KB. The same idea follows for semantic mode. The statistical mode is the point with the most number of hits. Likewise, semantic mode is the concept with the most number of hits.

To account for our altered KB structure, the semantic mean algorithm is modified slightly. It is applied to a tree in its first incarnation. Since our KB is not a tree, the algorithm has to manipulate the graph to simulate a tree like structure. It prunes the graph by simply omitting unused nodes. Since the graph is weighted, there is one most applicable path from a set of nodes to the top of the graph. This path is a tree to which the algorithm may be applied. In the case of the semantic mode, the algorithm can be applied to the graph without any significant changes. The hits are simply tallied up and the node with the most hits is located.

The mean algorithm has an inherent degree of optimization. Since the mean only takes into account the nodes in the applicable tree, it is not necessary to perform calculations on nodes that are not spanned by the tree. On the other hand, when calculating the mode, every entry in the KB has to be scanned for its hit number.

The semantic algorithms provides a valuable metric for semantic distance. Together, they provided insight into both the content and scope of a document. When applied, they provide a useful result and give the answer that an observer would expect. To see this, we provide an example from each of the two algorithms.

Suppose a document about computer network defense is scanned and we want to tag it with its hypothesis. From Figure III-3.3, we can see that the document parsed mentioned "UDP Alert" twice, "JTIDS" three times, "Log Events" twice, and "Network Scan" twice. The graph below shows what it would look like when mapped onto the KB. When the semantic mean algorithm is applied, the resultant concept node turns out to be "Network Scan". From the structure of the graph, it seems intuitive that a document containing these concepts would receive the tag "Network Scan." We see from this simple example that the result of semantic mean is intuitive.

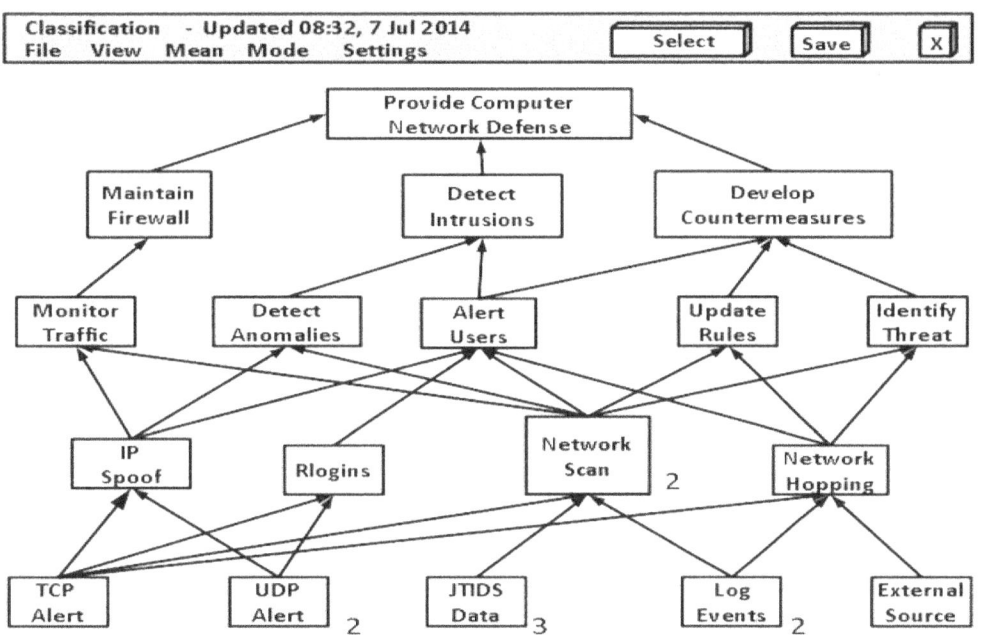

Figure III-3.2: Semantic Mean Example

Now, let's compute the mode for the previous example. We still have the same concept hits but we will perform a different calculation on them. We see from Figure III-3.2 that "JTIDS" has three hits while everything else has two hits. Since three is the maximum, "JTIDS" is the semantic mode.

The speeds of the two algorithms are directly affected by the attributes of the graph. The speed of the semantic mean is proportional to the number of edges in the graph. This is to be expected since the sum of the distances between hits must be minimized. It is only natural that every edge would have to be traversed to arrive at this estimation. The speed of the semantic mode is proportional to the number of nodes in the graph. It is necessary to compare the hit number from each node. The mode is located by finding the maximum of these hit numbers. Thus, it is necessary to traverse through each node in order to perform the calculation.

The effectiveness of the mean algorithm is dependent upon the weighting of the graph. Since a tree is produced in order to perform the classification, it is important that there are not too many applicable trees. If the weighting is an accurate estimation of the relationship between various concepts, the algorithm has no problem in determining the appropriate path. Since the semantic mode algorithm does not take into account the edges or their weights, this has no affect on the mode.

Both algorithms are needed to provide an accurate estimation of the content of a document. As in statistics, the mean is susceptible to fluctuations due to data spread. For example, if a piece of text contains information not directly relating to the topic, a concept might come up that really is not in its scope. Aberrations of this nature tend to skew the mean towards a more abstract answer. It is here where the mode can shed some light. Used together, the mean and the mode can give confirmation of content and also an indicator of degree of scope.

Classifier	Advantages	Disadvantages
Semantic	Based on meaning and context Can generalize hypotheses Intuitive Results	Requires an ontology Associates keywords
Bayesian	Gives best statistical match	Requires a training set Can't generalize Computationally intensive
Heuristic	Simple and fast Always gives desired answer	Requires exact keyword match Limited in applicability
Human	Provides "common sense"	Slow and inefficient

The implementation presented here defines several different methods for text classification. Independent of each other, the individual solutions offer some attractive features along with some severe limitations. Used together, the techniques enhance and complement one another. We will now discuss how the various methods can be used in conjunction with each other to provide a more robust solution than any one can give on its own.

The semantic classifier offers some useful contributions. First, the semantic classifier is based on meaning. When dealing with conceptual content, it is useful to define a metric that is based on semantics. This is precisely what the semantic classifier presents. It provides a way to measure meaning. Second, with the semantic classifier, we can get a degree of generalization. A concept does not have to be explicitly stated in the document in order for it to be tagged as the governing concept. This is a useful property that no other solution offers. Lastly, we can see that the semantic approach offers answers that appear intuitively correct. The two examples presented earlier illustrate this.

As attractive as the semantic solution seems, it has some limitations. Specifically, it requires a KB and attributes for associating keywords with entries in the KB. This can be prohibitive if little is understood about the domain. It is not necessary to have an overabundance of information, but a thorough understanding is required. As discussed previously, the relationships between various concepts must be described with weighted links. Some areas offer well-established connections between their components while others provide very little of such information. Without the connections, it becomes very difficult to make the explicit links this technique requires. In addition, the concepts themselves must be understood. We cannot create attributes for keyword matches if we do not have at least a basic understanding of the idea.

We now consider the Bayesian method of classification. This scheme is attractive because, historically, it produces the best results. However, it is extremely computationally intensive and it does not have the ability to generalize. The concept must be explicitly defined within the document in order for this classifier to recognize it as the documents hypothesis. In addition, it can only be applied to situations in which a lot of information has been collected and classified. It requires this collection for use as training data. For highly specialized or obscure domains, this information might be hard to come by.

With the semantic classifier, a thorough understanding of the domain is required. With the Bayesian classifier, on the other hand, a complete understanding is not necessary, but a certain amount of domain specific information must be present. Using both of these methods in conjunction provides a robust solution regardless of the domain. We can now work in domains regardless of their degree of uncertainty or obscurity.

The heuristic method for classification has a very specific boundary of usefulness. When applicable, it provides the fastest and most efficient scheme of classification. Since the rules are user defined, it provides the exact solution the analyst is looking for. The pitfall is that in order for this method to classify a document, an applicable rule must have been predefined. This severely limits the situations where this method is useful. When this scheme is applicable, however, it provides a very good solution. An example of one area where this might be useful is computer-generated reports. Some reports have a special class designed specifically for them. Instead of taking up CPU cycles by classifying these with bulkier methods, a rule may be defined to automatically place these in the correct class. Thus, both time and system resources are conserved while preserving the accuracy.

Classifying by hand is the last technique we allow. When human error or lack of understanding is not an issue, this method is correct in all cases, although it is both time consuming and inefficient. However, there are situations in which this offers the best solution. For instance, this method could be used for providing a training set for the Bayesian classifier or for defining more information about a domain that is not initially understood. Another situation where manual tagging might be useful is when the user is not satisfied with the machines solution. If the results achieved automatically are not satisfactory, the user might wish to manually tag the document with the desired hypothesis. We thus primarily define this method for bootstrapping and for granting the user greater control over the end results.

Using the techniques defined here, we now have a suitable solution for use in information synthesis. We can parse new information and correctly place it in the appropriate class. One such application where this might be useful is for the purposes of evidence fusion. Since our KB's represent concepts and they are cyclic graphs instead of trees, we can take our concept space directly from a superset of Dempster Shafer Belief Networks. This allows us to classify information that can then be fused to produce decision aids. One of the challenges of evidence fusion is assimilating the evidence and tagging it with the correct hypothesis. Our tagging process provides precisely this functionality.

Summary

This chapter presents innovative and robust text classification algorithms. We define three methods for classification that complement and enhance each another. In this way, we can avoid most of the pitfalls associated with each of the individual algorithms while simultaneously taking full advantage of their respective strengths. Two of the methods have been previously explored in some great detail while the semantic distance metric is a new idea. We provide a new way of linking the various methods and taking full advantage of each.

Chapter III-4
Threat Discovery

Our goal in this chapter is to find an upper bound on the number of threats. We can then make some predictions about the nature of our defense efforts. Based on probability, we can find an estimate of the number of enemies that we have not yet seen. Combining this estimate with the results from the Collector's Problem[18] provides a robust tool with which to make observations about the value of searching for more information, the value of sharing collected data between analysts, and the confidence that we have seen all threats. With this information, we can localize our defense efforts, achieve better threat recognition, and make the most of working together.

Introduction

The defense of any type of system requires some information about the threats. There are many tools to facilitate the discovery and processing of new information. We will look at some tools to analyze and quantify information that we do not yet have. It is easier to tailor defense efforts when we have an idea of what we are up against. With a bound on the number of yet unseen threats, we can optimize our defense efforts and gain a clearer understanding of our enemy.

The Collector's Problem is a statistical problem that inherently addresses the analysis of uncollected data. Simply stated, it asks how many items from a particular set need to be collected in order to guarantee with some certainty that the collector holds a complete set[19]. Taking attackers as the items to be collected, we find information such as the confidence that we have seen the entire set of threats, the effort versus benefits of probing for more information, and the general speed up gained in applying more analysts and sharing the collected information. The answers to these questions require a bound for the size of the set being sampled. Thus, we probabilistically determine the size of the set and then proceed to apply the analysis techniques given by the Collector's Problem.

The Role of Domain Size

We need to estimate the total number of attackers that pose a threat to our system. With this information, we can find the number of unknown attackers and tailor our defense efforts in order to maximize threat deterrence. The problem with quantifying the unknown is that it is uncertain. Experience tells us that we can make observations about what we don't know based on what we have seen thus far. We hope that previous threat discovery efforts can tell us something valuable about the overall number of attackers. As it turns out, the percentage of time that incoming information describes a new threat provides us with the information needed to find a bound on the number of unknown threats.

---

[18] https://www.math.ucdavis.edu/~tracy/courses/math135A/UsefullCourseMaterial/couponProblem.pdf, accessed 03/26/2015.

[19] http://www.spri,nger.com/mathematics/probability/book/978-0-387-97974-8, accessed 01/21/2015.

This bound is used in conjunction with the insight provided by the Collector's Problem to make some valuable observations about our domain. The expected number of samples needed to guarantee that all threats are known is easily estimated using a Monte Carlo (numerical sampling) approach. With this number, we compare how many samples we have already collected and determine the confidence that our collection describes a complete set of threats.

The classical problem assumes that samples are taken one at a time. We want to know the affect of taking more than one sample for every time step. In other words, what would happen if we have multiple analysts searching for information and collaborating? It is important to note that in addition to taking more samples per unit time, we take samples with replacement. The classical problem relies on the fact that samples are taken without replacement. Using the domain bound, we compute the expected number of samples expected to collect a complete set taking n samples at each time step. This can be interpreted as the expectation for n analysts to collect a complete set. Taking the ratio of the time for one analysts to collect a complete set to the time for n analysts to collect the same set, gives us a force multiplier. This multiplier is the relative speedup gained by using n analysts instead of one. This is helpful information in determining how many analysts should be assigned to data collection.

Approach

We have three distinct problems, the number of possible threats, the confidence that we know of all the threats, and the advantage gained by adding more analysts to the task. The latter two are modeled using both an analytic method and a Monte Carlo approach. Thus, using the result from the first problem, we find solutions for the latter two.

We solve the unknown threat problem in the following way. Let P_t be the set of all threats at time t. We are interested in knowing the size of this set. Let p_t be the set of known threats such that p_t is a subset of P_t. Now, suppose that we have a piece of data describing threat d. Notice that $P(d \in p_t) = |p_t| / |P_t|$. Thus, the size of the entire set of threats is found by: $|P_t| = |p_t| / P(d \in p_t)$. Then, the number of unknown threats is determined by subtracting the number of known threats from the number of total threats. Thus, we have a value for the size of the set of all attackers and the set of unknown attackers provided we have an estimate for the probability of finding a new threat. This information is easily estimated by tracking the number of times incoming information describes a new threat and the number of samples. The probability is the ratio of these two numbers.

With the newly acquired information, we use an analytic method to determine the confidence that our set of threats describes the complete collection. If there are N threats and we know about m of them, we simply simulate the incoming data using a random number generator and find the percentage of time that each piece falls into the set of known threats. This gives us the confidence that we have a complete set; that is, $P(m = N)$.

The speedup gained by applying more analysts to the problem is also found. To find the speedup, we require a method to find the expected time for *n* analysts to collect a complete set. The case where n = 1 is the classical collectors problem. For this simple case, we can write the probability of finding a piece of evidence describing a new threat as: #unknowns / #threats

Now, as the number of unknowns varies from 0...N, the expected number of times steps for one analyst to collect a full set is given by

E(time to collect N threats) = $1 + N/(N-1) + N/(N-2) + ... + N$

Now, we want to find the expected time for *n* analysts to collect the same set. We approach this in the following way. The number of analysts working on the problem is interpreted as the compression of the time it takes to collect a sample. For example, if one analyst can perform one task in one unit time, then five analysts can perform five tasks in one unit time. Thus, the probability of *n* analysts finding one of a piece of evidence describing a new threat is (#analysts) * (# of unknowns) / #threats

In addition, the expected time to complete a full set for n analysts is found by:

E(time to collect N threats = $(1 + N/(N-1) + N/(N-2) + ... + N) / n$

Knowing the size of the domain and the anticipated time to collect a full set of N threats for 1… n analysts, we find the relative speedup. If the time to collect a full set taking *n* samples is T*n*, then the naive relative speedup is found by taking the ratio of T_1 to T*n*. However, in real life, the actual trading of information will require some time. To compensate for this, we define Tc(*n*) as the time for *n* analysts to collect a complete set and Ts(*n*) as the time for *n* analysts to share the information collected. So, by definition, we have Tc(1) = T_1, Ts(1) = 0, and T*n* = Tc(*n*) + Ts(*n*). Let M*n* be the force multiplier (relative speedup). Then we have M*n* = T_1 / (Tc(*n*) + Ts(*n*)). Now, it will normally take longer for one analyst to complete a set than for *n* analysts to complete a set while sharing information. Thus, the M*n* will typically be greater than 1. However, as we will see, it is possible for the M*n* to be less than 1. This corresponds to the case where it would be more efficient for one analyst to perform the task than it would be for *n* analysts. This occurs because the amount of redundant information being shared becomes large. At this point, sharing information becomes useless. Using the behavior of M*n*, we gain some idea of the benefit of collaboration.

Note that Ts(*n*) will typically not be a constant valued function. There are many factors that affect this function. For instance, an analyst doesn't have to assimilate information that he has already collected. Thus, the time to share will depend on the average number of unknowns at the time. As this decreases, the time to share decreases because the need for information assimilation drops. Since information takes the form of knowledge instead of data, the rate of compression given by the representation also affects the sharing function. Knowledge is more compact than raw data. It is also easier for an analyst to assimilate. These factors must be taken into account when determining the time to share collected information.

Results

Provided that our known to unknown ratio is accurate, we find reliable values for the threat count even when the size of the system is small. The tabulated values closely approximate the actual sizes. The error is attributed to the estimation of the probability of finding a new threat. Each simulation is performed with about one tenth of the threats presumably known. The number of unknowns is found by subtracting the number of knowns from the estimated total number of threats.

Real, Estimated, and Known Threats

The second two problems are solved using a Monte Carlo method. This allows us to model the domain without the stringent time and space requirements of the analytical solutions. The analytical equations for the problems require too many resources to be realistically computable for any more than about twenty threats. With the Monte Carlo approach, we arrive at a reasonable answer in linear time and space.

The results from the confidence interval simulations gives some valuable information as to the usefulness of searching for data at each time step. We see in Figure III-4.1 that the confidence becomes asymptotic as more data is collected. At some point, the cost of collecting more data begins to outweigh the benefits gained. We call this the point of diminishing returns. Based on an effort bound, we specify exactly how much confidence we require using as much effort as the bound allows.

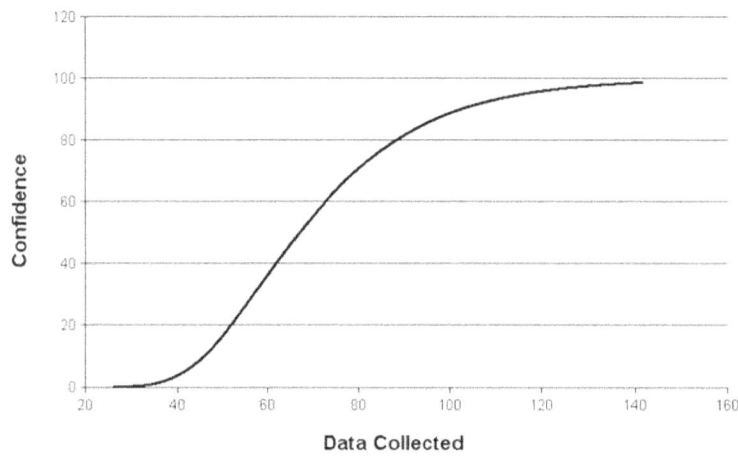

Figure III-4.1: Confidence in Having a Complete Set

Figure III-4.2 illustrates the behavior as more analysts are applied to the problem for varying number of threats. Notice the shape of the function. There is an obvious optimum. In the particular case where there are 500 threats, the highest returns are found at about 60 analysts. Once past this point, the benefit of adding more analysts is actually decreasing. In other words, after about 60 analysts, the efficiency starts to drop. This occurs because the time it takes to share the collected information begins to outweigh the benefits from doing so.

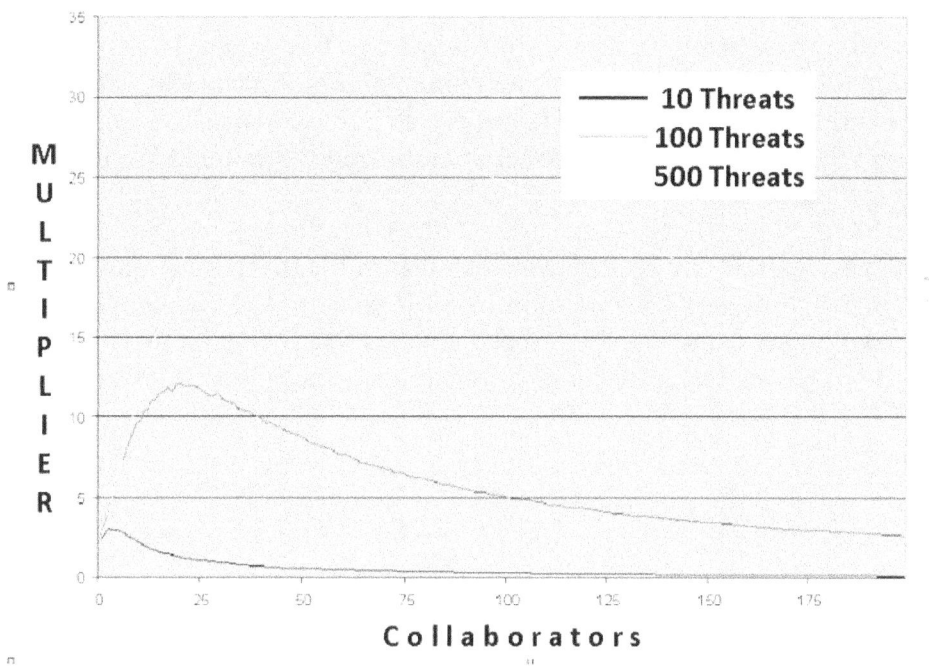

Figure III-4.2: Cooperation Produces a Force Multiplier

We are interested in the behavior when the set of threats is very small. It is quite possible that there are only a few attackers on a specific system. What is the value of cooperation when the threat size is very low. Suppose there are only five attackers. The system behaves in precisely the same way as a larger system. The maximum efficiency is found at about three analysts and efficiency tapers off as the number of analysts is increased.

For an over-worked system, the force multiplier finds a maximum and drops for a while slowly becoming asymptotic. However, the multiplier will eventually drop below one. This occurs between eleven and twelve analysts. Basically, after this point it is more efficient for one analyst to do the work than for twelve to be assigned to the task. After about twenty-one analysts, the efficiency is about half that of one analyst. In large systems this will not be an issue. There typically needs to be significantly more analysts than threats for this to occur. However, when the number of threats is small this will be an issue. One of the benefits of the methods presented here is the ability to recognize when this will occur and take steps to avoid it.

Issues

We previously assume that the set of threats does not change over time. It will never be the case that each and every threat is constant to our system. Threats will come and go. The results that we find using the methods described here are only valuable provided that the system do not change. We address this issue by assuming the system is constant for a certain period of time. In this window, we assume our analysis is an accurate estimation. However, once past the window, our data becomes useless. It is necessary to repeat the process in order to model the changing system.

For the purposes of preliminary analysis, we assume that threats are uniformly distributed. In reality, there are some threats that are quite common while others are extremely rare. The distribution is simply not known. In addition, threats do not attack in a uniform manner. An analyst may only be able to see attackers that are localized to his area. We model these effects by perturbing incoming data.

When some threats are rare, we see behavior like that in Figure III-4.3. The force multiplier is significantly higher than that of the uniform data. If some of the threats are rare, then adding more analysts will increase the discovery speed. Notice that the optimum is located around the same place. Thus, if optimal collaboration is the goal, then threat rarity will not affect the solution.

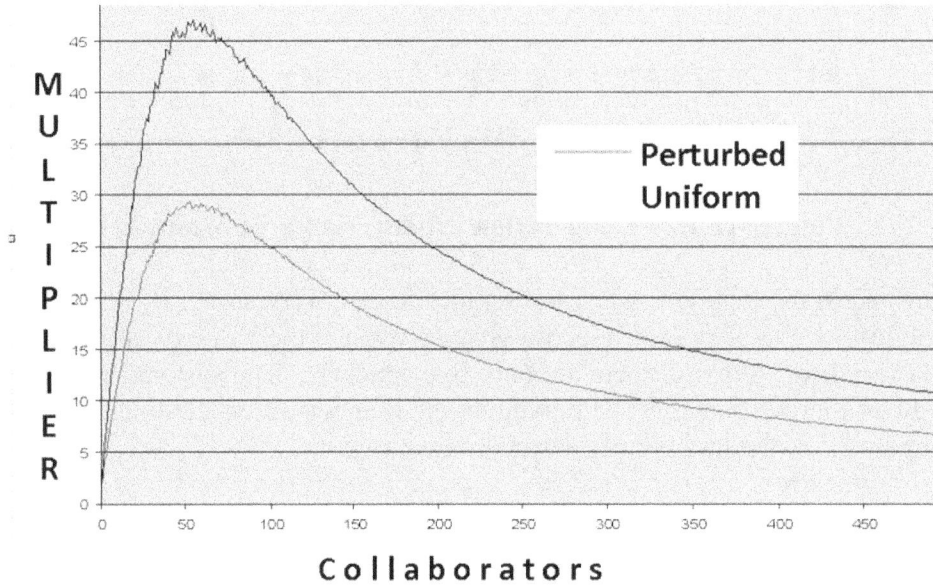

Figure III-4.3: Threat Distribution Perturbation Analysis

In contrast to the previous perturbation analysis, assuming a non-uniform attacking pattern decreases the threat multiplier. If an attacker is localizing his efforts, then simply adding analysts uniformly does not increase productivity. Thus, the force multiplier is significantly lower. Also note that the optimum has shifted. Attack distribution will affect collaboration optimization efforts.

Figure III-4.4 describes a system in which both distribution perturbations are taken into account. We see that the maximum force multiplier has not changed, but the location has. The function has been skewed to the right.

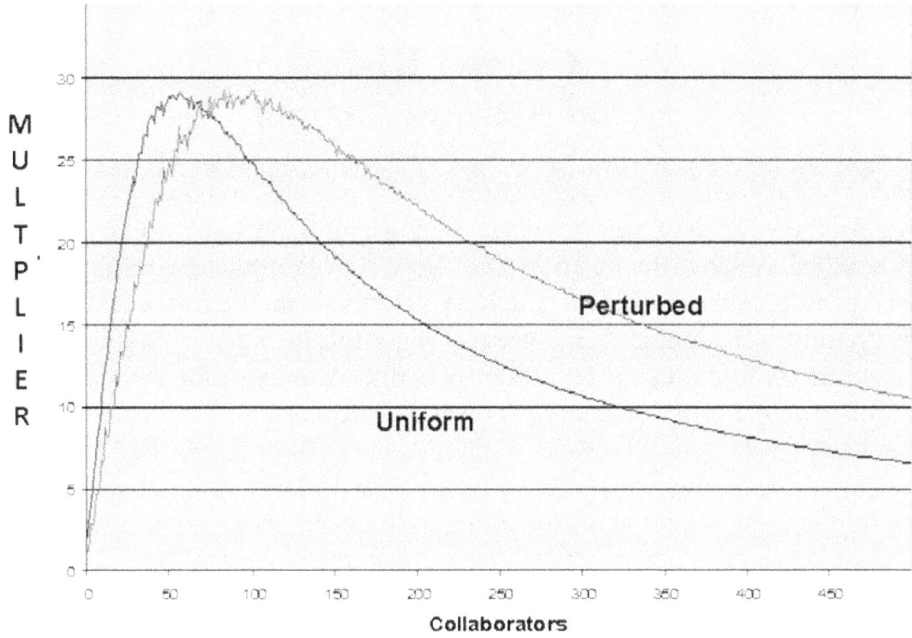

Figure III-4.4: Overall Perturbation Analysis

Summary

We find the size of the set of threats. From this, we calculate the number of unknown threats. Using the overall size, we find the confidence that we have a complete set, and the value of cooperation. These four pieces of information give valuable information about the domain and allow us to tailor our counter efforts to the set of attackers. We now have some very powerful tools to help us determine how many resources should be put toward threat discovery, threat deterrence, and threat analysis.

Chapter III-5
Social Network Analysis

A wide variety of situations are concisely described by a graph consisting of nodes and links. Nodes represent objects such as people, places, things, and concepts. Links are directed from one node to another to show connectivity and flow; for example, of influence, funding, or electrons. We quantitatively analyze a graph to determine metrics related to static network structure, uncertainty in node and link characteristics, and network dynamics. These network metrics are used to analyze an Internet forum network. A process for evaluating the network allows us to appropriately drill down to key nodes and links.

We need to identify specific network-related metrics for the computer network. These metrics provide input to operational metrics that give a mission perspective by allowing us to quantify adversary dynamics and the impact of friendly force interventions. From a network perspective, we need to represent the network topology and measure its structural properties; for example, the extent to which the network under consideration is robustly connected.

Our goal is to define network metrics for an adversary network. Tasks are to:

Task 1: Organize metric types in a taxonomy
Task 2: Identify metrics of potential interest
Task 3: Choose a set of metrics
Task 4: Provide rationale
Task 5: Define and implement algorithms for computing metrics

Rather than perform a trade study: evaluating candidate metrics against weighted selection criteria, ranking the candidates, and computing a score for each candidate, we identify candidate network metrics based on discussions with our subject matter experts (SME)s.

Selection Criteria

To decide on metrics to incorporate, we define criteria for discussing the relative usefulness of the potential metrics with our SMEs.

Criteria	Weight	Definition
Innovative	10	Provides thought leadership
Expressive	10	Accounts for important network behavior
Intuitive	10	Easily understood
Relevant	10	Ties network properties to networks encountered in practice
Suitable	10	Applies to our problem
Calculable	8	Easily computed, especially for large networks
Robust	8	Insensitive to errors and missing information
Scalable	8	Computation does not scale exponentially with network size
Acceptable	6	Commonly used in commercial and academic tools

| Independent | 6 | Doesn't overlap functions of other metrics |
| Global | 6 | Useful in static, dynamic, temporal, and uncertain networks |

The criteria are derived from discussions with subject matter experts, practical considerations, hands-on experience with (static) social network analysis tools and research of the literature. The static metrics are chosen from a long list of SNA metrics. The overwhelming response from our subject matter expert and a practitioner allowed the list to be trimmed. Many of these are already calculated in tools like Pajek[20] and UCINet[21].

Taxonomy

The metrics we choose are shown in Figure III-5.1. The remainder of the paper explains what they are and how they work. Focus is on network metrics, metrics computed with uncertain information (relevant for threat, dark, networks), and dynamic metrics that relate stability to network growth and interventions are the sub-categories of interest. Each of these is further detailed and examples are given.

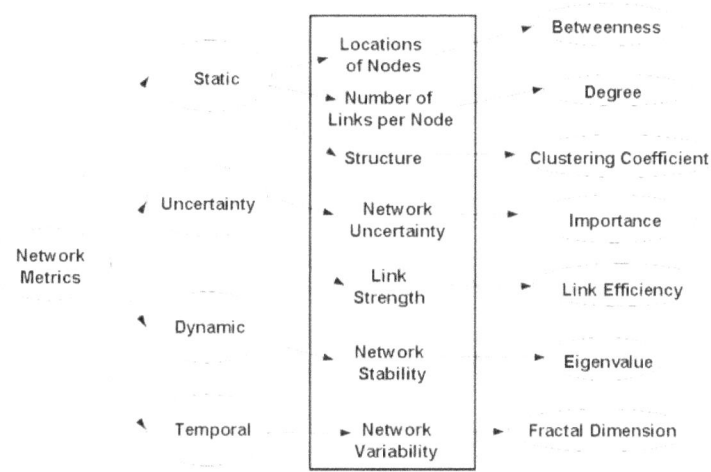

Figure III-5.1: Metrics Taxonomy

Metrics of Potential Interest are discussed by sub-category (static, uncertain, temporal, and dynamic) and each metric is briefly described. In some cases, static metrics have broad applicability: they provide the foundation for computing adversary Centers-of-Gravity.

Static Network Metrics

These measures provide information about the nodes, links, and overall structure of the network. The intent is not to provide an exhaustive list (which could number in the hundreds), but rather to identify those that are recommended by our SMEs, recur in the literature, are computed by popular SNA tools, and tend to support computing metrics for dynamic networks.

---

[20] http://pajek.imfm.si/doku.php, accessed 02/24/2015.
[21] www.lsu.edu/faculty/bratton/networks/ucinet.ppt, accessed 02/25/2015.

We first need to specify the number of nodes, number of links, whether links are directed, and the presence of loops and self-loops. These define the network, but are not considered network metrics. The most commonly studied property of a network[22] is the degree, followed by the clustering coefficient and the diameter. When compared with random graphs of the same size and average degree, real-world graphs are found to have a much larger clustering coefficient. This is an important point that is discussed in more detail under the dynamic metrics section: it drives our construction of appropriate graphs.

Uncertainty

Computing metrics under uncertainty is a topic of significant interest throughout this book. Two new metrics: importance & rank under uncertainty and link efficiency are presented. Sources of uncertainty that we may need to reckon with are;

1. Importance: of a person, institution, or posting
2. Influence: of links, based for example on the flow of idea from one node to another
3. Credibility: of evidence posted to nodes
4. Ambiguity: of group size and degree of interaction
5. Vagueness: in the roles of individuals and organizations
6. Redundancy: in the number of identities of individuals
7. Motives: multiple motives, ulterior motives, disguised motives
8. False reporting: in the evidence of events and activities
9. Deceptive behavior: in activities to deceive authorities
10. Evidence aging: over time, things change, credibility in evidence often decreases
11. Feedback: degree and motivation due to feedback is difficult to gauge
12. Missing data: forum interactions must be complemented by person-to-person E-Mail

Temporal

These metrics identify patterns that occur periodically, cyclically, or are otherwise evident over time. Although much work has been done in time series analysis, data for the analysis of temporal evolution of social networks has been scarce. With the advent of the internet and web forums, this data has become available in digital form and recent work[23] is of interest. A limitation of many techniques for time series analysis is the requirement that the time series be stationary. A *stationary process* is a stochastic process whose probability distribution at a fixed time or position is the same for all times or positions. As a result, parameters such as the mean and variance, if they exist, also do not change over time or position. For our purposes, the metrics become interesting when the process; that is, the Internet forum, becomes non-stationary! For this reason, our focus is on fractal dimension rather than standard deviation as a measure of dispersion.

Another topic, of future interest, is wavelet time series analysis, a "mathematical

---

[22] Stumpf, M. et al, Subnets of scale-free networks are not scale-free, www.pnas.org/cgi/content/full/102/12/4221, accessed 6 August 2007.
[23] Carley, K. Dynamic Network Analysis, http://privacy.cs.cmu.edu/dataprivacy/papers/socialnetworks/index.html, accessed 11 September, 2007.

microscope"[24], due to its ability to focus on weak transients and singularities in time series. In chaotic systems[25], wavelet analysis provides useful information based on the energy concentration at specific wavelet levels. The metrics of potential interest are the variance of the wavelet coefficients at various levels, and the Holder exponent[26]. An additional advantage is that wavelet coefficients can be tailored to destabilize a social network[27]. A non-parametric randomization test is discussed[28] that performs pattern recognition of time series (of interest rates for four countries) using time series.

Dynamic

Networks of interest continually change topology over time: internet forums grow and sometimes collapse. Nodes and links are intermittent. Forums increase and decrease membership rapidly. Postings are added and removed over relatively short periods of time. Our understanding of the roles of individuals also changes over time, given that we observe the dynamics of the network over an extended period. From a network perspective, we seek metrics that characterize the stability of a network and the volatility of network interactions.

Analysis of topology and evolution of complex networks is based on the field of statistical mechanics[29]. The focus is on understanding the interplay between topology and the network's robustness against failures and attacks, or conversely the vulnerability to intervention. Three concepts occupy a prominent place in contemporary thinking about complex networks – these are summarized, with related (static) metrics in parentheses:
- small worlds (path length)
- clustering (clustering coefficient)
- degree distribution (scale-free exponent).

Many algorithms are available for constructing large idealized networks with properties that match real-world networks to a limited extent. From earlier discussions, we are computing metrics on real-world data rather than doing theoretical experimentation.

As shown in the taxonomy, qualities that capture dynamic characteristics of a network and related metrics (in parentheses) are;
- stability (eigenvalues and eigenvectors of the adjacency matrix[30], Lyapunov exponents, and the effective coupling coefficient)

---

[24] Struzik, Z., Wavelet Methods in (Financial) Time-series Processing, http://db.cwi.nl/rapporten/abstract.php?abstractnr=627, accessed 6 August 2007.

[25] Murguia, J. et al, Wavelet analysis of chaotic time series, http://www.smf.mx/rmf/pdf/rmf/52_2/52_155.pdf, accessed 6 August 2007.

[26] Struzik, Z., Wavelet Methods in (Financial) Time-series Processing, http://db.cwi.nl/rapporten/abstract.php?abstractnr=627, accessed 6 August 2007.

[27] Wei, G. et al, Tailoring Wavelets for Chaos Control, Physical Review letters, Volume 89, Number 28, 31 December 2002.

[28] maharaj, E., Pattern Recognition of Time Series using Wavelets, http://www.quantlet.de/scripts/compstat2002_wh/paper/full/P_03_maharaj.pdf, accessed 6 August 2007.

[29] Albert, R. Statistical Mechanics of Complex Networks (2001), http://citeseer.ist.psu.edu/albert01statistical.html, accessed 8/8/07.

[30] Carley, K. http://privacy.cs.cmu.edu/dataprivacy/papers/socialnetworks/index.html, accessed 11 September 2007.

- volatility (fractal dimension), which doubles as a temporal metric.

The stability of a network is computed from an algebraic condition on its Lyapunov exponents[31]. The Lyapunov exponent is the average value of the logarithm of the derivative of the function.

The implication is that the structural properties of a network have some bearing on the dynamics taking place in it. A key concept is *network synchronization*, which refers to the ability of the network to efficiently propagate information to its nodes. Hence, a fragmented network in unsynchronized. The synchronization threshold is a dynamic property of the network that is measured from the network topology. Over time, each of the nodes must reach a stable point for synchronization to be achieved.

A sketch of the processing[32] follows: compute the eigenvalues of a Lapacian matrix (**L**) which is formed from the link adjacency matrix (**A**) matrix and the node degree matrix (**D**), so that **L** = **D** – **A**. The condition for stability is that for all non-zero eigenvalues of the Lapacian matrix, the corresponding Lyapunov coefficients are negative[33].

A recent paper[34] describes a more elegant way to assess the importance of a node or a link in a dynamical system. The dynamical importance of a link is defined as the amount ($-\Delta\lambda_{ij}$) by which the largest eigenvalue (λ) changes when the link is removed, normalized by the eigenvalue. Similarly, the dynamic importance of a node is defined as the amount ($-\Delta\lambda_k$) that the largest eigenvalue decreases when the node (k) is removed, normalized by the eigenvalue. A perturbation analysis provides approximations of these metrics. Although a Lyapunov exponent is not employed, these metrics do address the stability of the network.

Algorithms

The content in this section is based on analysis and a literature search to find fast-running algorithms for computing the metrics earlier identified. A summary of the computational complexity of the chosen metrics indicates that scalability ranges from linear to exponential.

Network Metric	Scaling Law	Explanation
Betweenness	$O(N * L * \log N)$	N = # nodes, L = # links
Degree	$O(N)$	N = # nodes
Clustering Coefficient	$O(N * <k>^2)$	N = # nodes, <k> = average degree
Dynamical Importance	$O(N^2 * \log N)$	N = # nodes
Link Efficiency	$O(N)$	N = # nodes
Eigenvalue	$O(N^{2.376})$	N = # nodes
Fractal Dimension	$O(N)$	N = # nodes

---

[31] Pecora, L. et al, Synchronization in Small-World Systems, Physical Review letters, Volume 89, Number 5, 29 July 2002, http://www.bg.ic.ac.uk/staff/barahona/SWsynchro.pdf, accessed 8/8/07.
[32] http://arxiv.org/pdf/1112.2297.pdf, accessed 02/25/2015.
[33] Xiang, L. Sync in Complex Dynamical Networks: Stability, Evolution, Control, and Application (2005), http://www.yangsky.com/ijcc/pdf/ijcc342.pdf, accessed 8/8/07.
[34] Restrepo, J. et al., http://arxiv.org/PS_cache/cond-mat/pdf/0606/0606122v1.pdf, accessed 11 September 2007.

Betweenness

We want to measure important characteristics of a social network. A metric identified by our SMEs is betweenness centrality, one of the most basic metrics in graph theory. We need to efficiently implement this metric because our networks may be large. Betweenness will allow us to define Centers of Gravity (COGs) in Internet forums as part of the Intelligence Preparation of the Battle Space (IPB) process. Nodes with high betweenness have prominence and function as "brokers". Degrading or destroying these nodes will have a high probability of degrading the forums.

Betweenness is a centrality measure for a node within a graph. Nodes that occur on many shortest paths (geodesics) between other nodes have higher betweenness than those that do not. Betweenness of a node is the proportion of all geodesics between pairs of other nodes that include it.

Social Network Example[35]: While a Moderator (Figure III-5.2) has many direct ties, the Webmaster has few direct connections, less than the average in the network. Yet, in many ways, she has one of the best locations in the network: she is *between* two important constituencies. She plays a 'broker' role in the network. The good news is that she plays a powerful role in the network; the bad news is that she is a single point of failure. Without her, Distributors and Originators would be cut off from information and knowledge in the Moderator's cluster. A node with high betweenness has great influence over what flows, and does not, in the network. A node like Webmaster holds a lot power over the outcomes in a network. As in Real Estate, the golden rule of networks is: location, location, location.

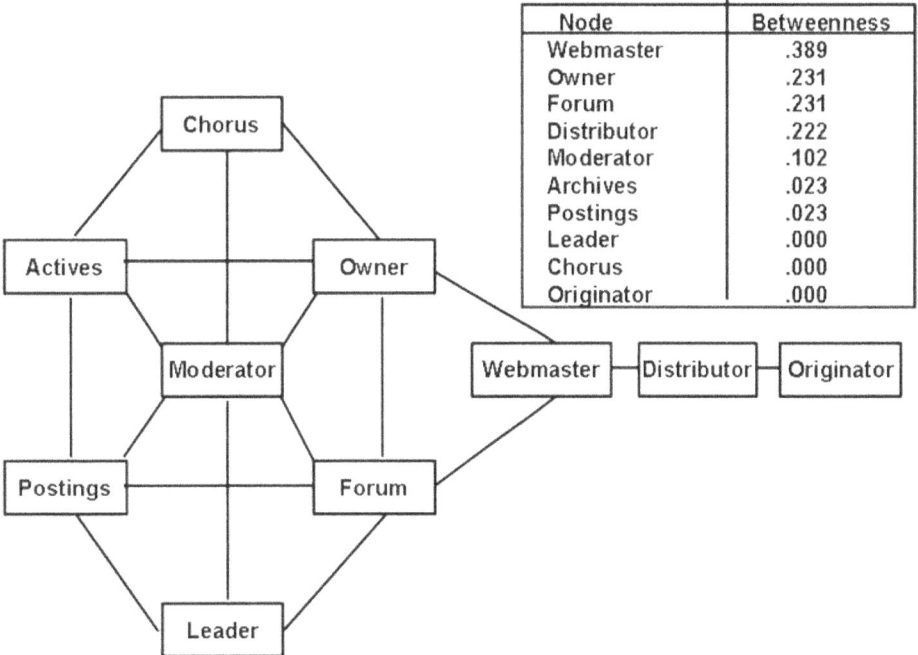

Figure III-5.2: Kite Network: Webmaster Has Highest Betweenness

---

[35] http://www.orgnet.com/sna.html, accessed 03/03/2015..

A Faster Betweenness Algorithm[36]: Betweenness calculations for nodes in a weighted graph scale as $O(NL + N^2 \log N)$ for time complexity and $O(N + L)$ for space complexity, where N = # nodes, L = # links. This means that the running time of the calculation is intense! Our networks will have a maximum of about 10,000 nodes, and about 50,000 links. This suggests a maximum number of operations of $10^4 * 5*10^4 + 10^4 10^4 * 4 = 5*10^8 + 4*10^8 + = 20*10^8$, or about 2 billion calculations. Previous to the publication of this algorithm, the fastest known algorithm was of $O(N^3)$, which would result in 1 trillion operations. Based on this scalability analysis, a faster algorithm is needed, but the calculation should be doable. In addition, "Results generalize to directed graphs (our case) with only minor modifications"[37].

Degree

We want to measure important static characteristics of a social network. Such a metric, identified by our Subject Matter Experts, is the degree. In graph theory, for a directed graph, the in-degree of a node is the number of links incident on the node. Similarly, the out-degree of a node is the number of links leaving a node. We need to efficiently implement this metric because our networks may be large. We deal only with directed graphs. We leverage existing network graphs on relationships between links and nodes. We also define an effective in-degree which accounts of the strength of influence of links.

Effective in-degree allows us to define Centers of Gravity in Internet forums as part of the Intelligence Preparation of the Battle Space process. Nodes with large in-degree are prestigious: many other nodes communicate to them. Degrading or destroying these nodes will have a high probability of degrading the forums.

Definition: The in-degree of a directed graph is a measure of the prestige of a node. For a node, the effective in-degree is the sum of the average belief and disbelief values of link strength. In-degree and effective in-degree are defined for a directed graph (Figure III-5.3).

Node	In-Degree	Out-Degree	Effective In-Degree	Effective Out-Degree
1	0	2	0	1.3
2	2	0	1.2	0
3	2	2	1.8	1.6
4	1	1	.9	1.0

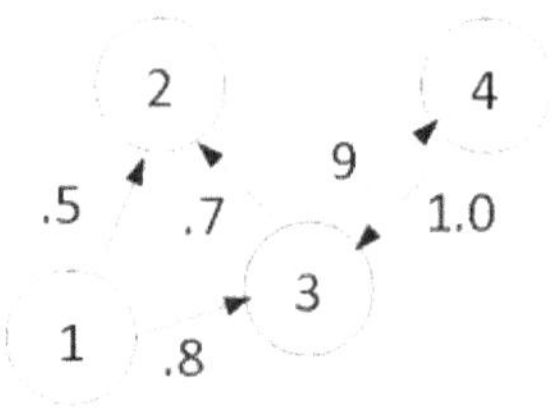

Figure III-5.3: Example Calculation of In-Degree and Effective In-Degree

Based on in-degree alone, nodes 2 and 3 are equivalent. However, when link strengths are considered, node 3 has a 50% higher effective in-degree and is therefore a more likely candidate for a Center of Gravity node.

---

[36] Ulrik Brandes, *A Faster Algorithm for Betweenness Centrality*, Journal of Mathematical Sociology 25 (2):163-177, 2001.
[37] ibid.

Clustering Coefficient

For measuring important characteristics of a social network, a metric identified by our SMEs is the clustering coefficient, proposed by Watts and Strogatz[38] in 1998. We need to efficiently implement this metric because our networks may be large. We also want to account for the fact that links among neighbor nodes may be of unequal strength, resulting in an effective clustering coefficient which provides more information for identifying an adversary Center-of-Gravity. We found code for an algorithm known from graph theory to have polynomial running time.

Clustering coefficients allows us to define Centers of Gravity in Internet forums as part of the Intelligence Preparation of the Battle Space process. Nodes with large clustering coefficients are highly connected: they are tightly organized clusters of people and things. Degrading or destroying these nodes will have a high probability of degrading Internet forums.

The clustering coefficient (C) is a measure to determine whether or not a graph is a small world network. For a node: C = # links among neighbors / # possible. For example, clustering coefficients are defined for three undirected graphs (Figure III-5.4). Note that the gray lines from the node for which the clustering coefficient is being computed (black node) are not counted! In the first, the number of links among the neighbors is 3 (black lines) and the number of possible links among the neighbors is 3 (black lines); therefore the clustering coefficient is 1. In the second example, the number of links among the neighbors is 1 (black line) and the number of possible links among the neighbors is 3 (black lines plus dashed lines); therefore, the clustering coefficient is 1/3. In the third example, the number of links among neighbors is 0 (no black lines) and the possible number of links is 3 (no black lines plus dashed lines); therefore, the clustering coefficient is 0/3 = 0. Rather than simply counting the number of possible links among the neighbors, we can use a formula which is based on the degree of a node. Here (blue node) there are three links (gray lines) between the blue node and other nodes, so the degree (k) is k = 3. The maximum number of links among the neighbors is k(k-1)/2 = 3(3-1)/2 = 3.

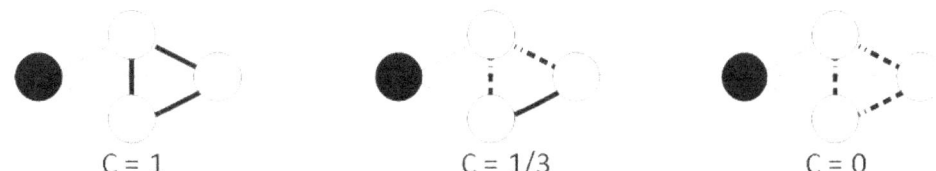

Example clustering coefficient calculation for black node. Thick lines connect neighbors, dotted lines are unused possible connections

Figure III-5.4: Example Calculation of Clustering Coefficients (Undirected Graph)

Our networks are directed, so we can count the number of links among the neighbors as before (without regard to the directionality of the link) and double the number of possible links (denominator). Alternatively, we set k = in-degree + out-degree and apply the formula k (k-1) to compute the denominator. The effective clustering coefficient is: C* = average strength of links

---

[38] http://en.wikipedia.org/wiki/Watts_and_Strogatz_model, accessed 02/25/2015.

among neighbors / possible strength.

Importance/Rank

We often need to quantitatively analyze a graph to determine metrics; for example, important nodes and the rank of nodes relative to one another. Importance and rank suggest solved problems (for example, Google page rank). This problem statement extends the challenge that the Google page rank algorithm addresses. It states an unsolved problem: computing metrics when the nodes and links are subject to various types of uncertainty.

The specific problem that motivated this idea is the need for a way to calculate adversary centers-of-gravity (COGs) in a terrorist social network for Intelligence Preparation of the Battle Space. COGs are important nodes in the graph that friendly forces focus on and attempt to neutralize to avert terror attacks.

What is New: We present an algorithmic procedure that robustly computes graph metrics where nodes are uncertain (objects with degrees of belief, ignorance and disbelief) and links have quantitative values that indicate variable degrees of influence.

- Sources of uncertainty include: relative importance, degree of influence, missing data, chaotic underlying processes, evidence credibility, data ambiguity, possibility of multiple voting, ulterior motives, false reporting, deceptive behavior, reporting redundancy, aging of evidence, and hedge words in feedback.
- Procedure for graph composition from sub-graphs through set operations such as graph nesting, graph union and graph intersection.
- Process for representing uncertainty in a directed graph structure that allows direct computation of graph metrics; for example, node importance, node rank, sub-graph embeddedness, and sub-graph cohesion. Because uncertainty is considered, classic social network analysis metrics, such as embeddedness and cohesion have a more general definition.
- Process for relating uncertainty in graphs to graph metrics to include, but not limited to, evidence, hypotheses, relationships, strength of association, knowledge bases, ontology's, distinguishing attributes, goals, priorities, patterns, iconic knowledge, geometric knowledge, metric maps, graphs, semantic networks, and inference rules.
- Processes (8 fusion algorithms) for fusing uncertain evidence that is dependent on the type of and relationships among hypotheses within a single belief network structure.
- Software that orchestrates and integrates the computation and organization of graph metrics computed under uncertainty.
- Displays that provide knowledge visualization of graph metrics.

What it Does: These processes and procedures provide a wide variety of metrics about graphs with nodes and links that are subject to multiple types of uncertainty. Metrics are computed based on graphs that are dynamically updated through uncertain evidence, user interaction, new nodes, new links, data aging, and dynamic composition of graphs through set operations such as graph nesting, graph union, and graph intersection.

How it Works: A graph consisting of nodes and directed links is composed. These graphs are highly cyclic and nodes can have self-cycles. This graph composition is either accomplished with a drag-and-drop display interface, or through a template-filling tool. The graph is then populated with attribute values for the constituent nodes and links. Nodes and links have attributes with values. Nodes have attributes which characterize underlying uncertainty; for example, a node has a degree of belief, ignorance, and disbelief about the hypothesis that a node represents. Links also convey uncertainty with distinct influence values for node belief and disbelief.

Uncertain evidence is posted to nodes, mathematically fused with existing evidence, propagated through the graph, and link analysis metrics are displayed to the user. Evidence is aged based on a persistence value and this is fused and propagated as a function of time. Metrics which deeply characterize the graph; for example, node importance, node rank, sub-graph betweenness, and degree, are calculated based on the uncertainty that is inherent in underlying node and link characteristics and associated values. Results are displayed intuitively to the user. The user can modify the graph and associated parameters for nodes, links, and evidence.

This is a new collection of algorithms for computing metrics for a graph that has uncertainty in its nodes and links. A solved problem that we're familiar with is web page ranking. Note that all node and link values are assumed known with certainty (100% belief). Importance calculation is a simplified version of Google's web page ranking algorithm. A recursive algorithm converges quickly to the answer.

Now consider the same problem, except that nodes are subject to belief, ignorance, and disbelief. For example, the degree of participation of a terrorist leader in a web forum may be uncertain. Links convey fractional amounts of influence (of node belief and disbelief) from one node to another. This paper provides algorithms for solving for the importance and rank of nodes under uncertainty. Other metrics that are readily computed are betweenness and degree.

A Progression of Algorithms: We begin with a problem and solution that is discussed on an episode of Numb3rs[39]. It is a "democratic" importance algorithm for nodes that has an analytical solution. The second and third are also analytic and account for node and link uncertainty. The fourth explores importance-based recursion. The fifth accounts for "dead ends" and "spider traps" to solve practical problems encountered with graphs, and combines uncertainty, practical extensions, and importance-based recursion.

Each of these algorithms has application. Rather than explore abstract examples, an internet forum is explored. In the matrix (Figure III-5.5), fractions arise when a node spreads its influence across more than one node; for example, the chorus (C) spreads its influence across eight nodes, giving $1/8^{th}$ to each.

Legend: O = Originator, D = Distributor, W = Webmaster, S = Site Owner, F = Forum, L = Leaders, M = Moderator, C = Chorus/Members, A = Active Members, P = Postings, T = Terrorist Acts

---

[39] http://education.ti.com/educationportal/activityexchange/Activity.do?cid=US&aId=7846, accessed 31 May 2007.

Importance Metric Definition

Each node has **importance**: the number of percentage points of importance attributed to that node, based on the values of incoming links. The **rank** (1st, 2nd,..) of a node is the relative position based on a sort of this importance metric.

Algorithm 1: Democratic Importance: To compute the importance that nodes have when each node has equal probability of providing influence, each component in the right-hand vector has a value of unity. Matrix multiplication produces importance values. Analysis of this result shows that the node rank (which is based on the number of incoming links), from most important to least important is: Forum, Moderator, Webmaster, Originator, Distributor, Chorus, Actives, Postings, Leaders, Site Owners, and Terrorist Acts.

Since this performance metric is based on connectivity, it makes sense that the Forum with four input sources and the Moderator with five input sources are ranked as most important. On the other hand, terrorists, leaders, and site owners get few inputs and are not very connected to the workings of the forums. The maximum value that a node can have is 11, which is obtained as the sum of the components on the left. These components are stated as a percentage of the maximum value for presentation to users, as defined above.

A disadvantage of this algorithm is that it does not account for the volume of activity incoming to the node. Also, it doesn't account for the idea that an input from an important node is more indicative of importance than an input from an unimportant node.

Algorithm 2: Node Uncertainty: In this case, we acknowledge that not all nodes provide the same volume of output, or support, to other nodes. Some nodes are active 24/7, others are not. As an aggregate, we de-rate each node (Figure III-5.5) to specify its relative effectiveness as a purveyor of importance to other nodes. The node values correspond to the hours/day that the node is effective.

Analysis of the resulting importance value for nodes gives a more reasonable sense of importance: as expected, the forum is still ranked highest. The chorus/membership is now rated 2nd highest, followed by the Leadership. Site Owners and Terrorist Acts are still rated as the least important nodes.

Algorithm 3: Node and Link Uncertainty: In this case, we also adjust link values to account for degree of influence, inefficient communication, volume of communication, possibility of redundancy, age of evidence, and hedge words in feedback. Implementation of these factors reduces the influence of links values.

Algorithm 4: Recursive Algorithm: The intuition[40] behind the matrix in Figure III-5.5, and its recursive solution by a technique called relaxation, is that each page initially has one unit of *importance*. At each round of iteration, each page shares whatever importance it has among those it links to, and receives importance from those pages linking to it.

---

[40] http://infolab/Stanford.edu/~ullman/mining/websearch.pdf, accessed 31 May, 2007

Eventually, the *importance* of each page converges to a limit, which happens to be its component of the principal eigenvector of this matrix. Recall that the set of eigenvectors for A is defined as those vectors which, when multiplied by A, result in a simple scaling λ of A.

In the Google domain, this *importance* is the probability that a researcher, starting at a random page, and following random links from each page, will arrive at the page in question after a long series of links. If the importance is high, he is more likely to arrive at it sooner.

The convergence criteria for the iteration are that the sum of the errors (differences between input and output vectors) for all components values is less than 10^{-6}. Convergence occurred in less than 30 iterations for all 11-node cases tested.

Importance Points

To														From
O	1.5		0	0	0	0	0	0	0	0	0	0	$\frac{1}{1}$	1/24	O
D	1.5		$\frac{1}{1}$	0	0	0	0	0	0	0	0	0	0	1/8	D
W	2.9		0	$\frac{1}{1}$	0	0	0	0	0	$\frac{1/8}{2}$	0	0	0	1/3	W
S	3.2		0	0	0	0	0	$\frac{1/4}{1}$	0	$\frac{1/8}{2}$	0	0	0	1/4	S
F	14.7	=	0	0	$\frac{1}{1}$	$\frac{1}{1}$	0	0	$\frac{1/4}{1}$	$\frac{1/8}{2}$	0	0	0	1	F
L	6.9		0	0	0	0	0	0	0	0	0	$\frac{1/2}{4}$	0	1/24	L
M	28.5		0	0	0	0	$\frac{1}{1}$	$\frac{1/4}{1}$	$\frac{1/4}{1}$	$\frac{1/8}{2}$	$\frac{1/4}{1}$	0	0	1/2	M
C	11.8		0	0	0	0	0	0	0	$\frac{1/8}{2}$	$\frac{1/4}{1}$	$\frac{1/2}{1}$	0	1/12	C
A	13.8		0	0	0	0	0	$\frac{1/4}{1}$	$\frac{1/4}{1}$	$\frac{1/8}{1}$	$\frac{1/4}{1}$	0	0	1/3	A
P	13.8		0	0	0	0	0	$\frac{1/4}{1}$	$\frac{1/4}{1}$	$\frac{1/8}{4}$	$\frac{1/4}{1}$	0	0	1	P
T	1.5		0	0	0	0	0	0	0	$\frac{1/8}{1000}$	0	0	0	1/3	T

Figure III-5.5: Importance Under Uncertainty

Algorithm 5: Problems with Real World Graphs: The basic algorithm doesn't account for a few circumstances[41] that arise in practical networks. The discussion again focuses on nodes in a network rather than pages on the Web.

- *Dead ends*: a node with no outgoing links has no successors to send its importance. Eventually, all importance will "leak out" of the network. Suppose a node has no

---
[41] ibid.

outgoing links. This is a dead end case and the result of the iteration is that all nodes eventually equal zero.

- *Spider traps*: a group of one or more nodes that have no links out of the group will eventually accumulate all the importance in the network. Suppose a group becomes inwardly focused and has links only to itself. In this case, all importance is vested in that group.

Google Solution to Dead Ends and Spider Traps: Instead of solving the matrix directly, "tax" each node by some fraction of its current importance, and distribute the taxed importance equally among all pages. We used a 20% tax. The Goggle Solution with uncertainty perturbations:

Recursion Equation, Algorithm 5

$$\mathbf{N} = [\mathbf{p} \circ \mathbf{L}] \times [\mathbf{q} \circ \mathbf{N}] + \mathbf{T}$$

where $\mathbf{N}$ = node vector, dimension (n)
$\mathbf{p}$ = link perturbation matrix, dimension (n,n)
$\circ$ = Hadamard (component-wise) product
$\mathbf{L}$ = link matrix, dimension (n,n)
x = standard vector and matrix product
$\mathbf{q}$ = node perturbation vector, dimension (n)
$\mathbf{T}$ = tax vector, dimension (n)

The algorithm is programmed and evaluated for a wide variety of "taxes". The convergence is stable, typically iterating to the same solution as Algorithm 4, but with a different "total importance" which grows slowly and linearly with the tax rate.

User Interface: For this proof of concept, the network graph and its associated characteristics is driven by our choice of Pajek[42] as the graph visualization tool. A side-by-side comparison of the results obtained from these algorithms indicated intuitive results. For all but the democratic algorithm, Moderators are ranked as the most important node. Forums are always ranked either 1st or 2nd. Active members and postings are ranked 3rd in the recursive algorithms, with double the importance percent over non-recursive algorithms.

Node	Algorithm 1	Algorithm 2	Algorithm 3	Algorithms 4 & 5
Originators	9.1	8.2	9.2	1.5
Distributors	9.1	1	1.2	1.5
Webmasters	10.2	3.4	3.6	2.9
Site Owners	3.4	0.5	0.4	3.2

---

[42] www.indiana.edu/tutorials/pajek/, accessed 03/01/2015.

Forums	21.6	17.8	19.7	14.7
Leadership	4.5	12.4	3.5	6.9
Moderators	17	30.4	33.8	28.5
Chorus	8	14.7	16.2	11.8
Actives	8	5.7	6.3	13.8
Postings	8	5.7	6.1	13.8
Terror Acts	1.1	0.3	0	1.5

Node Importance for Various Algorithms

The visualization of this graph is easily accomplished in Pajek. The layout has nodes sized according to importance and link values. Colors are useful so node labels don't always need to be shown. The visualization clearly shows important nodes (larger in size) and influential links (numerical weights, thickness and gray scale can be added).

An example of a Pajek graph (Figure III-5.6) shows an energy-based graph representation (Kamada-Kawai, Free), with minor node repositioning for clarity.

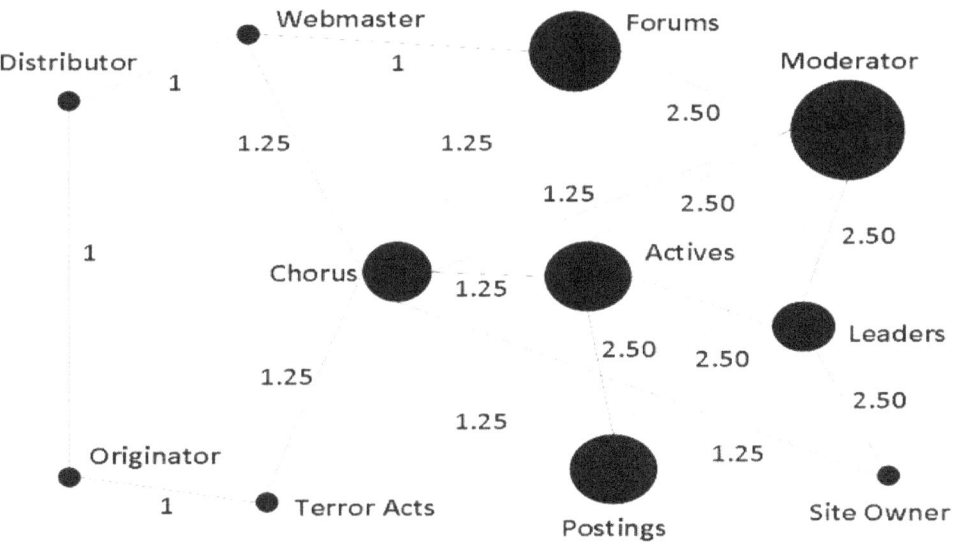

Figure III-5.6: Kamada-Kawai Visualization

Conclusions and Recommendations for Importance and Rank Algorithms: The algorithms give intuitive results. All are stable under large perturbations; for example, tax = 50%. No iteration is required for the 1st three algorithms. Iteration is fast for recursive algorithms and results for this small case are immediate. Scalability per iteration is linear, and algorithm scalability overall should be less than n * log (n), where n = number of nodes in a network. The networks modeled can be directed or undirected. Cycles are accommodated, including self-referenced nodes. Extension to compute other measures of social networks is straight-forward.

Operationally, this algorithm can help to identify a Center-of-Gravity (COG) as the node with the most importance. Here, importance is based on the number of inputs nodes to a node, the importance of those input nodes, the volume of input provided by the input nodes, and the strength of influence of links from input nodes.

Link Efficiency

As an example, the density (D) of a network is defined as the number of actual links divided by the number of possible links. For an asymmetric network (directional links), the number of possible links is n*(n-1), where n = number of nodes. Suppose there are 28 links connecting 11 nodes. The density is:

$$D = (\text{\# links} / \text{\# possible links}) = (\text{\# links}) / (n * [n - 1]) = 28 / 110 = \sim.25 \text{ or } \sim 25\%.$$

Suppose the link values (L) total 11. A generalized metric, the Effective Density (D_E) is defined as:
$$D_E = L / (n * [n - 1]) = 11 / 110 = .1 = 10\%.$$

The Link Efficiency (L_E) is then defined as the ratio of the Effective Density and the Density:

$$L_E = D_E / D = L / \text{\# links} = 11 / 28 = .39 \sim 39\%.$$

This generalization provides additional information: the graph is structurally connected with 25% density and the efficiency of these links is 39%.

Eigenvalue

The matrix calculator[43] computes inverses, eigenvalues and eigenvectors of up to 5 x 5 matrices, multiplies a matrix and a vector, and solves the matrix-vector equation **Ax = b**. An example (Figure III-5.7) using five network nodes is instantiated. Connectivity shown is accomplished using a connectivity matrix to represent the network.

Networks are explored and results are tabulated. The dynamic stability metric is the difference between the largest negative original and perturbed eigenvalues divided by the original eigenvalue. Large negative differences indicate less stability, as shown in F => B. It is the difference between the fully-connected network (F) and the baseline network (B) which is less stable.

---

[43] https://www.math.ubc.ca/~israel/applet/mcalc/matcalc.html, accessed 9/25/2014.

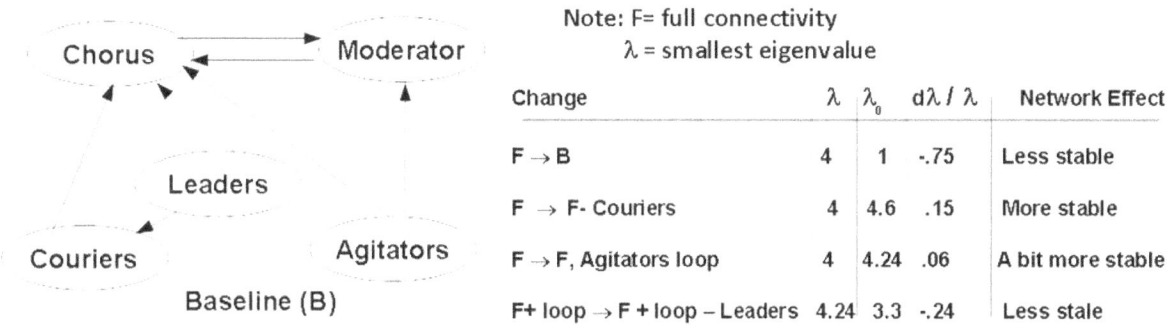

Figure III-5.7: Sample Eigenvalue Calculation

Volatility: the statistical standard deviation is a measure of the variation, or dispersion, of a parameter from its mean value. A more robust measure of the variation is the fractal dimension. A fractal is a geometric pattern that is repeated at ever smaller scales to produce irregular shapes and surfaces that cannot be represented by classical geometry. Fractals are used especially in computer modeling of irregular patterns and structures in nature. The most intuitive way of measuring the fractal dimension is the box-counting dimension (D_B)[44]. The fractal dimension is defined as D_B = log (# of self-similar pieces) / log (magnification factor). For time series, it has a value between 1 and 2. With frequent up-down fluctuations, D_B increases (Figure III-5.8), whereas there is no distinction between smooth and rapid changes with the standard deviation. Another advantages of the fractal dimension is that, unlike a standard deviation, it does not require stationarity.

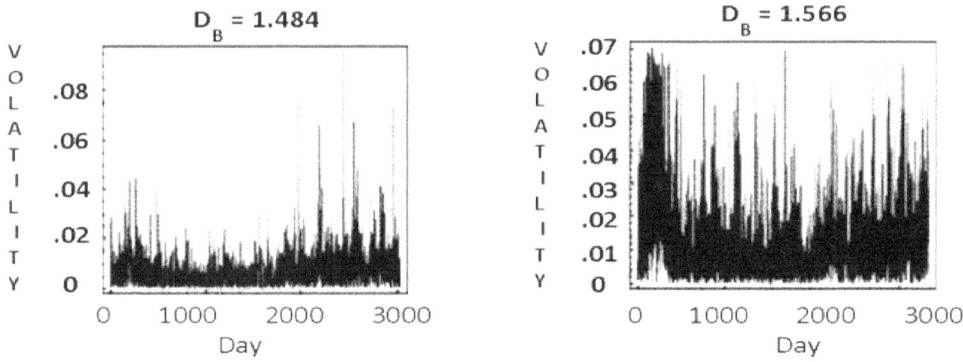

Figure III-5.8: Fractal Dimensions for Two Time Series

Chapter Summary

Network metrics have been defined, formulated, and implemented, based on link weights and node belief values, as appropriate.

---

[44] Ming, C., Statistical properties of volatility in fractal dimension and probability distribution among six stock markets. (2002). http://129.3.20.41/econ-wp/em/papers/0308/0308002.pdf, accessed 8/8/07.

Chapter III-6
Case-Based Reasoning

Case Based Reasoning is a feasible method for military planning and plan updating. In today's hectic world, the time available for generating plans, and updating those plans, is growing increasingly shorter. Therefore, methods must be developed to ease the use of semi-automated planning aids that assist decision-makers and planners. One goal of this project is to study new ways of using Case Based Reasoning algorithms to help the decision process of selecting a Course-of-Action, or plan, for a given situation from among preplanned options.

We describe prototypes of Case Based Reasoning algorithms[45] to solve the problem of selecting Courses-of-Action in four different situations: an optimum locator that determines where to put resources based on constraints; a terrain reasoner that determines how to navigate a patrol from the landing area to the defended area based on terrain, enemy, logistics, and time constraints; a collaborative planner that translates the broad objectives from a high-level directive into Course-of-Actions; and a plan update application that updates detailed plans based on best match to previously stored cases. Results are promising – based on our experience, case-based reasoning appears well suited as an algorithm for a wide variety of planning applications.

Introduction

New missions dictate new policies, weapons, and planning strategies. Such policy shifts have wide-ranging mission planning and execution implications for military commanders. Semi-automated aids are needed to optimize the planning and plan update cycles in future operational centers.

Many strategic planning organizations have been downsizing personnel since the late 1980s, in response to reduced budgets and rapidly changing technology goals. Mission planning, execution, and assessment are now assigned to smaller crews, who must handle more data — in fact, overwhelming amounts of data — in less time. They are responsible for planning and executing missions of increasing diversity and complexity — for example, just-in-time merchandising, computer network defense, and time-critical targeting of terrorists. In addition, because missions are executed in an increasingly dynamic environment, objectives change as the plan unfolds and the original plan fails. To ensure success, the planner needs semi-automated help in developing plans, monitoring execution as the mission progresses, and updating plans in light of continuous mission assessment.

Because of the complexity and size of the planning space and the number of possible combinations of options and constraints that must be considered, manual planning methods and most current automated methods are inadequate. They are narrowly focused, inflexible, time consuming, and non-scalable. As options or constraints are added, the plan complexity increases

---

[45] http://www.cs.cmu.edu/~sycara/cabins.html, accessed 04/08/2015.

several-fold, as does the time necessary to develop the plan.

Fortunately, new technologies and algorithms enable users to plan complex missions successfully, despite lack of time and manpower. Semi-automated, automated, and autonomous decision support tools reduce decision-making time. Unfortunately, few have been implemented in systems for operational users. To be useful, they must be integrated in operational planning systems with which the decision maker is familiar and in which he or she has confidence.

Many organizations, both government and civilian, are currently investigating the implications of Computer Network Operations[46] (CNO) systems for defending computer networks. Our interest is in two major objectives: (1) to investigate the feasibility of predicting possible threats before they occur, giving the decision-maker time to put preventative measures in place; and (2) to define a spectrum of operational concepts for CNO centers.

A major sub-objective is rapid mission planning and plan updating. This paper focuses on the development of an aid for generating Courses-of-Action. We exploit the latest technological innovations in system analysis, software engineering, and artificial intelligence to produce a decision toolkit of fast, flexible, semi-automated aids for planning, execution monitoring, and mission assessment. We focus on developing CNO planning support tools that can be incorporated into both routine and crisis planning environments.

Case Based Reasoning Basics

A Case Based Reasoning (CBR) system makes feasible decisions from preplanned options, rules, and uncertain reasoning. It also adapts solutions as the information or situation changes and learns from each experience.

Following a common type of human reasoning, reasoning by analogy, CBR algorithms "remember" stored data, both helpful and hindering, from past situations to develop a solution to a current problem. CBR accomplishes this task by having a database of past events organized into cases containing information about each scenario. Each case is thoroughly described by a list of defined index values.

Effective and desired index variables contain a value generic enough to judge the usefulness of a case but complex enough to discriminate amongst cases. CBR algorithms implement index values in the search for past cases that show usefulness in developing a solution to the reasoner's current problem. The discovered useful past cases are the basis for creating a problem solution. The solution strategy for this project is summarized in Figure III-6.1.

Upon the selection of legitimate past cases, a CBR algorithm advances through four basic tasks found in all CBR applications. After the creation of an appropriate case base, the process is reduced to these four tasks to solve and learn from a problem. First, retrieve the most similar case, or cases, to the problem at hand. Upon retrieval, use the rules specified to extract the best information and knowledge in that case, or cases, to solve the problem.

Once a possible solution is found, revise the proposed solution to better fit the current problem. Finally, retain the new solution, and other parts of the experience, for future problem solving.

---

[46] http://en.wikipedia.org/wiki/Computer_network_operations, accessed 04/08/2015.

The general process shown above retrieves the most similar cases by assigning weights to selection criteria and computes the score as a weighted sum of the selection criteria. The process is similar to an engineering trade study that is best described as finding several possible solutions to an identified problem, weighing each possible option as to its benefits and drawbacks as a solution and, finally, selecting the best solution based on a weighted sum. Similarly, the CBR algorithm takes an identified problem, searches a case base for similar past cases that show promise, and chooses the case that best matches working solutions from the past. The CBR approach goes beyond a trade study in that modifications to the selected case are allowed to create better options[47], and that the modified case is stored for future reference when similar situations arise. This implementation of the case based reasoner allows a user to modify the selection criteria, that is, change the problem to assist in finding the best possible solution.

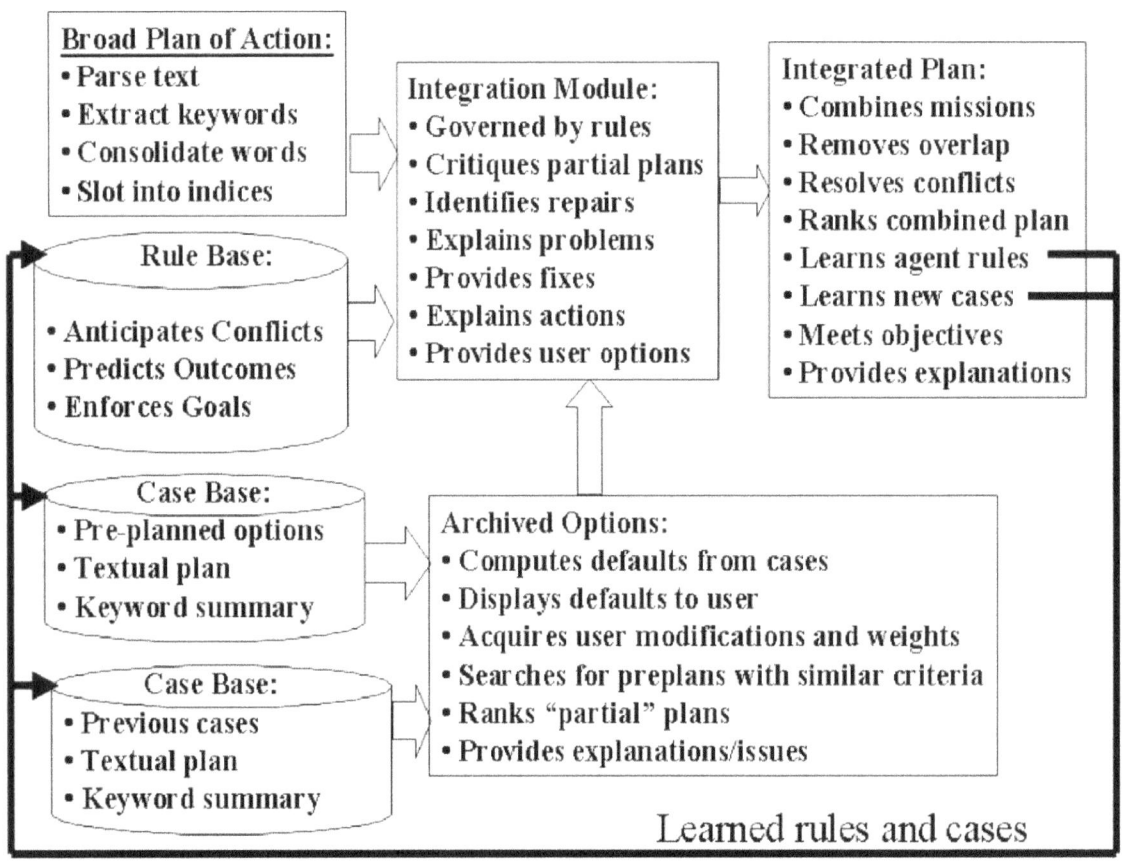

Figure III-6.1: Case-Based Solution Strategy

Approach

An effective study of the usefulness of any CBR algorithm as a decision-making technique requires an understanding of the process and the development of a prototype to demonstrate the

---

[47] J. Kolodner, *Understanding Creativity: A Case-Based Approach*, Georgia Institute of Technology, Paper, 1993

process. The prototype implements the basic process of CBR, allows further study, and creates a CBR reasoning tool for decision-making.

The implementation of any CBR application requires the creation of a legitimate case base. A case base contains organized information about past occurrences of a particular type of situation. Case bases differ among problems since it is a representation of acceptable, possible, and modifiable solutions to the problem. We focus on military planning because the process is well-established and is easily related to other domains. Due to the multitude of recorded military experiences and the highly structured nature of military planning, it is not difficult to create a reasonable case base for any current military situation. This ease of gathering past information to modify a past solution to a current problem makes CBR a logical choice of study for a decision-making and planning tool.

For example, for the Terrain Reasoner application described below, the case base contained "created" past Army Patriot Battalion missions grouped in cases that describe possible paths of travel and information critical to the mission, troops, and equipment. Each case includes fields of interest such as the mission and goal, route traveled, intelligence, fire support, mobility, air defense, and logistics. The case base is in essence the creation of routes presented as Course-of-Actions (COAs) to the Patriot Battalion Commander by his staff. Multiple cases are created to form the case base that is to be searched, allowing the selection of the "best" COA.

Included with the description of the case is the route information that consists of military intelligence data, terrain type, slope of the land, and probabilities of mission failure due to difficulties with personnel, equipment, weather, and communications. Also contained in a case is information on personnel and equipment replacement as well as data concerning the morale and welfare of the personnel. Theoretically, the number of possible cases is combinatorially explosive; however, CBR limits cases by "spanning" the space of possibilities with a limited number of feasible plans that can be "mixed and matched".

After determining the necessary information to include in the case base, an important aspect of CBR is to generically label, and provide values for, indices that define each case. Based on the information stored in the cases, index values are defined to describe each field of the case. Each of the index values describing the fields are defined to have discrete or continuous values. These values provide a general description of the information in that field about the case, but sufficiently separate it from other cases. These values specify under what circumstances a case is useful and retrievable for observation, manipulation, and selection.

Once index values are created, numerical weights are assigned to each possible value of each index variable. Based on a scale of 0.0 to 1.0, a value determined as best for a case is given a 1.0, while a value not considered good receives a 0.0. For example, under an index variable "terrain trafficability", a value of "flat" is the best (1.0), while the value "alpine" is assigned a 0.0. These weights are assigned to the index variable's values for every past case and the values describing the problem the prototype is to solve.

After creation of the cases, each with indices and associated values, the prototype allows the user to enter the values describing the conditions of the current situation. These conditions are described by assigning values and weights to the same indices used in the case base. To determine the cases closest to the desired final case, the weights of the index values describing the cases and the index values of user-entered criteria are compared. Point totals are assigned to a past case based on how the index values compare to the user-entered criteria. Higher point scores declare a case's similarity to the user-entered criteria.

Results

During the course of our studies, we developed four prototypes in differing military domains:

- Optimum Locator: places antennas at locations based on soft (logistics, terrain) and hard (satellite passes/day) constraints.
- Terrain Reasoner: navigates a route from initial area to final area based on terrain, adversary, logistics, and time constraints.
- Collaborative Planner: translates broad objectives into Courses of Action based on best match to previously stored cases.
- Plan Update: updates detailed plans based on best match to previously stored cases.

Optimum Locator

The Optimum Locator is a case-based planning algorithm that supports planning in a communications mission. The resource to be positioned is an antenna mounted on a HMMWV (HumVee or Jeep) that exploits high bandwidth data downlink from satellites. The goal of the application is to find the best location for the antenna based on kinematic considerations such as antenna and over-flight patterns of satellites and logistic constraints such as the availability of communications support.

Three types of operator selection criteria (Figure III-6.2) are defined: Resources and Measures of Merit, Satellite Accessibility, and Constraints. The input display allows the operator to start with defaults stored in a Case ID, choose ground sites and required communications connectivity, and define a figure of merit (here, image quality) and its desired value. The analyst also edits default values for selected satellites, stipulates desired number of downlinks supported per day, and special geometries (for example, spot beam). Finally, constraints such as mission support, basing, timing, keep-out zones, and supportability criteria are set.

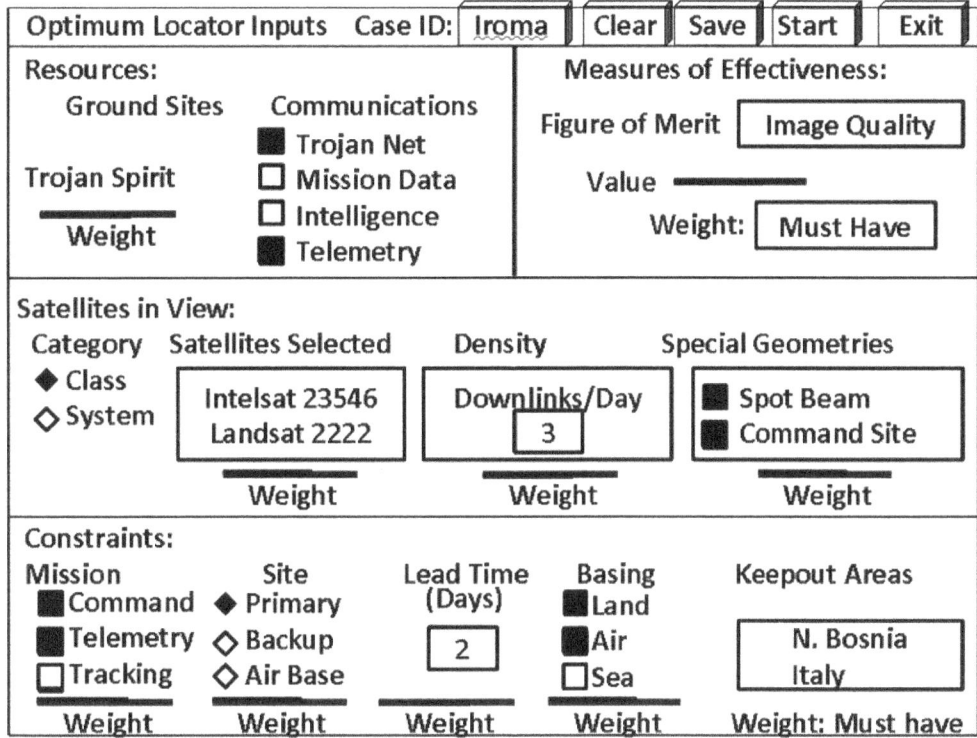

Figure III-6.2: Optimum Locator Input Display

The algorithm relies on a similarity metric that provides a quantitative comparison between operator selection criteria and site criteria for each of the cases in the case base. It produces pareto-optimum[48] "best fits" and explanations.

A geographic view (Figure III-6.3) provides a powerful visual explanation: icons show the optimum location that is within a required spot beam, antenna pattern (large circle) and meets other criteria as well. The other icon is not within the spot beam and are therefore don't score as well, while one icon is outside both the spot beam and the antenna pattern and scores poorly.

The detailed output (Figure III-6.4) consists of a visual characterization of how well each of the candidate sites met the selection criteria and a listing of the best sites along with the scores that they achieved. Clicking on a colored box produces an explanation.

The optimum locator is coded and has prototype quality software. A database of over 300 military installations in Europe and their characteristics is provided. Antenna patterns for 22 satellites are also provided. Astrodynamic and communications link analysis algorithms are integrated into the algorithm to compute satellite link geometries and bit error rates. A customer uses the code to determine where in the European Theater to place resources. Feedback from a customer indicates that the application efficiently sorts through the characteristics of military installations and the dynamics of satellite pass geometries to find the best locations for resource placement.

---

[48] http://www.businessdictionary.com/definition/Pareto-optimum.html, accessed 02/25 2015.

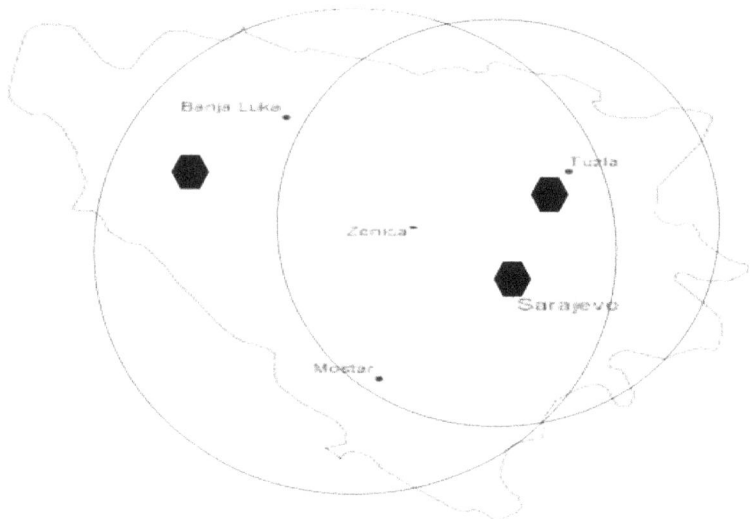

Figure III-6.3: Optimum Locator Geographic Display

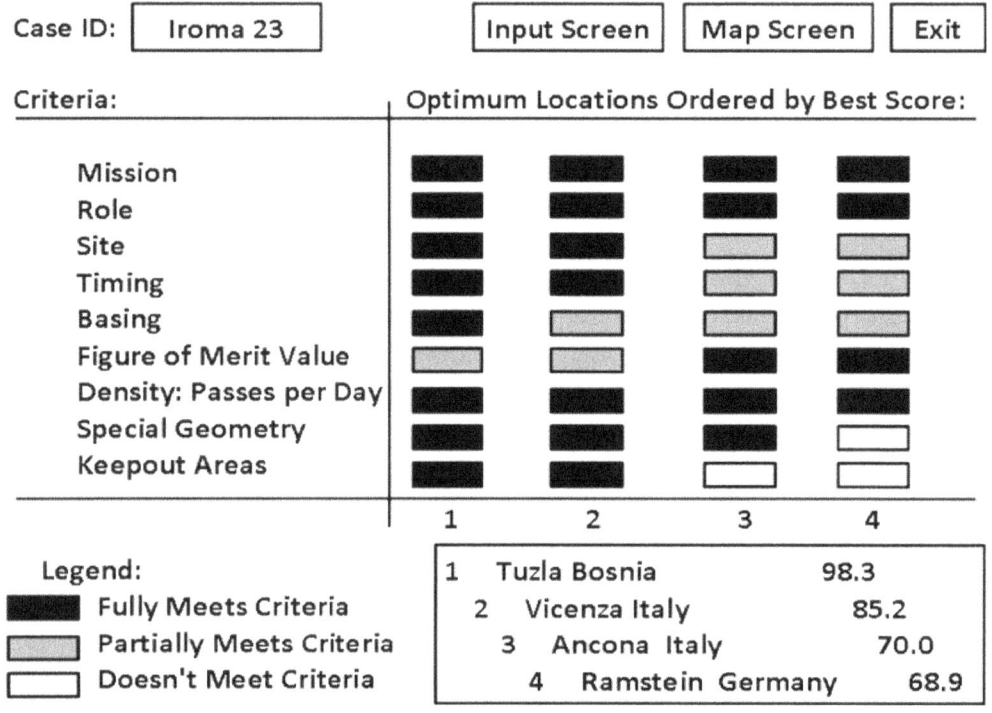

Figure III-6.4: Optimum Locator Output Display

For business applications, such as choosing the best product, or deciding where to locate a new facility, we configure an Excel spreadsheet to perform case-based reasoning.

Terrain Reasoner

The task is to develop a tool for the selection of a best route for ground troops to be deployed through varying terrain and in the face of obstacles such as enemy deployments. Depending on the need of the user and the development of the application, case-based reasoners provide the ability of acting in either of two modes - as a planning assistant or as a decision-maker. This demonstrates that CBR has the ability to provide various levels of user interaction. At one extreme, the user is allowed final say in all decisions and the reasoner is only used as a selective data provider. At the other "fully automated" extreme, the reasoner is autonomous and has complete control over manipulation of data and all decision processes..

For Terrain Reasoning, we implement CBR as a reasoning tool replacing a real person in war-gaming simulations. Ideas come from computer gaming[49]. The user has control over what type of route is desired and is allowed to manipulate index values to simulate last minute changes, but has no control in selection of the appropriate route. This simulates a military commander's use of subordinates to collect information and design plans, while retaining autonomy and responsibility for making the final decision based on an overview of each plan. For this project, a person simulating a commander's subordinate creates the case base of routes for a specific area and stores the information. A user then specifies the type of route needed by selecting the appropriate index values to describe the characteristics of the new situation or mission. The selected index values are used to search the case base for matching or similar cases.

An important aspect of the project is providing a visual method for determining whether the system behaves as predicted or desired. To that end, we develop a graphical display that shows Google-like maps with routes drawn to present the results of the CBR planning tool during a Terrain Reasoning planning session. As an example, the CBR algorithm chooses and displays three routes and associated similarities to weighted indices. At this point, the user decides which route is the most appropriate.

Collaborative Planner

The task is to develop a tool for the selection, based on the broad objectives contained in a high-level directive, of a set of Course-of-Actions from previous cases, simulated or actual, with similar situations.

The characteristics of this decision are: a decision-maker is given broad objectives that result in an ill-posed problem, exponential explosion of possible solutions, many dimensions or criteria, and difficulty in quantitatively rating the worth of a strategy. When faced with these circumstances, people often attempt to 1) recall a course-of-action that worked in previous similar circumstances, 2) caucus to arrive at a consensus; that is, group-think; or 3) because of cognitive overload, identify a dominant criteria and assemble a strategy around it. We implemented and tailored a case-based planning tool that provides computer-assistance for the first option. It retrieves preplanned options, computes scores based on weighted selection criteria, and presents options for the decision-maker to edit. It helps an operator produce

---

[49] www.gamasutra.com, accessed 01/15/2015.

Courses-of-Action (COAs) for inclusion in a Commander's Estimate. Advantages of this approach are that it mimics the human decision-making process, produces a finite set of cases, computes a score, doesn't "jump to conclusions", and provides an explanation.

Again, we provide a graphical display as discussed earlier to present the results of the CBR planning tool during a Course-of-Action planning session.

The 'n' closest matching cases, where 'n' is a user selected parameter, in the database are highlighted (similar to Figure III-6.4) across the lower right of the display. Positioning the mouse over a "Case n" button will display the case name. Clicking on a button will select that case and show the match criteria of that case versus the incoming directive.

Plan Update

The task is to develop a tool for the selection of incremental plan repairs from a database of previously used and saved cases. This approach is referred to as ***revision-based*** plan update. It starts with a precomputed plan and incrementally "fixes" that plan in response to unanticipated events that take place during plan execution. The advantage of this approach is that the plan has a guaranteed monotonic increase in quality after each update. The effect of the unanticipated events on the current plan is determined using our Dempster-Shafer Belief Network mission assessment algorithm that provides a level of belief that the plan is, or is not, meeting expectations.

The construction of a cognitively or computationally cheap sub-optimal plan that is then incrementally updated to meet unanticipated events is preferable in practice, because one can interrupt a plan that is no longer appropriate or performing satisfactorily and use a new, updated approach.

Using a CBR approach has the advantages of dealing with uncertain data (not just numbers such as performance metrics), acquiring user knowledge in complex domains (from experience, the user knows that a given plan update is expected to provide the "best" results for this situation), and expending less effort in knowledge acquisition than would be necessary for a rule-based (that is, expert system) approach.

Since it is impossible to judge beforehand the effects of a planning decision on the mission objectives, a plan update must be applied to the plan and its outcome evaluated in terms of the resulting effect on those objectives. Therefore, having a single update action as a case provides advantages in terms of focus and trace-ability of the planning process.

A case represents application of an update action to one activity of the plan, the update context, an indication of how well the update worked in the past in similar situations, and a suitable update action. CBR allows capture and reuse of this knowledge to dynamically adapt the planning procedure and differentially bias planning decisions in future similar situations.

As with all the other implementations, we developed a graphical display (Figure III-6.5) to present the results of the planning tool during a Plan Update session.

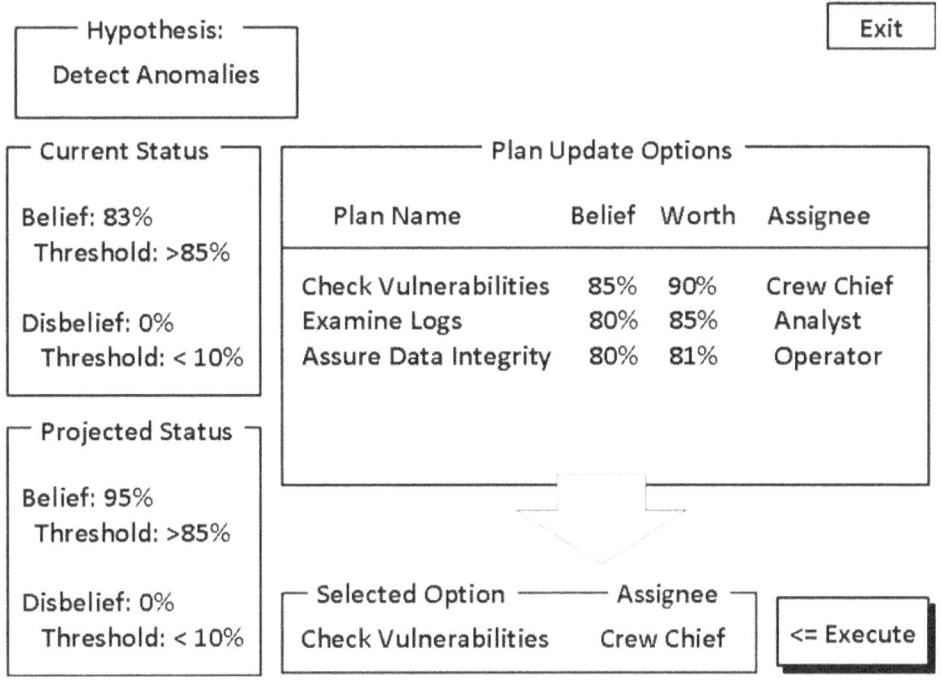

Figure III-6.5: Plan Update Strategy Selection Screen

This display shows all the possible strategies applicable to updating the plan at this point, the figure of merit (belief) for the plan updates and the "worth" of the plan update. The worth is a parameter value that is learned based on the percentage of times that the plan update either worked successfully or was shown to work successfully in a simulation.

Relevance/Significance of Results

As with all practical artificial intelligent applications today, the effectiveness of a CBR process is due to the restricted domain of the problem. Current CBR systems are domain specific and the designs for each problem reflect index values and past cases that fit strictly in that defined domain. Creating a case base and the needed index values to define information and solve a problem in a more general domain is a complex task. Our applications are useful primarily because they exploit a narrow domain.

CBR is not a general-purpose algorithm for military decision-making. While it is well suited to static planning and plan update, it does not compete with rule-based reasoning for decisions based on doctrine, or belief networks for uncertain reasoning with incomplete data. For this reason, upon completing the one defined goal of a single CBR system, an entire new CBR application would have to be developed and implemented to plan and decide the next task. For example, this project implemented a simple CBR algorithm to decide the best route choice for a route to travel. Finding the path of travel for a ground unit is the first and most basic step of a planning process. An entire new CBR or other artificially intelligent systems would need to be developed for every goal a ground unit must accomplish. Nonetheless, several CBR applications exist that exhibit successful planning and "learning" algorithms and research continues on the creation of broader CBR processes. For example, Hammond[50] (1986) describes a computer program that successfully created tasty, new recipes from a case base of simple stir-fry recipes and ingredients. Also, CBR systems are being developed that have access to multiple case bases and indexing schemes allowing the planning and solving of multiple tasks.

In the world of decision-making, intelligence is dynamic information and is always changing and sometimes incorrect. To alleviate this problem, a user must see the best matching possible cases found to match the desired criteria and be allowed to manipulate the data. Operator interaction allows updating cases due to changed real-time data. In addition, viewing cases before making the final decision allows a human user to check for errors. In the natural world, information constantly changes or is incorrect so this aspect needs to be available. Also, cases are usually planned well in advance of their actual need, so it is important to understand the natural situation of changing data and allow case modification before final decisions are made. Once a user has completed modifications, the cases' indexes are again computed and re-compared to final case criteria. The best match is selected as the final case to be used by a military commander for producing or updating the plan.

Finally, the goal of this project is to evaluate CBR as a method of augmenting high-level military decision-making. After developing a trade study, studying effective commercial and academic CBR applications, and prototyping several CBR applications, CBR appears well suited as an algorithm for a wide variety of planning. See Chapter V-11 for an application of case-based reasoning to longevity prediction. Depending on the domain of the problem, a developed CBR application could provide cognitively and computationally cheap, automated planning. As technology develops, especially in the artificial intelligence field, CBR algorithms may be developed to achieve goals in a variety of related decision processes and, combined with other artificial intelligence techniques, create a simulated "thinker" capable of human-like planning decisions in situations with dynamically changing plans driven by uncertain and incomplete information.

Related Directions

We have algorithms for information extraction into evidence templates. This technology allows automatic extract information from free text and semi-structured reports to automatically index cases, thereby building cases from text.

The literature cites work in creativity theory[51] that uses a case-based approach. Can case-based reasoning be leveraged to endow the computer with the ability to identify creative hypotheses and to find creative solutions?

Another source of creative solutions is to use our existing genetic algorithms to find optimal solutions; that is, evolve cases to obtain highest case-based score by combining indices from multiple cases for a new solution. Novel solutions may be obtained by experimenting with the "diversity" parameter in the genetic algorithm code. See Chapter III-7.

---

[50] http://alumni.media.mit.edu/~jorkin/generals/papers/Kolodner_case_based_reasoning.pdf, accessed 02/26/2015.

[51] http://www.cs.indiana.edu/~leake/projects/swale/, accessed 02/25/2015.

Commercial applications of case-based reasoning include video games. Gamasutra[52] employs this technology to "enemy forces" in first-person shooter games to make the enemy interesting and predictable. The ability to simulate a responsive, intelligent enemy is lacking – it is usually scripted - in current command and control simulations.

Summary

Exploiting emerging computer science technologies and algorithms, we have developed an integration test bed for studying issues related to military mission planning. Specifically, we used the test bed to prototype and demonstrate Case Based Reasoning algorithms for Course-of-Action planning and updating. More generally, we have used it to develop domain expertise in Computer Network Operations mission planning to meet hacker threats that affect banking, retail, and government domains.

The CBR process can be conceptualized as a trade study involving the selection, manipulation, and learning of previously saved cases. Using the trade study idea of comparing the weights of effectively defined and indexed cases and problems produces a reasonable solution in a large number of situations, using a simple Excel spreadsheet.

Powerful results are achieved by combining human skills (such as creativity and common sense) with computer skills (such as number crunching, searching, and remembering) using Case Based Reasoning algorithms.

---

[52] http://www.gamasutra.com/, accessed 03/30/2015.

Chapter III-7
Genetic Algorithms and Plan Optimization

We explore genetic optimization to investigate the implications of optimizing strategic, theater, and tactical plans. We choose the military planning domain because the planning process is well-defined and highly structured. However, the reader will appreciate the extension of genetic optimization to business, transportation, computer network defense, and other domains. The environment within the Department of Defense is changing. The advent of Missile Defense signals a shift from an "offense only" policy for deterrence to a weapons arsenal containing both strategic and theater offensive and defensive systems. The same is true of the computer network operations domain. A good offense may be the best defense. Wide-ranging mission planning and execution implications for decision-makers are apparent. The need is therefore to identify the benefits associated with semi-automated help in planning[53] in strategic and tactical offensive and defensive systems in future operational centers.

Mission planning, execution, and assessment are being handled by smaller crews which are subjected to shorter timelines and overwhelming amounts of data. The diversity of missions including both offensive and defensive options that must be planned, and the complexity of executing these missions, is increasing. For these reasons, the mission commander needs semi-automated help in developing plans, monitoring execution as the mission progresses, and updating the plan based on continuous plan assessment. One focus of this research project is the development of a planning aid for optimizing detailed plans. We use a genetic algorithm to choose combinations of response options (actions) and targets that best satisfy an operator specified goal. Results are scalable: a search space of eight trillion possible combinations is efficiently searched to produce an optimal solution.

In this chapter, we also discuss two helper algorithms: the Replan Trigger that provides an optimal policy to tell an operator when to replan, and a Planning Horizon that computes the practical plan-ahead time for detailed plans, based on the cumulative effect of uncertainties on a plan.

Introduction

The objective of this algorithm prototype is to produce a fast, flexible, semi-automated decision toolkit that provides aids for mission assessment, planning, and execution monitoring by using the latest technological innovations in the fields of system analysis, software engineering, and artificial intelligence. This Chapter discusses the underlying technology of a prototype implementation of a weapon-target pairing algorithm and its integration into this decision toolkit.

Objective

Mission planning, execution, and assessment are being handled by smaller crews which are

---

[53]Lugar, G.F. & Stubblefield, W.A., "Artificial Intelligence – Structures and Strategies for Complex Problem Solving", http://www.abebooks.com/book-search/isbn/0805311963/, accessed 04/08/2015.

subjected to shorter timelines and overwhelming amounts of data. The diversity of missions that must be planned; for example, Offense/Defense Integration (ODI), Missile Defense, Computer Network Warfare, and Time Critical Targeting, and the complexity of executing these missions, is increasing. In addition, the environment in which the missions take place is increasingly dynamic, thereby causing the objectives to change as the plan unfolds and the original plans fail. For these reasons, the mission commander needs semi-automated help in developing plans, monitoring execution as the mission progresses, and updating the plans based on continuous situation assessment. For concreteness, we address ODI.

The obstacles described above are large and complex in the areas of strategic planning. Attempting to develop a mission plan in which both offensive and defensive options are available, and can be used to complement each other, increases the plan complexity several fold. If tactical missions are added to the equation, the complexity becomes overwhelming, especially if the planning must be done manually.

Fortunately, new technologies and algorithms are available to overcome many of the impediments described above. Unfortunately, few of these technologies have been implemented in multi-strategy reasoning systems. Today, there exists no integrated decision toolkit to satisfy the required functionality.

These semi-automated, automated, and autonomous decision support tools are not only needed for reducing the decision making time, they must also be included in planning systems with which the decision maker is familiar. Hence, the effort of this study is directed toward incorporation of proven optimization tools into planning environments. Specifically, the planning process includes detailed plan development, monitoring, and updating. It is toward this goal that the algorithms described here are aimed.

We should, at this point, define what a detailed plan means in the context of this study. A detailed plan is defined as a set of 1-to-n {Response Option, Target} pairs that produces a desired effect, where a Response Option is an action taken against a target. In order to limit the complexity and size of the data used in this study, we limited the number of ROs to 14 and the number of targets to 50. The implementation of the decision aid, however, is limited only by the amount of memory and computing power available.

Genetic Algorithm Basics

Genetic Algorithms (GAs) are patterned after the processes underlying evolution: shaping a population of individuals through the survival of its most fit members. Three distinct phases are applied in solving a problem using genetic algorithms:

- Define the objective function: a fitness function that judges which individuals are the "best" life forms – that is, most appropriate for the eventual solution of the problem. These individuals are favored in survival and reproduction, thereby shaping the next generation of potential solutions.
- Define the genetic operators: mating and mutation algorithms, analogous to the sexual activity of biological life forms, produce a new generation of individuals that recombine

features of their parents.
- Define a representation: the individual potential solutions of the problem are encoded into representations that support the necessary variation and selection operations.

Eventually, a generation of individuals is evolved from the original population as the solution for the problem.

To use a genetic algorithm to solve our optimization problem, we represent a single solution to the problem in a single data structure. Defining the appropriate representation is the single most difficult (and important) aspect of using GAs and is still considered an art rather than a science. The representation chosen must represent any solution to the problem, but cannot represent unfeasible solutions.

For the implementation described here, we use two arrays of integer numbers (one array composed of the integers 0-13 and the other composed of the integers 0-49) representing the Response Option and the target, the combination of which (taken "n" at a time) defines the makeup of each individual of the population.

After evaluating each candidate, the algorithm selects pairs for recombination using ***genetic operators*** to produce new solutions that combine components of their parents. There are a number of these operators, but the two most common are ***crossover*** and ***mutation***, both of which are used in the implementation described here. Crossover takes two candidate solutions and divides them, swapping components to produce two new children that have genes that are swapped (crossed over) from their parents. Mutation takes a single parent and randomly changes some aspect (gene) of it to create a new child. Mutation is important in that the initial population may not contain some essential component of an optimal solution. Mutation is needed to introduce this component in later populations. It has the practical advantage of driving the solution to a global optimum, while avoiding local minima.

Approach

An important prerequisite to the plan optimization process is determining the objective function; that is, the metric to optimize. We quantify the potential for ODI to provide a "force multiplier" based on a new performance metric - the Ratio of Damage Expectancies (rDE). This metric is derived from the ratio of Damage Expectancy (DE) of Blue-on-Red to Damage Expectancy of Red-on-Blue (throughout this paper, "Blue" means friendly forces and "Red" means enemy forces). DE is a concept obtained from the strategic deterrence community; that is Intercontinental Ballistic Missile analysts, and modified for our needs. It is palatable to the Missile Defense community because it naturally includes the concept of threat "leakage": Red-on-Blue DE is proportional to leakage. Further, pre-launch destruction of Red resources, a factor in Blue-on-Red DE, is also directly proportional to leakage.

Damage Expectancy is composed of three probabilistic factors: Probability of Pre-launch survivability (SV), Probability of Arrival (AR), and Probability of Damage (DM). The rDE is determined as outlined below.

a = SVbr = Pre-launch survivability (Blue on Red)
b = ARbr = Probability of Arrival (Blue on Red)
c = DMbr = Probability of Damage (Blue on Red)
d = SVrb = Pre-launch survivability (Red on Blue)
e = ARrb = Probability of Arrival (Red on Blue)
f = DMrb = Probability of Damage (Red on Blue)

$$rDE = (a \blacklozenge b \blacklozenge c) / (d \blacklozenge e \blacklozenge f)$$

Default values for (a,b,c,d,e,f) are available from a database. These factors are based on rules of thumb obtained from subject matter experts. They do not depend on target class, instance, or value. Representative values are based on past experience, best estimates, and system analysis. The "(C)" and "(N)" indicators following the actions in the table refer to the weapon type used: conventional or nuclear. Other types may be used if desired.

Action	SVbr	ARbr	DMbr	SVrb	ARrb	DMrb
Strategic Defend (C)	0.8	0.9	0.7	1.0	0.1	0.7
Tactical Defend (C)	0.8	0.9	0.8	1.0	0.1	0.9
Strategic Preempt (N)	1.0	0.9	0.8	0.2	0.9	0.7
Tactical Preempt (N)	1.0	0.9	0.9	0.1	0.9	0.7
Strategic Preempt (C)	1.0	0.9	0.3	0.5	0.9	0.7
Tactical Preempt (C)	1.0	0.9	0.7	0.4	0.9	0.7
Strategic Destroy (N)	0.6	0.8	0.7	0.4	0.9	0.7
Tactical Destroy (N)	0.6	0.9	0.8	0.2	0.9	0.7
Strategic Destroy (C)	0.6	0.8	0.7	0.8	0.9	0.7
Tactical Destroy (C)	0.6	0.9	0.8	0.6	0.9	0.7
Strategic Retaliate (N)	0.5	0.8	0.9	1.0	0.9	0.7
Tactical Retaliate (N)	0.5	0.9	0.7	1.0	0.9	0.7
Strategic Deny (C)	0.9	0.9	0.4	0.6	0.9	0.7
Tactical Deny (C)	0.9	0.9	0.5	0.4	0.9	0.7

We use data mining techniques to assure that the heuristics underlying the estimated table values are consistent and complete. The tool we use is a public domain tool called Weka that uses a rule-induction tree algorithm, the J48 Classifier[54], to visually identify outliers and portray relationships. We find that the rule tree highlights the hierarchical structure of the data and clearly identifies "outliers" for further processing. Specifically, when we insert the data from the above table into the rule induction algorithm, it initially provides a very asymmetric rule tree with many "weak" branches that point to inconsistent numerical values. After modifying these values and re-executing the rule induction algorithm, the tree is pruned of inconsistencies and gets smaller because fewer rules are necessary to capture the parameter values. A few iterations are necessary to "catch" all inconsistencies. The insight we gain is that the data mining "tree" discovers structure in the data – inconsistencies, omissions, and conflicts are explicitly shown.

So, after defining the objective function, we tackle the problem of formulating a suitable representation and solution strategy: Given the desired number of actions ("n") and a desired

---

[54] http://weka.sourceforge.net/doc.dev/weka/classifiers/trees/J48.html, accessed 02/25/2015.

Ratio of Damage Expectancy (rDE), we determine a detailed plan, which consists of "n" {Response Option, Target} pairs, that most closely matches the desired rDE. Solution of this problem consists of searching through all possible combinations of pairs. Due to the size of the search space, an exhaustive search is out of the question. We decide that using a genetic algorithm (Figure III-7.1) is the best approach.

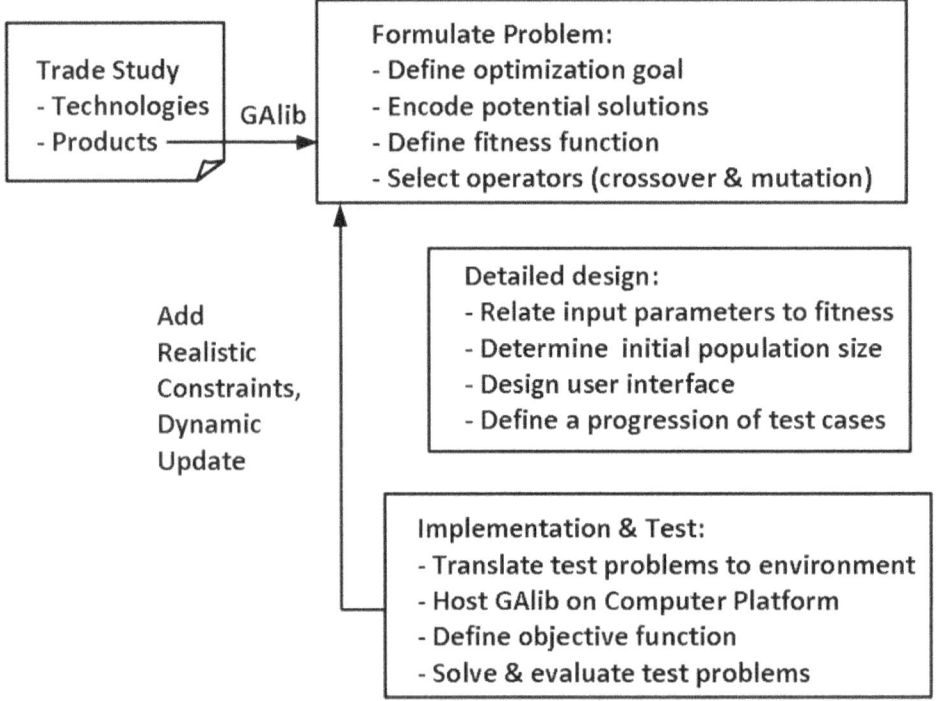

Figure III-7.1 Approach to Genetic Optimization

Since the objective of this study is to develop a decision tool using GAs, as opposed to implementing the underlying algorithms, we choose to use an available genetic algorithm implementation. The results of our trade study show that the GAlib product developed by Matthew Wall at MIT[55] provides the features and capabilities that best fit our needs.

The instantiation of the representation of the problem consists of using arrays of integers to represent both the Response Option and the target. This allows us to use the integers returned by the GA as indexes into tables of data associated with the Response Options and the targets. Each Response Option consists of the following fields:

- Name – Assigned by user for the user's reference
- Weapon – Type of weapon used (Conventional or Nuclear)
- Arena – Area in which action takes place (strategic or theater)
- Type – Type of action (Preempt, Defend, Deny, Destroy, Retaliate, etc.)
- Factors – Six Response Option factors making up the rDE
- rDE – Preomputed rDE before constraints

---

[55] http://lancet.mit.edu/ga/, accessed 01/20/2015.

A related planning tool included in the prototyping test bed is the Target Assessment tool which allows the decision maker to examine the current attributes of targets contained in a database for selection of the most appropriate targets for use in the target part of the weapon/target pairings.

Each target consists of the following fields:
- Name – Assigned by user for the user's reference
- Coordinates: latitude, longitude, altitude
- Locale – Area/City in which target is located
- Weather conditions : dynamic update required
- Type – Classification of target (Airbase, Sub port, Chemical facility, etc.)
- Value – Importance of target to the mission
- Hardness – Hardness of target
- Mobility – Mobile or Fixed target

In genetic algorithm terms, the {Response Option, Target} pair is referred to as a genome and in our implementation is composed of two genes – the Response Option (represented by an integer in the range of 0–13) and the Target (represented by an integer in the range 0-49).

The solution strategy is to use a genetic algorithm that starts from an initial population randomly generated by GAlib and then, using crossover and mutation operators, evolve new genomes for each individual in the population consisting of new combinations of genes. As each new population is generated, a user-supplied "objective" function (as described above) is called to determine which individuals should be kept and which should be dropped for the next generation.

Our initial implementation imposes one constraint on the objective function: target values must be above a specified number. As more constraints are added and the number of ROs and targets increase, the number of possible combinations and the complexity of calculating rDE increases exponentially – certainly beyond the capability of finding the optimal solution manually, or even with automated exhaustive search techniques.

The objective function calculates an rDE using data from the Response Option table (indexed by one gene integer) and the target table (indexed by the other gene integer) according to a combination rule described below. The rDE is returned to the GA and when it matches the criteria set by the initialization parameters, the "optimal" solution has been found.

Combination Rule Algorithm:
For each factor (for example, SVbr)
Find all actions on same target instance
- if not dependent, use OR rule P = { 1 - (1 - p_1) (1 - p_2),.. }
- if dependent, use AND rule: P = $p_1 * p_2 * ...$)

Find all actions on different target instances
- look up for each target

- normalize so the sum of the values is unity => weights
- apportion factors according to weights

Results

Since this is a proof-of-concept project having the ultimate goal of being integrated into an operational system, an important aspect of the project is providing a visual method for determining whether the system is behaving as predicted or desired. To this end, we developed a graphical display (Figure III-7.2) showing the results of a GA-based optimization during a detailed planning session. On the X-axis of the graph are the targets and on the Y-axis of the graph are the Response Options. The results shown here are for a detailed plan composed of three {Response Option, target} pairs. The lines connect the 'n' pairs that comprise the plan (three pairs in this case, resulting in two lines). The lines are significant in that they visually show patterns: a vertical line indicates Response Options against a single target while a horizontal line indicates a single Response Options against multiple targets

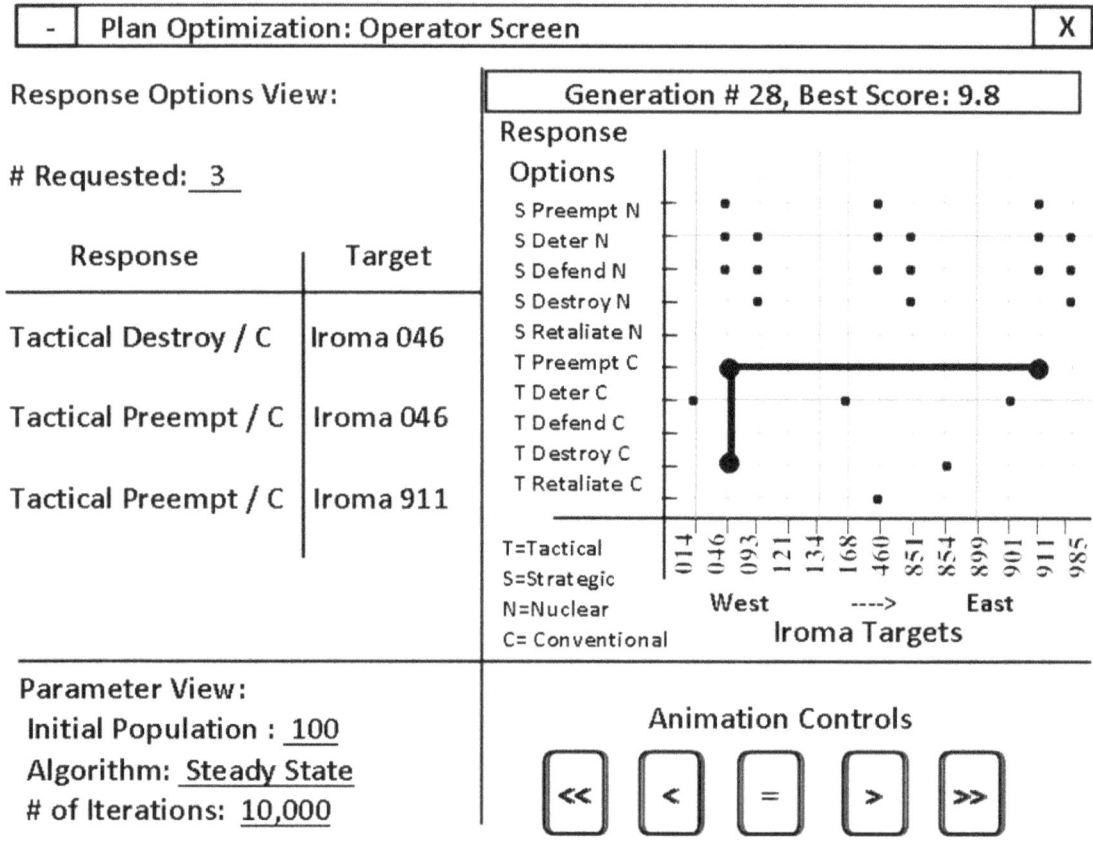

Figure III-7.2 Genetic Optimizer Output Screen

The population size is 250 individuals and the optimal solution is found at generation 21 with a best score of 0 (that is, the desired rDE exactly matches the calculated rDE for that generation). The 'Response Option View' area shows the three {Response Option, target} pairs making up the detailed plan. For this example, they are:

- **Tactical Destroy with Conventional weapons against Target 046**
- **Tactical Preempt with Conventional weapons against Target 046**
- **Tactical Preempt with Conventional weapons against Target 911**

The 'Parameter View' area shows the values of the genetic algorithm-specific parameters, including the initial population size, the GA algorithm used, and the maximum allowable number of iterations.

Decision Algorithms test bed

One of the overall objectives of this effort is to implement and demonstrate a prototype decision tools test bed that allows us to explore the use of a variety of planning tools in the context of command and control systems. This Chapter describes one of several of the tools implemented.

As stated earlier, to gain the most benefit from the advantages that semi-automated planning tools can provide, they must fit into an existing operating environment and application. As illustrated in Figure III-7.3, we integrate the GA-based planner into our prototype test bed – it receives broad planning objectives from the Course-of-Action Planner and the Target Assessment applications and provides an optimum detailed plan to the decision maker.

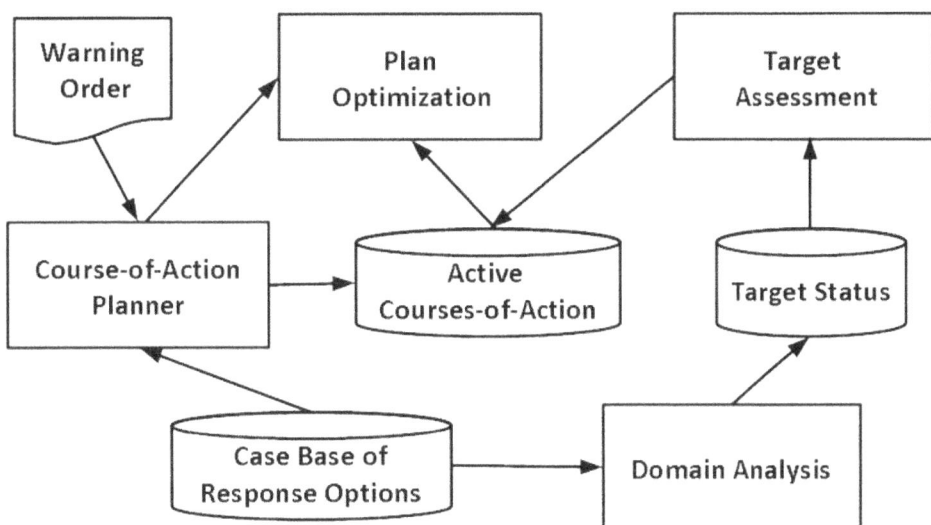

Figure III-7.3: Course of Action Flow Diagram

Future Research Directions

Here are ideas to enhance the implementation to provide additional options and capabilities:

- Find the most efficient parameter settings and benchmark our implementation
- Apply temporal constraints:
    - $T_{preempt} < T_{defend} < T_{retaliate}$

- Tpreempt < Tdestroy < Tretaliate
- Tpreempt < Tdeny
- Apply weapon constraints (for example, weapon reserve, firing doctrine, fratricide,…)
- Apply temporal constraints for a detailed schedule; for example, response options have duration, minimize overlap
- Customize crossover, mutation, representation, and objective function

A particularly difficult problem for military decision-makers is to decide when to either repair or replace a plan, given that it is going (or predicted to go) bad. When faced with this circumstance, people often: 1) abandon the plan and continue the mission in an ad hoc fashion; 2) continue with the plan for too long a period of time; 3) cannot decide whether to repair or replace the plan; and 4) generally don't seem to have a cognitive strategy for coping with the problem.

Replan Trigger

Another tool planned for inclusion into the toolkit is a "Replan Trigger". This tool monitors a Probability of Mission Success metric to assess the "goodness" of the currently active detailed plan to order to help a decision-maker decide whether and/or when to repair or replace a plan. It is based on an optimal policy algorithm that computes the "best" time, "now versus later", to repair a detailed plan. The decision point is based on a long-term strategy that minimizes the ratio of planning time to execution time. It is a function of mission time, probability of success of the current plan, the time required to repair or replace the current plan, and the anticipated success of the new plan. We found a optimization problem[56], called "Controlling a Production Process" that we translate to the planning domain. We mention this because, in almost all cases, a decision support tool that finds applicability in one domain will have equal applicability in many other domains. Although the optimization law complex, it is easily stated in a single equation:

$$C/R = [\lambda t e^{-\lambda t} + e^{-\lambda t} - 1] / (1 - e^{-\lambda B})$$

where C/R = Cost to repair at time "t" versus at breakdown
 λ = Rate at which the current plan is failing (% /hour)
 t = Repair time (hours)
 B = Breakdown time (hours)

Suppose the projected probability of success for the plan an hour ago was 90% and the projected probability of success now is 80%. At his rate (λ = 10%/60 minutes = .1333 %/minute) the plan will be less than 70% successful in another hour and be deemed to have broken down (B = 60 minutes) . The equation is evaluated for t and solved for the "cost now / cost later" ratio (C/R). Once the decision has been made to repair the plan, the planner uses the same Target Assessment tool to select new targets and the GA-based detailed planning tool to determine a new detailed plan.

---

[56] http://www.amazon.com/Applied-Probability-Optimization-Applications-Mathematics/dp/0486673146, accessed 02/15/2015.

Planning Horizon: A companion to the repair algorithm is the Planning Horizon. Specifically, this application addresses detailed planning. Broad, strategic, goal-oriented planning, in contrast, can be projected out for years. In practice, the question, how far ahead should I plan, often arises. The question has less to do with available time to plan than with the realization that uncertainties will grow over time and make the detailed plan meaningless.

Three examples illustrate the need to establish a planning horizon:
- **Climate Change**: we see projections of the detailed effects of climate change for the next 100 years. Really? How could these simulated results possibly be worthy of study, given that climate is chaotic, experimental data is sparse and uncertain, and partially understood c, and likely chaotic, climactic mechanisms (tipping points) are at work. Climate predictions would be more credible if a thorough analysis of all sources of uncertainty were undertaken and if these were "rolled up" into a planning horizon for climate predictions.
- **Space Object Collision**: typical near-earth space object tracking results in a ellipsoidal error volume that is approximately 10 km. x 1 km. x 1 km. This error ellipsoid grows over time, and new observations are continuously required to update orbital parameters. Any predictions of collisions two weeks in the future are therefore suspect, when we know that orbits are updated multiple times per day. Space object collision predictions are overwhelmed by uncertainty with so many possible collisions cited that satellite owners and operators typically ignore the warnings. This is a fairly easy problem to address because the uncertainties, though large, are known.
- **Anti-Satellite Engagement Windows**: this was the problem we tackled. The questions was, how far ahead can detailed anti-satellite engagement windows be planned, given a myriad of uncertainties, including, target location predictions, target probability of maneuver, anti-satellite readiness uncertainties, and even the possibility of intervening political truces..

Our technical approach was to use a little-known two-dimensional Monte Carlo sampling theorem[57]. Suppose we have a data sample with time values between 0 and 100 and plot it as a distribution by counting the number of data points where a disruptive event did not occur between 0 and 10, 10 and 20, 20 and 30, until we reach the 90 to 100 bin. Plotting the bin versus the number in the bin produces a frequency distribution. We then calculate the mean and the standard deviation. The most common failure rate distribution is the exponential distribution, $R(t) = \exp(-\lambda t)$, where $R(t)$ = survival probability, λ = failure rate or average time to failure, and t = time.

However, if we integrate the exponential distribution over time, we get probabilities of when the failures are likely to occur.

$$\int R(t) = \int \exp(-\lambda t)\, dt = Ln(t) / \lambda \quad , \text{where Ln is the natural logarithm}$$

Monte Carlo sampling proceeds by drawing a uniformly distributed number from a time interval; for example, [0 – 100] as suggested above, and comparing the value with $1 - Ln(t)/\lambda$. If it is more than this value, a disruptive event is declared.

---

[57] http://helper.ipam.ucla.edu/publications/matut/matut_5898_preprint.pdf, accessed 02/13/2015.

As a proof-of-concept, we identify six sources of uncertainty, some with uniform distributions (all values equally likely), exponential distributions with failure rates, and a more exotic Poisson distribution[58]. Implementation is straightforward. We draw many random samples from each distribution to form the probability that each would occur in a time interval, stack the probability of occurrence for each uncertainty source and provide a display, with a cutoff time as shown in Figure III-7.4.

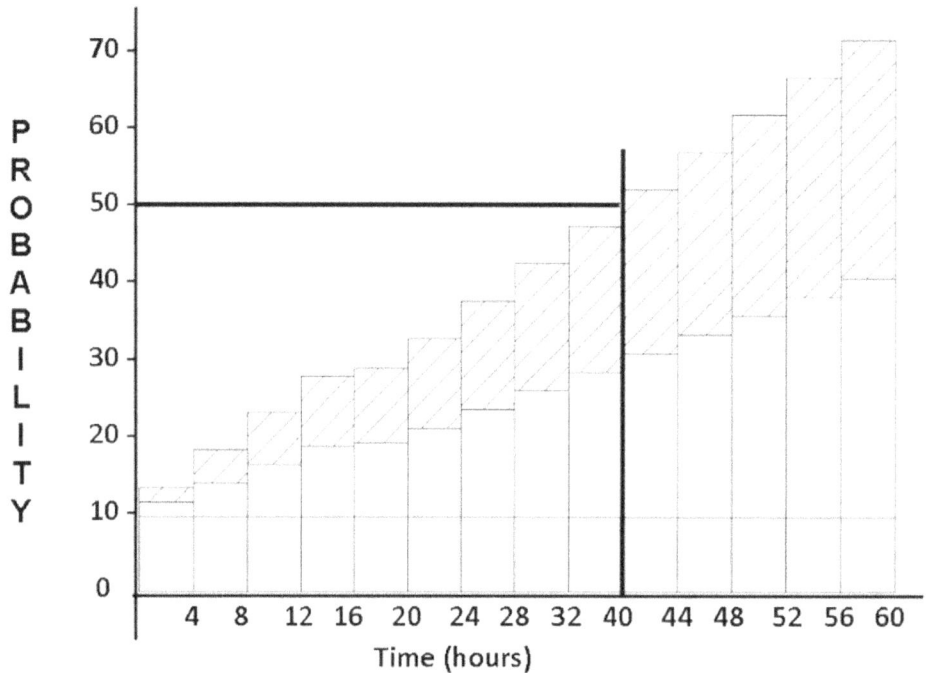

Figure III-7.4: Probability of a Disruptive Event Versus Time

Summary

Our research results in the development of an integration test bed that is used to study issues related to plan optimization using emerging technologies and algorithms in the computer science community. Specifically, we used this test bed to prototype and demonstrate the use of genetic algorithms to optimize detailed planning, and introduce the Replan Trigger and Planning Horizon as helper algorithms.

Another area where genetic algorithms fit extremely well is Time Critical Targeting (TCT). A TCT plan consists of pairing Red targets with Blue assets (usually aircraft). Each of the multiple targets and multiple available assets have numerous parameters and constraints that affect which target should be paired with which asset. The time constraint in determining the plan is also extremely short, making it almost impossible to determine the "best" plan in the time available either manually or using automatic exhaustive search techniques given realistic numbers of TCTs and available assets. Although many weapon/target pairing algorithms exist

---

[58] http://www.phys.utk.edu/labs/modphys/poissonstatistics.pdf, accessed 02/13/2015

(most based on Dykstra Search[59], A* algorithm[60], etc.), the beauty of the genetic algorithm is that it avoids local minima and maxima while allowing flexible specification of the objective function.

An area of potential use of GA-based detailed planning is Space Control, which deals with how to protect space assets to avoid what the Space Commission refers to[61] as a "Pearl Harbor in Space". This mission will clearly involve a tight coupling of Intelligence, Surveillance, and Reconnaissance actions and defensive actions to protect space assets. Depending upon the threats and the defense concepts, this could lead to mission planning considerations for other systems dependent upon these space assets - including missile defense, commercial broadcast systems, and the Global Positioning System for navigation.

Two companion, or "helper", algorithms, the Replan rigger and the Planning Horizon, were also discussed. These algorithms extend the prototype test bed from classic artificial intelligence algorithms to optimal policy and Monte Carlo statistics.

---

[59] http://en.wikipedia.org/wiki/Dijkstra's_algorithm, accessed 02/25/2015
[60] http://theory.stanford.edu/~amitp/GameProgramming/AStarComparison.html, accessed 02/25/2015.
[61] http://www.dod.mil/pubs/spaceintro.pdf, accessed 02/25/2015.

Chapter III-8
Evidential Reasoning

This chapter describes an evidence fusion engine for assimilating uncertain evidence, propagating evidence to form indicators, and predicting intent. Data fusion is a difficult, multifaceted problem with many unsolved challenges, the major one being that the data fusion problem is ill posed - data fusion means many things to many people (Figure III-8.1, where the arrow shows where evidential reason fits in the taxonomy). Fusion of evidence provides a focused solution in cases where a decision maker is faced with discrete bits of information that are inherently uncertain. The Dempster-Shafer Belief Network algorithm is the basis for our approach to data fusion. Advantages are that it: 1) provides intuitive results; 2) differentiates belief, ignorance, and disbelief; and 3) resolves conflicts. Insights into the behavior of the belief network are obtained from an analytical study. Mathematical and empirical properties are explored and codified.

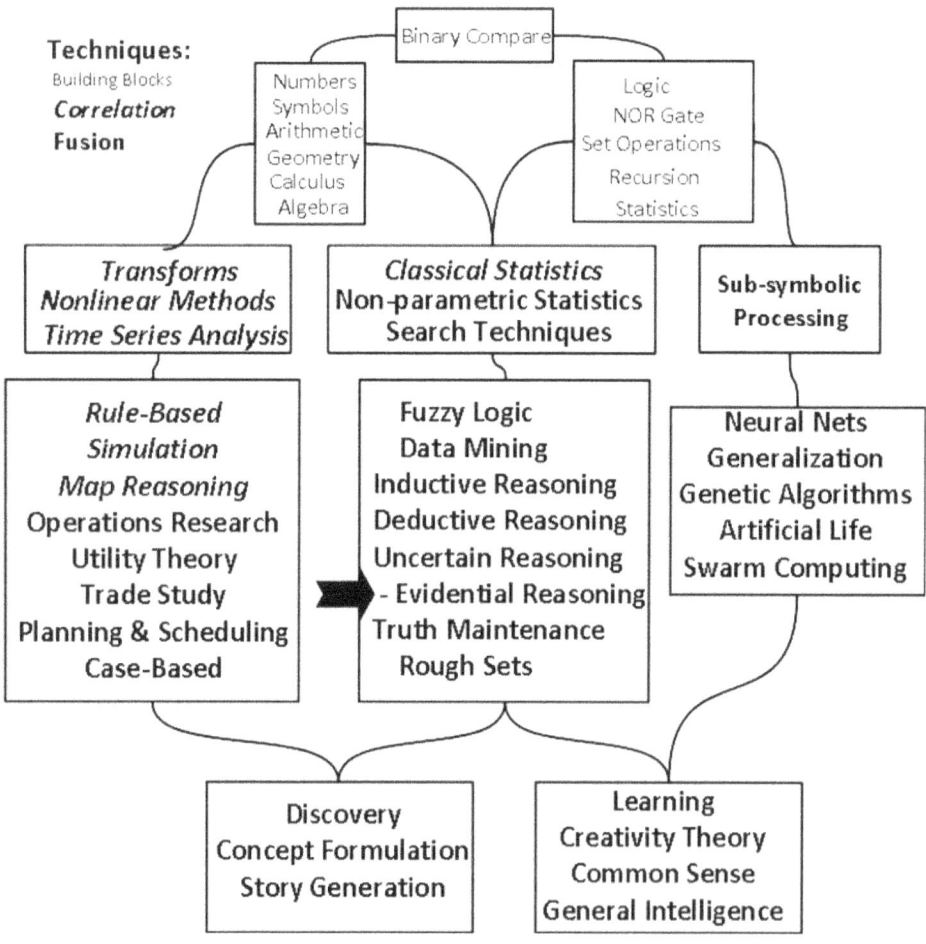

Figure III-8.1: Data Fusion Hierarchy

A novel feature of the test bed is that the user is allowed to override the belief or disbelief associated with a hypothesis and the network will self-adjust the appropriate link values – or learn - by instantiating the override. The back-propagation algorithm from artificial neural network research is used to adjust the links. Alternatively, an analytical solution to fusion equations allows the tasking agency to forecast quality and quantity requirements when tasking for future data. Our algorithm has much in common with deep belief networks[62].

Introduction

This chapter describes our experience with the tailoring of a general-purpose fusion engine to the task of fusing evidence in a wide variety of domains, including individual longevity predictions, contract award likelihood, weather, climate change, terrorism, time critical targeting, intelligence operations, and anti-satellite operations. The tool provides semi-automated reasoning support for decision makers faced with risky decisions and unknown intents based on assessment information that is inherently uncertain, incomplete, and possibly conflicting. Decision makers often find it difficult to mentally combine evidence - the human tendency is to postpone risky decisions when data is incomplete, jump to conclusions, or refuse to consider conflicting data. Those versed in classical (frequentist[63]) statistics realize it doesn't apply in situations where evidence is sparse. A data fusion engine is needed.

Data fusion is a difficult, multifaceted problem with many unsolved challenges. A contributing factor to these challenges is that the problem is ill posed – data fusion means many things to many people. The taxonomy shown in Figure III-8.1 attempts to organize and differentiate mathematical building blocks, correlation schemes, and "true" data fusion methods. Clearly, there are many types of data fusion, and the difference between fusion and correlation is tenuous. At the top are the fundamental building blocks, which are then differentiated into analytical, statistical, and sub-symbolic techniques. Uncertain reasoning, the topic of this paper, is among the statistical approaches.

Related techniques, such as Bayesian Networks[64] and Rough Set Theory[65], are assessed for applicability. The evidential reasoning approach[66], which relies on the Dempster-Shafer Combination Rule, is chosen because it provides intuitive results, differentiates ignorance and disbelief (sometimes described as "skeptical" processing), and performs conflict resolution. Bayesian networks also provide intuitive results, but are better suited to causal reasoning. Rough sets differentiate between what is certain and what is possible and are of potential future interest – a truth maintenance system appears necessary to track the validity of hypotheses as evidence is accumulated.

Insights into the behavior of the belief network are obtained from an analytical study. Mathematical and empirical properties are explored and codified. Discovery of a representation for an identity and its inverse reveal fascinating properties with practical application - for example, fusion equations are soluble in the same way that matrix equations are.

---

[62] https://www.cs.toronto.edu/~hinton/nipstutorial/nipstut3.pdf, accessed 04/08/2015.
[63] http://simplystatistics.org/2014/10/13/as-an-applied-statistician-i-find-the-frequentists-versus-bayesians-debate-completely-inconsequential/, accessed 02/25/2015.
[64] http://people.math.umass.edu/~lavine/whatisbayes.pdf, accessed 02/25/2015.
[65] http://www.nit.eu/czasopisma/JTIT/2002/3/7.pdf, accessed 04/08/2015.
[66] http://www.aaai.org/Papers/AAAI/1988/AAAI88-037.pdf, accessed 02/25/2015.

A novel feature of the technology is that the user is allowed to override the belief or disbelief associated with a hypothesis and the belief network self-adjusts the appropriate link values – or learns - by instantiating the override. The back-propagation algorithm from artificial neural network research is used to adjust the links. At the forefront of machine learning today is the deep belief network that does deep learning[67] by training the network a layer at a time. Our belief network can train one or more layers at a time and has many characteristics of a deep belief network. In addition, the belief network constructs explanations of how outcomes are obtained – this is important in risky decision making environments. The work described in this paper has broad applicability to risky decision-making in circumstances where evidence is uncertain, incomplete, possibly conflicting, and arrives asynchronously over time.

Belief networks are constructed for six mission areas, leading to the characterization of our prototype test bed as a general-purpose data fusion engine[68] for which we have received a patent. The most sophisticated of these networks consists of a pair of six-layer networks that map evidence and assessments to high-level terrorist and anti-terrorist objectives that are overlayed to predict terrorist intent and derive best anti-terrorism "moves".

Approach

The Dempster-Shafer Combination Rule (Figure III-8.2) for fusion of evidence is the core algorithm. Node values in the network are represented as evidential intervals with values from the set of real numbers ($0 <= n <= 1$). Three parameters specify each node: a "belief" (B), an "unknown" (U) and a "disbelief" (D). The words "unknown", "ignorance", and "don't know" are used interchangeably throughout this chapter. The "unknown" parameter is computed as: $U = 1 - B - D$. New evidence is fused with existing evidence according to the equations in the figure. Although the theory of evidential reasoning, which is a "skeptical" brand of reasoning, allows "m" states for each node (producing computational complexity 2^m-1), we find that specifying nodes with a single state and a set consisting of three values {B,U,D} is satisfactory. Disbelief (D) is the complement of the more obtuse "plausibility (P)" parameter ($D = 1 - P$) encountered in the literature.

Having discussed how evidence is fused at nodes, we now discuss how nodes are combined with links to form a network. We use "hierarchical directed a-cyclic graphs[69]" exclusively, since more complex representations are not required. The networks are composed of layers of nodes, each having values {B, U, D}, connected by links with constant values {L}. Three types of nodes form the networks: input or evidence nodes, multiple layers of intermediate or hypothesis nodes, and an output layer consisting of outcome nodes. The function of the links is to convey the influence, or impact, that one node has on another. Link values, either supporting or detracting, are input for each network based on domain expertise or back-propagation.

---

[67] http://en.wikipedia.org/wiki/Deep_learning, accessed 02/25/2015.
[68] https://www.google.com/patents/US20040019575, accessed 02/25/2015.
[69] We have demonstrated uncertain reasoning with graphs that have cycles. This is useful in social network analysis.

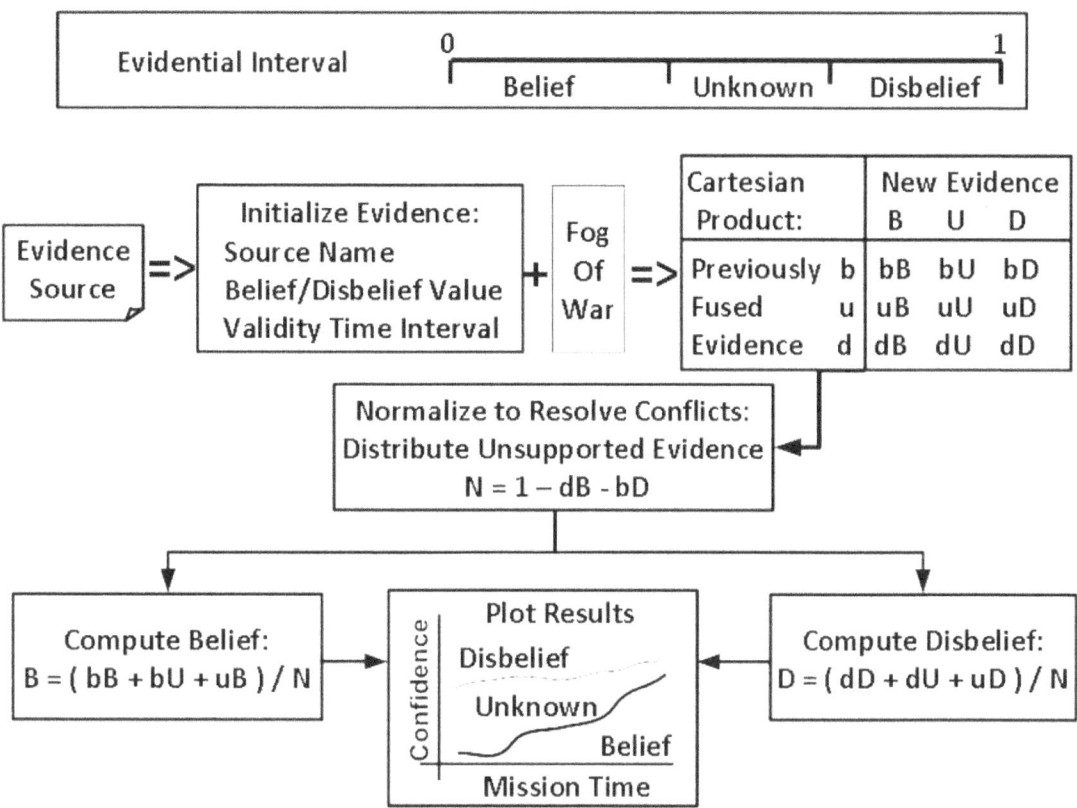

Figure III-8.2 Dempster-Shafer Combination Rule

As a practical matter, determining link values turns out to be the most difficult aspect of implementing belief networks. We reasoned that it's difficult or impossible to determine link values for artificial neural networks, which are equivalent sub-symbol representations of networks. Since link values in neural networks are learned automatically from training data using a back-propagation algorithm, we adapt this powerful approach to our use. That is, we allow a user to simultaneously change multiple node values to override existing values and use the back-propagation algorithm to correctly adjust node and link values. Constraints are imposed on which links are adjusted. If links that influence the preceding layer of nodes are adjusted, that layer requires modification. Typically, link values are constrained to $[0 <= L <= 1]$, but link values outside this interval are found to have practical significance; that is, in some applications, $L > 1$ suggests a multiplier effect, while $L < 0$ is interpreted as a contrary effect.

We exploit the ability of belief networks to provide the user with explanations. We believe that it is very important to explain the meaning of node labels; for example, "clouds" means: "what is the belief or disbelief in evidence stating that clouds will not be a problem?". The value associated with links and nodes are explained in plain English sentences. Concatenated phrases relevant to the node or link form explanations. For example, the following explanation is computer generated to explain a link value of "1.0" connecting a hypothesis node and an outcome node: "Belief in the hypothesis that assessment is complete has a "certain" (1.0) impact on the outcome that the target is neutralized". Numerical values are replaced with linguistic variables - for example, if the belief associated with a node is between 0.47 and 0.53, the phrase "it may" is furnished as part of the explanation.

The algorithm (Figure III-8.3) combines nodes, links, and explanations and executes as follows:

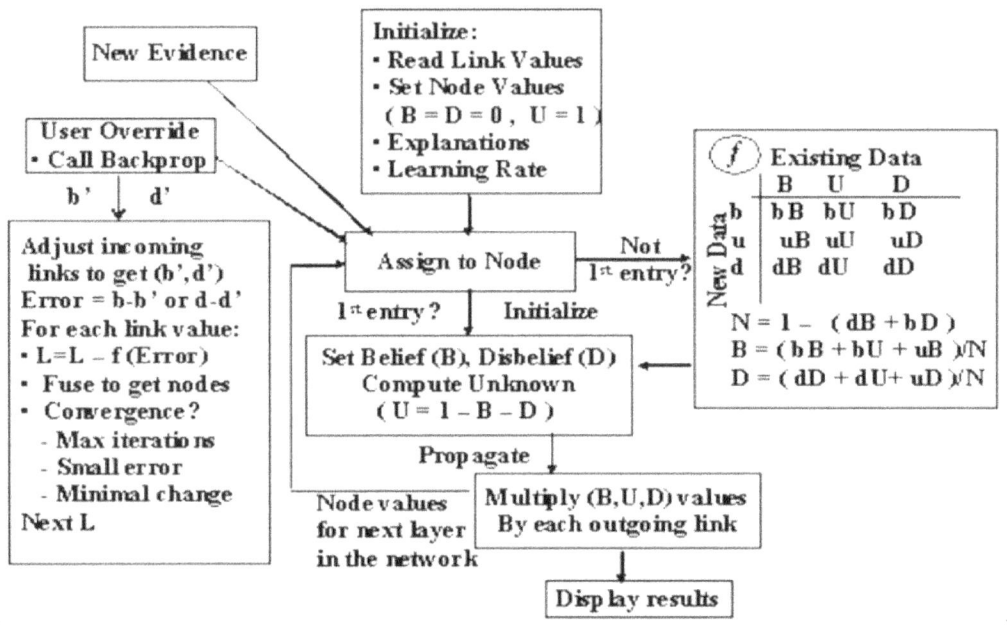

Figure III-8.3 Data Fusion Flow Diagram

Following initialization, new evidence is received.

- If a node value is unknown {0, 1, 0}, the evidence initializes the node, otherwise, it is fused with previous data.

- Node values are multiplied by link constants for each outgoing link to produce node values for the next layer in the network.

- The process continues until evidence is propagated to the final layer and results are displayed.

- The user can request an explanation for node or link values.

- The user has the option of overriding any or all node values and the back-propagation algorithm adjusts links and (optionally) nodes in previous layers to correct, or reconcile, the network.

Results

Experimentation with the prototype fusion engine allows us to confirm the mathematical properties of the Dempster-Shafer Combination Rule and to derive useful computational properties, summarized in the tables below.

Mathematical Properties	Importance
• Stable	No numerical instabilities; for example, divide by zero
• Symmetric	Belief and disbelief are computed equivalently
• Bounded	Belief + Ignorance + Disbelief = 1
• Commutative	Same result, regardless of fusion order
• Unit Triad	Fusion with [0,1,0] leaves the previous result unchanged
• Inverse Exists	Can solve fusion equations added belief needed
• Generalizes Bayes	Any Bayesian model can be replicated
• Conflict Resolution	Explicitly provided by normalization factor

Derived Properties	Importance
• Scalable	Scales linearly with evidence, # nodes, # links, # times
• Intuitive	Produces commonsense results rooted in evidence
• Comprehensive	Meaningful results produced in all cases
• Slight belief ineffective	Make belief either zero or greater than .5
• Reversing Trends	Reverse with strong belief or rapid evidence aging
• Polarization	Ignorance is quickly reduced favoring strong belief/disbelief
• Network Connectivity	Make sparse as possible to avoid cascading
• Link Values	Make link value for belief and disbelief equal

Mathematically, the algorithm is stable, except for the case where the denominator vanishes (1 - bD - dB = 0). If this condition, which is interpreted as reversal of belief or disbelief, is about to occur, the new evidence is merely perturbed by an infinitesimal amount to avoid division by zero. That the combination rule is symmetric, bounded, commutative and associative is shown by a derivation. This derivation elegantly produces the Dempster-Shafer Combination Rule and therefore seems worthwhile to include. Let $\{b,u,d\}$ and $\{B,U,D\}$ be two pieces of evidence to be fused. Use the property

$$1 = b + u + d \text{ and } 1 = B + U + D$$

to obtain:

$$1 = (b + u + d)B + (b + u + d)U + (b + u + d)D$$

then combine terms to get:

$$(bB + uB + bU) + (uU) + (dD + uD + dU) = (1 - dB - bD)$$

These terms correspond to the new belief, unknown, disbelief, and normalization factors, respectively! The Combination Rule (Figure III-8.2) is obtained.

In matrix theory, the identity matrix multiplied by any matrix leaves the matrix unchanged. The equivalent for Dempster-Shafer theory is the $\{B, U, D\}$ triad: $I = \{0, 1, 0\}$ corresponding to complete ignorance. The inverse of a matrix, multiplied by the matrix, produces the identity ($A \oplus A^* = I$), where $\oplus$ denotes the fusion operator. For Dempster-Shafer theory, this inverse is determined. Fused with the corresponding evidence triad, the inverse triad produces the identity triad $\{0,1,0\}$. The criteria for deriving the inverse is:

$$0 = (bB + bU + uB)/(1 - dB - bD)$$
$$1 = uU /(1 - dB - bD)$$
$$0 = (dD + dU + uD)/(1 - dB - bD)$$

Solving these equations simultaneously for {B, U, D} in terms of {b, u, d} yields,

$$U = 1/[\, u - bd \{\, 1/(b+u) + 1/(d+u)\}]$$

Similar expressions result for B and D. The existence of an inverse means that fusion equations can be solved arithmetically. Given triads A and C, we can find triad B such that A ⊕ B = C. In other words, it is possible to determine how much additional evidence is required to push existing evidence over a decision-making threshold; for example, given fairly strong belief {0.75, 0.10, 0.15}, what additional evidence is required to reach a decision threshold of {0.95, 0.05, 0.0}?

Derived properties are obtained by executing the code parametrically to span the computational space. We fuse certain, moderate, weak, and null belief and disbelief to determine if the results are intuitive. We also vary the degree and strength of link connectivity. The algorithm produces common sense results in all cases and also suggests rules-of-thumb to "get what we want, not what we asked for". Four heuristics are discussed in detail. Examples give results based on belief – they are also true for disbelief because belief and disbelief are mathematically interchangeable.

Slight Belief (or Disbelief) is Ineffective - It is not effective to state mild belief in the face of strong disbelief (or vice versa). For example, moderate disbelief (0.7) compounded four times gives strong disbelief, even in the face of mild belief (0.3). To make conflict meaningful, its value must be greater than ½, otherwise it may as well be zero. Another intuitive result (which is not obtained from Bayes Rule) is that equal belief and disbelief, compounded four times, yields equal belief and disbelief!

Trend of Evidence - It is not easy to reverse the trend of evidence. Compounding a moderate belief (0.7) produces a strong belief. Fusing strong disbelief with this only slightly decreases earlier belief. The strategy to reduce a trend is to decrease belief in evidence over time (we used a linear decay rate), and delete previous evidence that is "overtaken by events" or redundant based on new evidence.

Polarization of Assessments - Oscillations in belief based on addition of new evidence rarely occurs. Belief tends to grow, even in the face of weakening belief and mild disbelief. Compounding belief above ½ also produces strong belief.

Cascading - When network connectivity is dense, belief in outcomes (last layer in the belief network) rapidly becomes strong, even if link values are not large. When link values are large, the cascading effect is accelerated. To avoid the phenomena, only define meaningful links and avoid large link values when many links contribute to a node value.

Display Interface

An important aspect of the test bed is providing a visual method for determining whether the prototype is behaving as predicted or desired. To this end, we develop graphical displays showing the results of the data fusion process.

Displays that present abstractions of belief networks are misleading. The assumptions inherent in the model are not explicitly shown. This is a dangerous over-simplification in belief networks because the user relies on the graphics to make potentially risky decisions. We designed the displays to show explicit belief and disbelief for all nodes. Links are color-coded with variable thickness to show the degree to which they influence nodes. Automated explanations for nodes and links are also available.

In addition, given two linked networks with the same topology, how effective one network is compared to the other can also be shown. The result of the comparison is the color-coding of the nodes based on the value of the ratio of the beliefs at the corresponding nodes (for example, red $\Rightarrow$ <1.0, yellow $\Rightarrow$ 1.0-10.0, green $\Rightarrow$ >10.0).

Network Instantiations

Terrorist/Anti-terrorist: Terrorist attacks have historically come without warning. In retrospect, there has usually been enough "shreds" of evidence to suggest that the attack is about to occur, but these indicators reside in people's heads, on paper, and in multiple disparate databases. There has been no way to combine these various bits of information to provide timely warning. The initiative addressed here directly addresses the need for an integrated tool set to conceptually cluster information, fuse evidence, predict terrorist intent, and prioritize anti-terrorist actions.

Evidence of terrorist activity is what drives the fusion engine. The content of this evidence is critically important to the formulation of the belief network, which is the representation that the fusion engine works on. We reason that a format for collecting evidence that captures answers to the "Reporter's Questions" (who, what, when, where, why, how, and how certain) as well as the original text is a good start. We take a "learn-by-doing" approach. A newspaper article[70] that contains about 100 pieces of evidence is the basis. We parse each piece of evidence into the Reporter's Template, and concurrently add fields to accommodate practical aspects of the data set. Specifically, we add fields for how severe, how certain, and unplaced keywords to produce a workable data structure.

Beyond providing a simple way to organize evidence of terrorist attacks and anti-terrorism actions, the completed template proves useful in deriving a belief network and automatically producing the evidence to drive it. For example, the "Who" block indicates to which belief network the evidence relates - if Threat content is filled in, it goes on the terrorist belief network, while information in the Friendly section means it is associated with the Anti-terrorism belief network. Another finding is that the content of the "What" block provides

---

[70] The Case Against Osama bin Laden, Colorado Springs Gazette, 5 October 2001, which contained the full text of Tony Blair, Prime Minister of the United Kingdom

events, activities and tactical objectives for the belief network, while the "Why" block provides content for the operational objectives and strategic objectives in the belief network. The "When" block provides a key for chronologically sorting the evidence. Finally, the persistence entry is used in the belief network to degrade the belief in the data over time, if appropriate; for example, weather predictions are not persistent but anti-terrorist intents are.

The evidence templates are analyzed and become the basis for deriving the Terrorist Assessment belief network (Figure III-8.4).

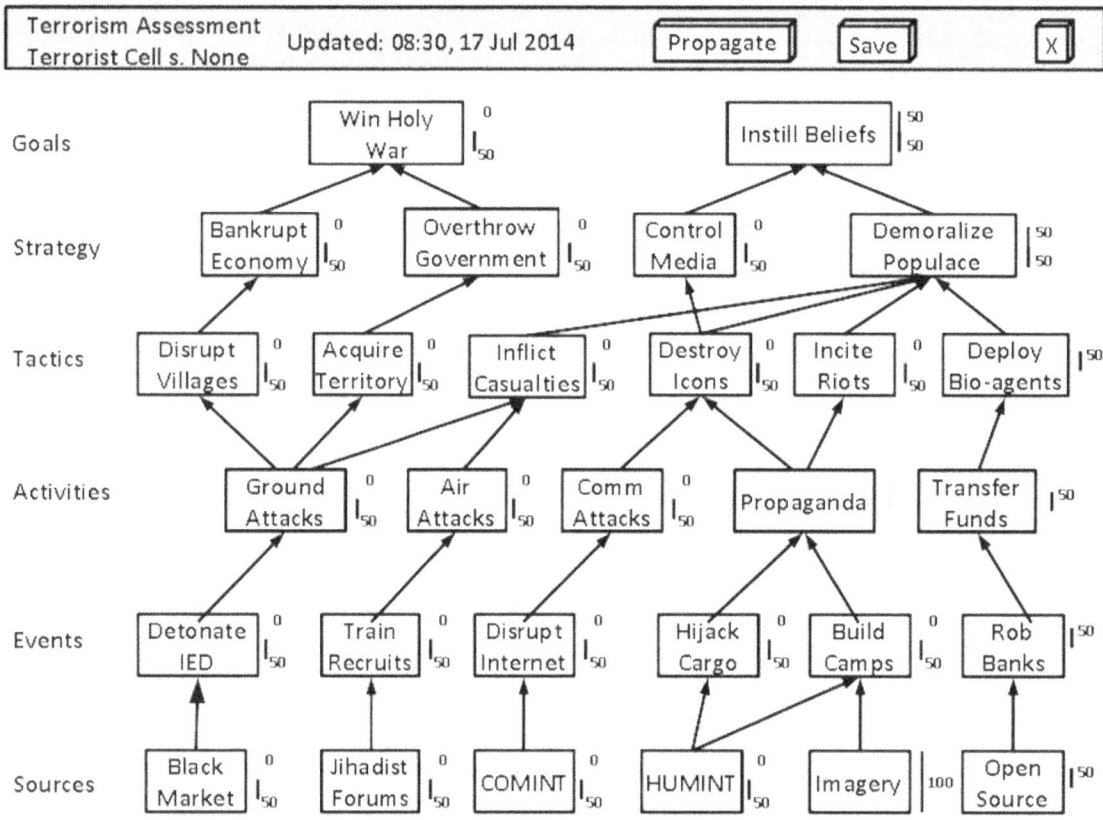

Figure III-8.4: Terrorist Threat Assessment

Nodes are hypotheses, with the information sources row showing low level hypotheses, the middle layers providing intermediate level hypotheses, and the last row showing high-level goals – also called intents. Although the algorithmic flow of evidence is from low-level input to high-level intent, what is of most interest in this problem is predicting (low-level) terrorist events! We found that evidence may be introduced at any level in the hierarchy. To accommodate this reality, we allow evidence to be associated with any hypothesis. Evidence is then fused with existing evidence for the hypothesis and the back-propagation algorithm is invoked to reconcile the network. The evidence templates and the terrorist belief network forms the basis for the Anti-Terrorism belief network. In many cases, our activities, tactical objectives, and strategic objectives relate directly to anti-terrorist nodes. In other cases, asymmetries are evident: for example, air superiority is an anti-terrorist operational objective that is virtually unchallenged by terrorists.

The Anti-terrorism and terrorist belief networks are then overlaid. The question is: what strengths and weaknesses are the terrorists or the anti-terrorism forces likely to exploit. Algorithmically, we compute the ratio of anti-terrorism belief to terrorist belief for each node and use it as a score. A ratio of unity is defined as parity, while a ratio greater than 10 corresponds to an anti-terrorism strength, and a ratio less than 1.0 corresponds to a terrorist strength. A ratio is tabulated for each node, and portrayed using color-coding on an effectiveness display. This effectiveness algorithm is analogous to assessing the relative worth of chess moves using a (one ply) 'minimax' algorithm, although our implementation is much simpler.

Other Examples

In addition to the Terrorist/Anti-terrorist instantiation, we test the belief network in a number of other mission areas in order to show its versatility. These include:

- Individual Longevity Predictions: Chapter V-11
- Combat Assessment
- Weather Prediction Impacts
- Kill Chain Belief Network
- Space Weather Fusion
- Foreign Missile Launch Assessment (with & without Fog-of-War Effects)

Combat Assessment - How do we fuse battle damage assessment reports, intelligence information, and operational knowledge over a period of weeks to determine if strategic, operational, and tactical objectives are being met? The combat assessment instantiation shows how weapon-target-assessment information influences belief about task success, which in turn influences beliefs about tactical, operational and strategic objectives.

Weather Impact Predictions - How will the 24-hour forecast impact tomorrow's Air Tasking Order for sorties that employ laser-guided munitions? This instantiation shows if weather conditions obscure an optical path between a laser designator and a target.

Kill Chain Belief Network - For Time Critical Targets (TCTs), such as mobile missile launchers, a variety of information sources provide information on how well targets are: found => fixed => tracked => targeted => executed =>assessed (this is referred to as the kill chain). This instantiation allows TCT kill-chain evidence to be posted, organized, and coherently represented in real time.

Space Weather Fusion - Over a period of hours-to-days, a satellite control officer may be confronted by a bewildering variety of information, including communications outages, reports of ionospheric scintillation, radio frequency interference reports, and indicators of threat activity. This instantiation allows varied evidence to be organized and fused.

Foreign Launch Assessment - This early demonstrator is based on the need to fuse assessments from space sensors, radar, and intelligence reports to determine, as a function of time, whether a possible foreign missile launch is hostile, deliberate, and whether it is an intercontinental ballistic missile (ICBM) or a spacecraft launch.

Fog-of-War Effects - A software module that perturbs data and human cognitive performance is integrated with the Foreign Launch Assessment belief network. Examples of the factors that we allow the user to specify are: the frequency and severity of lost data, data overload, human confusion, and false assumptions, among others. We find that the result of applying Fog-of-War perturbations to the evidence is significant. See Chapter V-3 for more detail.

Discussion

The work described in this paper has broad applicability to decision making in circumstances where evidence is uncertain, incomplete, possibly conflicting, and arrives asynchronously over time. The Dempster-Shafer theory of evidence has very useful mathematical properties - in particular, an inverse has been discovered that allows fusion equations to be solved arithmetically. In addition, the derived properties of these belief networks collectively suggest intuitive application of the technique as a general-purpose fusion engine for uncertain reasoning. A novel feature of our implementation is the addition of a back-propagation algorithm that allows the user to override fused beliefs and disbeliefs in nodes or link values. The back-propagation algorithm adjusts precursor node and link values to reconcile the network. Thus, the network learns from user training data in the form of overrides. Additional detail is given in the Sidebar.

Sidebar: Belief Network Inference Requirements and Combination Rules

Input

These requirements specify the content of the evidence and the components of the belief network graph.

- **Evidence Content.** Evidence shall consist of a hypothesis, a degree of belief ($0<=B<=1$) in the hypothesis (optional), a degree of disbelief ($0<=D<=1$) in the hypothesis (optional), a mission time (YYYY:DDD:HH:MM:SSS) that indicates when the initially is valid (optional - default id current time), and a duration (HH) that indicates the length of time the evidence is valid (optional – default is forever). We provide a pointer to an explanation.

- **Obsolescence.** Belief and Disbelief in evidence shall be linearly de-rated from the initial time to initial time plus the persistence time.

- **Graphical Node Input.** Graphical display interaction (nominally a two-sided horizontal slider associated with each hypothesis node) shall allow manual modification of the belief and disbelief in a hypothesis.

- **Graphical Link Input.** Graphical display interaction (nominally a two-sided vertical slider associated with each link) shall allow manual modification of the influence a hypothesis has on another.

- **Graph Structure.** The belief network graph shall consist of hypothesis nodes and influence links. Graphs shall be acyclic. A graph shall consist of up to 10 nodes per layer. A graph shall consist of one or more layers that need not be connected.

- **Link Values.** Links shall indicate the (constant) degree of influence, which may be supporting or detracting (-100% <=L<=100%) that a hypothesis has on the hypothesis to which it is linked by an arrow.

- **Automatic Generation**. The belief network graph shall be (optionally) automatically generated by appropriately filled-in templates from the Strategy Development Tool[71].

- **Graph Format.** Whether manually generated or built using the Stratgy Development Tool, the graph shall be stored in .xml format: this requirement is a constraint that allows us to leverage existing software.

- **Manual Node Input**. Nodes shall have means (for example, pull-down boxes) to allow evidence to be added, modified or overridden. All manual input shall constitute new evidence.

- **Manual Link Input**. Links shall have means (for example, pull-down boxes) to allow link values to be added, modified or overridden.

- **Logging.** Evidence shall be logged for reference.

- **Evidence File**. Evidence shall be introduced to the nodes through a formatted input file

Processing

These requirements define the properties of the inference engine, known as the general purpose evidence fusion engine (GPEFE).

- **Fusion Rules.** At each node, incoming evidence shall be mathematically fused based on a re-settable combination rule.

- **Manual Rule Selection**. Rules shall have means (for example, pull-down boxes) to allow combination rules to be modified.

- **Types of Combination Rules**. Each hypothesis shall allow evidence fusion based on the following combination rules:

    1. **Replace:**
       B = latest external input B
       D = latest external input D
    2. **Average:**
        For all external and incoming inputs:
           Add all the B's together, divide the sum by the number of B's

---

[71] http://www.amazon.com/Effects-Based-Operations-using-Strategy-Development/dp/B00NYGVRU4, accessed 03/28/2015.

Add all the D's together, divide the sum by the number of D's
3. **Optimistic:**
   Loop thru external inputs and parent inputs
   Get the greatest B value and corresponding D value
   Where multiple inputs have the same B value,
   Get the lowest corresponding D value
4. **Pessimistic:**
   Loop thru external inputs and parent inputs
   Get the greatest D value and corresponding B value
   Where multiple inputs have the same D value,
   Get the lowest corresponding B value
5. **Multiplicative:**
   Looking at the external inputs and parent inputs
   Multiply all the B's together
   Multiply all the D's together
6. **Dempster-Shafer:**
   Loop thru external inputs and parent inputs
   Fuse all of the (B, D) pairs together
7. **Fuse (used by Dempster-Shafer and Bayes):**
   norm = 1.0 - BxD' - DxB'
   fused[B] = (BxB' + BxU' + UxB') / norm
   fused[U] =   UxU'   / norm
   fused[D] = (DxD' + UxD' + DxU') / norm
8. **Bayes:**
   Loop thru external inputs and parent inputs
   Fuse all of the (B, 1-B) pairs together
9. **Screening:**
   Looking at the external inputs and parent inputs
   Multiply all the B's together and Compound the D's
   (1- [1-d1]*[1-d2],..) all the D's together

- **Forward Propagation.** Evidence shall be fused at evidence nodes (bottom layer), derated by links, fused at functional nodes (middle layer), de-rated by links, and fused at operational nodes (upper layer).

- **Backward Propagation.** Based on analyst modification of evidence belief or disbelief in one or more nodes via vertical sliders, the difference between current and desired node values shall be propagated backward to modify link values. Iteration shall continue until resulting forward propagation produces the override value(s). Current software uses the back-propagation technique for artificial neural network theory.

- **Evidence De-Rating.** At a user-selectable time step, de-rated values of all evidence shall be recalculated and re-propagated.

- **Scalability.** Propagation shall retain linear scalability with respect to the number of nodes and links at each propagation time-step.

- **Default Node Values.** The default for nodes shall be Belief = 0 and Disbelief = 0. This results in Unknown = 1 – Belief – Disbelief = 1.

- **Default Link Values**. The default value for links shall be a link weight of 100%.
    - Note that for these default node and link values, the graph is mathematically correct: Belief + Unknown+ Disbelief = 1 for every node at every time-step.

Output

These requirements define the properties of the output, based on displays.

- **Network Output**. In order to leverage existing investments, the output of the belief network application shall be the belief network editor (BNE) visualization and a file in .gml format. An example of the visualization is shown in Figure III-8.4.

- **Network Source**. The BNE output shall be obtainable manually composed or automatically imported from the Strategy Development Tool.

- **Manual Network Output**. Drag-and-Drop interaction with the BNE screen shall allow a belief network to be created, infused with evidence, and propagated to produce a result.

- **Time History**. An x-y plot of confidence (belief, unknown, disbelief) versus mission time for any node shall be available by clicking on the node.

- **Stored Graphics**. BNE shall output displayed data to a .gml file.

Chapter III-9
Inductive Reasoning: Generalizing from Data

Inductive reasoning produces general rules from a set of specific instances. It is one of the most powerful data mining techniques. Arguably, the rule induction algorithm is the "killer application" that got the field of data mining started. One of the earliest algorithms is Quinlan's[72] ID3 rule induction algorithms. We use a later version, dubbed J48[73], and available from the Weka data mining library. The algorithm automatically discovers rules in structured data sets. It is a supervised learning algorithm: the user specifies the variable that is the result of the rule.

Inductive reasoning, in the form of rule induction, and a clustering algorithm are used in the computer network intrusion domain. Both rule induction and clustering discover interesting patterns in multiple data sets with this domain. Data is collected from two sources, Knowledge Data Discovery, in the form of a pre-cleaned comma separated file, and a University of Massachusetts data set, in the form of raw packet log files. The data sets are cleaned and converted to an intermediate format for easy transformation into the file formats required by Weka. We "mine" the data using these techniques and the output is analyzed. Weka produces valuable results, giving information about how to predict what is going on in the network based on the data that is received. Weka provides information in support of predicting network attacks, which is an end goal of the research.

Introduction

This chapter documents several data mining techniques that we use to find interesting predictive patterns in data for a computer network intrusion data sets. The tools help to analyze past records of network traffic in an attempt to define a set of rules or patterns that predict whether or not future computer network traffic constitutes intrusive activity. We examine a broad range of techniques in data mining to identify the best for experimentation as a part of the predictive analysis subtask of the research project.

We use existing techniques; that is, Weka rule induction, and search the Web for new data mining tools. The techniques we find most valuable are clustering (in both a supervised and unsupervised mode) and rule induction. We rely on the clustering algorithms and graphing functions native to the Weka Knowledge Explorer. Rule induction is also included in Weka in the form of the J48 classifier.

We identify two data sets to test these techniques; one from the Knowledge Discovery and Data-Mining Cup[74] of 1999 and the other from the University of Massachusetts (UMass) at

---

[72] https://wwwold.cs.umd.edu/class/fall2009/cmsc828r/PAPERS/fulltext_Quilan_Ashwin_Kumar.pdf, accessed 02/25/2015.
[73] http://www.academia.edu/4375403/Decision_Tree_Analysis_on_J48_Algorithm_for_Data_Mining, accessed 02/25/2015.
[74] http://kdd.ics.uci.edu/databases/kddcup99/kddcup99.html, accessed 02/25/2015.

Lowell[75]. Both of these data sets consist of TCP Dump data gathered from live sites. We then clean the data and execute each of the techniques with these data sets as input. Weka produces valuable results: it gives information about how to predict what is going on in the network based on the data that is received. Weka provides information in support of predicting network attacks, which is an end goal of the research.

Approach

To decide which tools to use, we conduct a trade study on tools with the required functionality. A heavily weighted factor in assessing the tools is experiences of other researchers as reported in the literature[76]. Another important factor is actual use of the tool for network intrusion analysis. At first we look at all tools that fulfill the minimum requirements for each technique we test in data mining. We continue to narrow the field based on factors such as price and source code availability.

Two data sets are downloaded and loaded into Microsoft Access for easy manipulation of the files, which contain comma separated fields. The KDD data set is pre-cleaned, so it is ready to be inserted into the Data Mining algorithms. The UMass data set is in raw packet form and requires conversion to the same format as the KDD data set. Importantly, this format is a general view of the entire conversation, rather than a sequential collection of the individual packets. A conversation is defined as the entire sequence of packets starting and ending with a Three-Way-Handshake. We perform the conversion by taking the data out of one access table; processing the data through a Visual Basic for Applications (VBA) script that looks for the Three Way Handshakes, and performing calculations to find attributes such as duration, bytes transmitted, bytes received and service connections. When this is complete, the new data file is loaded back into Access, and a second VBA script looks for more general information such as the number of connections to a service per unit time and the percentage of connections to different services on the same computer per unit time.

The final data form of the Umass data closely matches the formatting of the data from the KDD data set. On a Pentium III 600 Mhz with 128 megabytes (MB) of random access memory (RAM), this reformatting of the Umass TCP dump data to a transitional state takes at least a day for each sub-data set. After both data sets are extracted, cleaned, and formatted we do a final conversion (minor re-formatting for compatibility) of the data files to be imported into each of the applications.

Algorithm

Decision tree algorithms are based on a divide-and-conquer approach to the classification problem. They work in a top-down manner, seeking at each stage an attribute to split on, that separates the classes best, and then recursively processing the partitions resulted from the split. An alternative approach is to take each class separately, and try to cover all examples in that class, at the same time excluding examples not in the class. This is the so called, covering approach, because at each stage a rule is determined that covers some of the examples.

---

[75] http://traces.cs.umass.edu/index.php/Network/Network, accessed 02/25/2015.
[76] http://www.predictiveanalyticstoday.com/top-15-free-data-mining-software/, accessed 02/25/2015.

Covering algorithms operate by adding tests to the rule that is under construction, always trying to create a rule with maximum accuracy. Whereas a decision tree algorithm chooses an attribute to maximize the separation between the classes (using information gain criterion), the covering algorithm chooses an attribute-value pair to maximize the probability of the desired classification. The algorithm below depicts the basic rule induction method[77]. We use the Weka rule induction algorithm, called the Prism method, cited in Witten and Frank.

Basic rule induction method:
  For each class C
      Initialize all examples of E
      While E contains examples in class C
          Create rule R with an empty left-hand side that predicts class C
          Until R is 100% accurate (or there are no more attributes to use) do:
              For each attribute A not in R, and each value V
                  Consider adding the condition A=V to the left-hand side of R
                  Select A and V to maximize accuracy and covering of the A-V pair
              Add A= V to R
          Remove the examples covered by R from E

Rule Induction generates only 'correct' rules, measured by an accuracy formula. Any rule with accuracy less than 100 percent is incorrect, since it assigns examples to the target class in question that do really belong to the target class. The Prism method continues adding clauses to each rule until the rule is correct. The outer loop iterates over the classes, generating rules for each class in turn, re-initializing each time to the complete training set of examples. The algorithm always starts with an empty rule which covers all examples, and then restricts it by adding new conditions (attribute-value pairs), until it covers only examples of the desired target class. At each stage of adding the condition to the rule, the best attribute-value pair in terms of accuracy is chosen. If there are more attribute-value pairs with the same value of accuracy, then the one with greatest coverage is chosen.

Results

The tool package chosen is Weka, a freeware package from the University of Waikato in New Zealand. We obtain results for two data sets, and two data mining techniques. In the process, rule trees, clusters, hierarchical structures, and affinity diagrams are obtained. Results obtained from the tools are now discussed in more detail.

UMass Data Set: The first attempt to "mine" the UMass data set is based on a paper about the SAS[78] tool to cluster the data and look for meaningful results. We initially use Quinlan's classifier algorithm on the individual packets. This produces no interesting results for several reasons. First, the data set (4 GB) is too large for 128 MB of RAM. Second, while reading through the paper produced by the earlier group, we realized that it is not the individual packets in the conversation that are being mined. Instead, it is entire conversations, the number of

---

[77] Actually the Prism method, described in the book by I.H. Witten and E. Frank "Data Mining: Practical Machine Learning Tools and Techniques with Java Implementations"

[78] http://www.sas.com/en_us/insights/analytics/data-mining.html, accessed 04/09/2015.

conversations preceding a conversation, the number of conversations connected to the same service, and the amount data sent and received in each conversation. Our first data representation is quickly scrapped and a transition data set is created. The transition data set is an overall view of the connection. Instead of looking at each individual packet that is transmitted across, this data set is constructed so that one record contains all the packets that go back and forth for one data transaction. This new data set cuts the amount of data going into the Classifier Algorithm by 90%. The data is now input into the program, but valuable information isn't extracted. At this point a second data set is sought in hopes that it is just the poorly created transition data set that is causing all of the problems.

KDD Data Set: We process the Knowledge Discovery and Data-mining data set next. This data set is already preformatted to show the overall data transaction, rather than individual packets, and is quickly input into the Classifier. Due to extremely large size of the data set (745 Megabytes), a random sample is taken from data set so that the Classifier algorithm is not overwhelmed by the immense size of the data. The sample includes nominal or benign data as well as equal amounts of data from several attacks. This sample composition is important because the Classifier is shown in earlier attempts with the UMass data to be strongly discriminatory against smaller groups when trying to classify it against other larger data sets. Our results (Figure III-9.1) indicate which fields the Classifier algorithm found important in distinguishing normal network traffic from the communications containing attacks. These fields are:

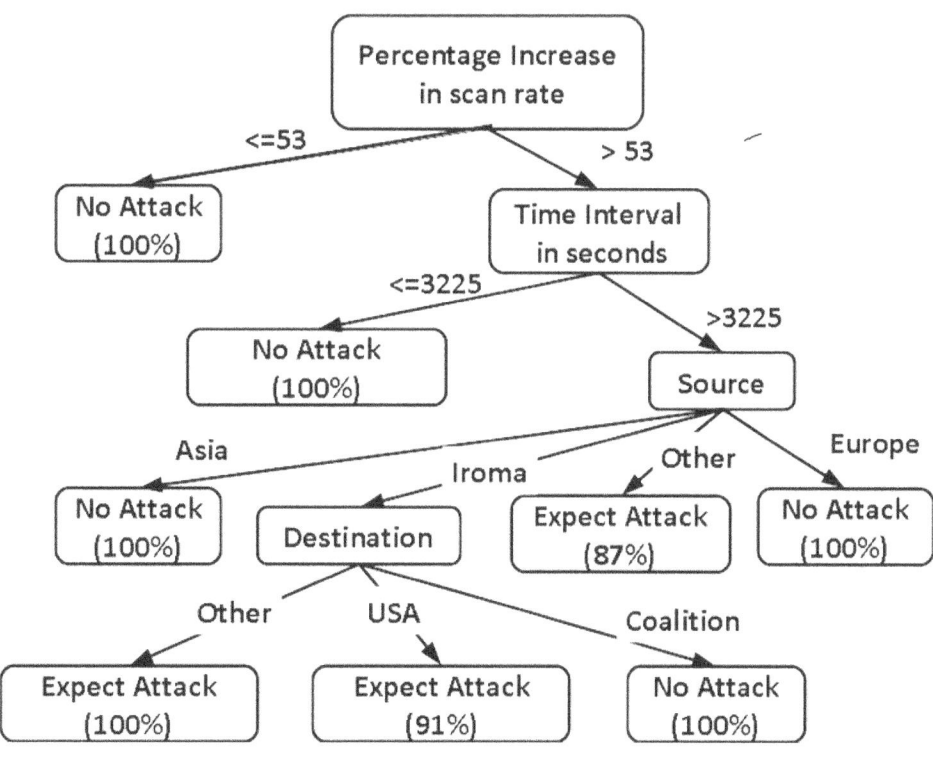

Figure III-9.1: Rule Tree

- Source Bytes: number of bytes sent by the source the computer,
- Server Count: percent of transmissions going to the same service on the same computer

in the past two seconds,
- Server Rate : percent of transmissions going to different services on the same computer in the past two seconds,
- Count: number of connections to the same service on the same computer in the past two seconds, and protocol: transmission type.

UMass data set Revisited: The information in this data set identifies fields in TCP dump data that predict threat activity. The UMass data set was restructured and transformed to match the fields that are found to be useful in the KDD data set. Once we have the data set reconfigured, we test the rules created by the other data set. We set the result field of the UMass data set to normal and then run the UMass data set as a test data set to the KDD data set.

The results of this method are quite favorable - we are on the right track with this procedure. We obtain an accuracy of 97% on the UMass data set; that is, 97% of the data set is correctly classified as being normal.

Clustering: We also use a Weka clustering algorithm that provides unsupervised learning. For the KDD data set, we find signatures (Figure III-9.2) of attacks that clearly differentiate them from normal network traffic. As shown, normal traffic is characterized as having a very small number of connections in a two-second interval that "tail off" smoothly, whereas Satan and Smurf attacks have very high connection rates that do not "tail off".

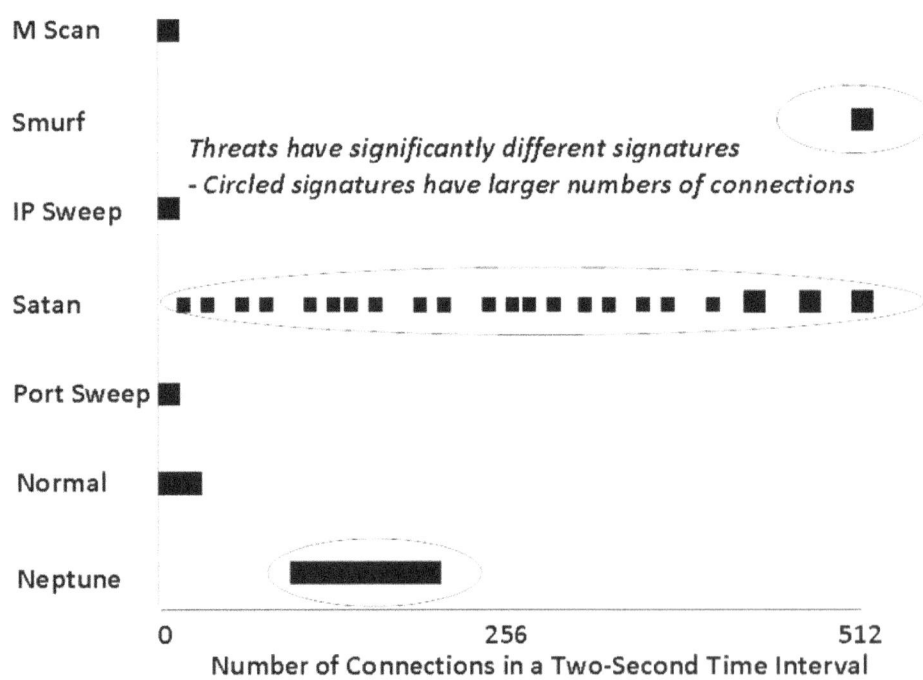

Figure III-9.2: Weka Clustering

Rule Induction Study: the READI Force Management Database

Issue: The READI[79] database contains readiness data for forces, equipment and personnel. READI presents this data to the user in the form of stoplight diagrams. To obtain these stoplight diagrams the data must be highly filtered and integrated. Thus, while stoplight diagrams provide ease of interpretation, they forfeit meaningfulness and traceability.

Objective: Our goal is to identify patterns in the data that can be used to report military force readiness. These patterns or rules are described with decision trees that are easily interpreted and leave a clear path to the data that caused that particular state of readiness.

Approach: We use the data mining application WEKA 3.0.2 to evaluate the Details table from the READI database. Weka's classifiers require the user to specify which column to treat as the dependent variable (class); all other data fields then become the independent variables. Weka's classifiers develop a rule tree that specifies the sets of constraints for the independent variables that lead to all possible outcomes for the dependent variable. In READI, the data field that best appears to reveal the readiness for a particular squadron is the SORTS Overall C-ratings. Therefore, we use SORTS Overall as the class for the Weka J48 classifier algorithm. If the user finds no compelling patterns, input characteristics (fields to be used) can be modified.

Result: Weka requires that all data be in the .arff format which is not a standard database format. Hence, getting the data to such a format is a tedious process. We accomplish this by writing small VBA programs in Access to manipulate the data into the correct form; that is, no spaces, periods, or single quotes in data entries and only integer or text types. After exporting the derived data set, we fill all empty values with question marks and generate the header for the .arff format that specifies all possible values for all data fields.

Metrics: The rule induction algorithm provides two pattern metrics: the number of instances of a rule and the number exceptions to a rule. These metrics provide accuracy and support. The goal is to form an evidential interval from the rule induction metrics (accuracy and support). Accuracy provides the mean (m) and the number of samples provides (via a simple equation) the dispersion (d). For example, in a survey, a candidate may receive 35 % +- 3% margin of error of the endorsements: m=35, and d=3. The evidential interval [0,1] is the set of fractional values for Belief (B), Ignorance (I), and Disbelief (D) such that $B + I + D = 1$. These are computed from the mean and dispersion as follows: $B = m - d$, $I = 2 * d$, $D = 1 - B - I$.

Belief, ignorance, and disbelief are presented to the user as the worth of the solution. Belief provides the degree of certainty, ignorance indicates "don't know", and disbelief indicates conflict in the solution.

Conclusions: WEKA 3.0.2, an academic application, doesn't provide any tools for getting the input data formatted or for displaying the results in an easily understood manner. This may be

---

[79] http://www.stratcom.mil/files/foia_requests/FOIA%2011-023%20Response%20pgs%20225-259.pdf, accessed 02/25/2015.

improved in the graphic user interface (GUI) of the new version. It does, however, allow for experimentation with many different data mining algorithms. Weka also has open source code, which allows for the complete understanding of how a certain algorithm works. Some results obtained from evaluating the snapshot from the READI database are trivial (to be expected); however, two meaningful rules were discovered.

For network anomaly detection, both the rule induction and the clustering algorithms produced results that helped predict the activity of the network. The clustering algorithms and the Quinlan's Classifier algorithm differentiated between normal network traffic, and traffic that contained attacks. Unsupervised clustering using Weka produced the most meaningful patterns of threat activity. The Weka rule induction algorithm provided historical patterns of threat activity as a basis for future predictions.

Future: Possible areas for further investigation include: Weka's clustering algorithms, the Weka GUI, Weka's association algorithms, test bed integration, larger READI snapshot, comparisons between Weka's numerous classifier algorithms, adding a stoplight column that would weight different attributes accordingly, and ease of integration into an Offense/Defense Integration (see Chapter V-2) test bed.

Chapter III-10
Deductive Reasoning: Rules and Logic

Next generation computer simulations are being developed for analyzing future ballistic missile and air defense systems. An important component of this simulation is the representation of a realistic, automated decision-making player, the "Simulated Commander." This paper describes our work toward simulating the concepts of "Rule Based" decision-making. The Simulated Commander receives the same information as the real world commander it represents. Our initial prototype uses the Fuzzy C Language Integrated Production System[80], called Fuzzy CLIPS, to implement the required Rule Based decision-making. These rules are perturbed by various reasoning algorithms to simulate lost inputs, latencies, modified priorities, degraded confidence, operator confusion, and misplaced outputs. A test bed is constructed to prototype a "thinker" module, a missile launch scenario, and related decision-making. A graphic user interface is implemented and connected to a Belief Network that provides input. Results to date are reasonable and provide a proof of concept for the "Simulated Commander. Rule Based Reasoning is further enhanced using a new tool, G2[81], that provides more flexibility and maintainability.

Introduction

In the conduct of a war-game, a considerable number of players are required, and many travel to the war-game facility from distant locations. Allowing the "players" to participate from their home locations by using remote terminals and established communication nets substitutes human travel for purchased hardware and communications costs. In some situations, this is cost effective; however, many people are still required to devote considerable time for each war-game that we conduct. This is especially true if the war-game is focused on some particular role, then the other "players" are needed to keep the test realistic but benefit little from participating. Selectively replacing human participants with Simulated Commanders is an approach that is intended to minimize monetary and human costs while still providing realistic "players" for a wide range of war-game functions.

Requirements are for fully automated command decision-makers that realistically represent human commanders at any level in the command hierarchy of a joint services Ballistic Missile Defense war-game. An important component of this simulation is a realistic automated command decision-making process, or a "Simulated Commander" (SC). This paper describes the concept of using rules for the SC decision-making process. These rules are affected by various factors such as the influence of real-world uncertainties, conflicts in information, and attention deficits that degrade human decision-making. The SC rules represent the rules and thought process that a real commander would execute. A test bed is constructed to prototype this "thinker" module using a missile launch scenario, and probabilistic decision-making for concreteness. Fuzzy CLIPS is used for implementation of the rule set for the initial prototype SC. After the initial success of creating the SC with Fuzzy CLIPS, a new tool, G2, is chosen to

---
[80] http://en.wikipedia.org/wiki/FuzzyCLIPS, accessed 02/25/2015.
[81] http://www.gensym.com/, accessed 02/25/2015.

replace Fuzzy CLIPS. G2 uses a natural language code that results in code more easily understood and maintained.

Approach

The overall approach to model the rule-based decision-maker is to begin with a missile defense scenario that requires a limited rule set. The rules are derived through research and interviews with commanders that are in charge of making command, control, and tactical decisions. The rule set is required to make credible decisions based on high level aggregated orders, construct and generate orders related to the SC position, and make decisions based on predefined rule sets. These rules are implemented to represent the chain-of-command for three commanders and are representative of the rules each commander are required to follow. A Belief Network (Chapter III-8) and a Fog-of-War module (Chapter V-3) are added to demonstrate the characteristics of lost or delayed communications, degradation of information, uncertainty, and ambiguity on the input to the SC.

The SC receives data from the Battle Manager to reflect predefined default values at the start of the simulation. Various other data such as Battle Plans, Rules of Engagement, Defense Engagement Authority, Readiness Posture and Mission Objectives are some of these preset values. The Battle Manager also relays information reflecting situation data, air picture, decision aids, and checklists to help the SC in his decision-making. These processes are represented in Figure III-10.1.

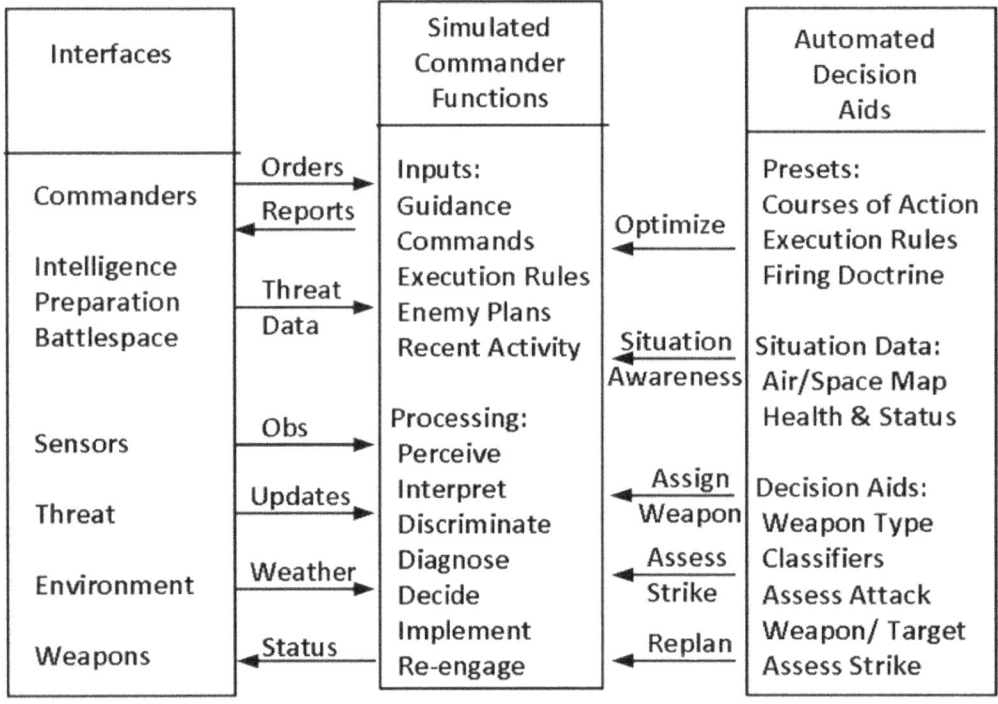

Figure III-10.1: Decision Making Process

The objective of this task is to prototype various decision-making techniques and demonstrate flexibility, including uncertainty conditions (Fog-of-War), and providing an interface to a wide range of software platforms. The Fuzzy CLIPS software package developed by National Aeronautics and Space Administration (NASA) is selected to implement the Fuzzy Logic component of the SC. A Decision-Centered Methodology is chosen to ensure an orderly development of the SC as discussed in Part I.

Mission Requirements are first examined. The Functional and Quantitative Requirements are derived from the expected scenarios and the operations the SC is expected to perform. The Allocation to Process function provided the hierarchical structure for dividing the Mission Requirements among the various SCs. This allow a chain of command to be emulated.

The Operator Decisions are derived from the requirements and processes that provide the basis for the displays and algorithms. The Decision Paradigm employed is "A decision is a state transition from a current state to a future state." This transition from state to state follows an established plan. This is where the major efforts for the SC occur. Six specific command decisions are implemented to change Defense Condition level, Battle Plans, Rules of Engagement, Defense Engagement Authority, Readiness Posture and Mission Objectives.

The Data Visualization function is accomplished using the XForms[82] Graphical User Interface (GUI) builder. Results from the Operator Decisions are displayed for the purpose of aiding the decision-maker. These displays are designed to emphasize the features the decision-maker needs. These displays provide vital information such as SC Identification (ID) numbers, DEFCON, RP, and DEA. Intelligence data including the date and time, the source of the intelligence information, and the associated weight and persistence factors are available for display.

SC deductive reasoning is driven by algorithms that simulate uncertain input data. are driven by the need to process data for other decision algorithms and Artificial Intelligence tools. The SC Controller has display buttons to turn these algorithms on or off. Algorithms included are Belief Networks and "Fog-of-War".

Results

The SC is designed to execute on various platforms. This keeps the decision-making process of the SC independent of exact particulars of any given system. The SC is a proof of concept model showing that a commander's decision-making process modeled as a rule-based "thinker". Additional influences on these rules are demonstrated through the use of Belief Networks and "Fog-of-War" decision-making algorithms.

Some rules require data fusion for the chain of command hierarchy that are represented by the message flow between the Battle Management Director, Command and Control Director, and the Fire Units Director. Each SC is supplied with rules applicable to the position they are simulating along with a database of corresponding associated with that role. These rules and databases are constructed by interviewing domain experts in these areas and are not expected to be absolutely complete allowing the Fuzzy Logic decision process to cover some of the gaps.

---

[82] http://en.wikipedia.org/wiki/XForms, accessed 02/25/2015.

Commander	If	Then
Rules of Engagement	The threat situation changes	Propose change of ROE to the other commanders
	ROE is changed	Notify other commanders
Acknowledge	Accept acknowledgements	Reply message received
	Clear acknowledgments	Final acknowledgment clear
Battle Manager	if	Then
Readiness Posture	There is a change in readiness	Propose a change in RP to the other commanders
	Readiness Posture is changed	Notify other commanders
Mission Objectives	There is a change in mission objectives	Propose change of MO to the other commanders
	Mission Objectives are changed	Notify other commanders
Defense Engagement Authority (DEA)	There is a change in DEA	Propose a change in DEA to the other commanders
	There is a change in DEA	Notify other commanders
Defense Condition (DEFCON)	The threat of attack is normal	Notify other commanders that DEFCON level is 5
"	There are signs of indications of increased threat activities	Notify other commanders that DEFCON level is 4
"	There is an unannounced launch of a missile	Notify other commanders that DEFCON level is 3
"	Initial direction of missile is toward North America	Notify other commanders that DEFCON level is 2
"	Predicted impact point is in Continental United States	Notify other commanders that DEFCON level is 1

Summary

The Simulated Commander prototype demonstrates the ability to model the human decision-making process for a war-gaming scenario. A rule base set is established to represent the SC decision process for a missile defense scenario. These rules are designed to allow other algorithms such as Belief Networks and "Fog-of-War" to affect the decision-making process.

Each SC module has its own set of facts, rules, and agenda. The SC's share different data through the hierarchical chain. The SC prototype is built to make six specific command decisions. The SC receives all the data the human commander receives and the data the human commander sends out to the Battle Manager. This allows switching between the SC active or inactive status without requiring the simulation to pause for the SC data to be updated when switched from inactive to active. The challenge is for the SC to continue to make decisions based on all data even though the decisions will not be used unless the SC is reactivated.

Chapter III-11
Abductive Reasoning to the Best Explanation

Abductive reasoning is reasoning to the best explanation. Unlike deductive and inductive reasoning, abductive reasoning is not guaranteed to provide correct inferences. Early work was done by Peirce[83]. We use the Subdue[84] abductive reasoning tool from the University of Texas at Arlington. The algorithm produces new nodes and links from structured data based on machine learning; specifically, minimum description length principles.

Subdue

To find interesting patterns in network traffic based on Abductive reasoning, or reasoning to the best explanation, we use the Subdue Discovery Algorithm to create a create substructures that separated the attack data from the nominal traffic. We use the supervised learning mode, which allows for two graphs to be input into the algorithm, one positive graph and one negative graph. In unsupervised learning mode, the algorithm then attempts to compress the positive graph into substructures, looking for the one substructure that compresses the graph the most. In Supervised Learning mode the same thing is accomplished, except the algorithm also makes sure that the substructure that is found to compress the positive graph, does not also compress the negative graph as well.

The data sets are easily converted into the file format that the Subdue algorithm requires - a list of vertices and edges. We again use a VBA script, and the structures of the graph are created as a tree of depth 1, where the top node contains a generic label "Connection". The children of this node hold field attributes, and the edges hold the field names. We use this structure for the records because it is said to provide the best results. Subdue processes this graph by searching for substructures (Figure III-11.1) that compress it, performing the compression, and repeating the process until there are no more substructures that compress the graph.

Subdue works well to find new structure (nodes and links) when semantic networks from the knowledge base are input and new evidence is added. This idea is used to advantage to find "unknown unknowns", or "things we didn't know that we didn't know". The idea of automating the discovery of unknown unknowns (Chapter IV-4) works best, in our experience, with abductive reasoning.

Summary

Abductive reasoning is the last of a wide variety of reasoning algorithms that this part of the book focuses on. Evidential reasoning uses concepts from probability and statistics, and implemented belief networks. Deductive reasoning makes specific inferences from general data. Inductive reasoning provides rules that are automatically discovered from structured content: it reasons from specific to general. Abductive reasoning provides the best explanation. Analogical reasoning is based on similarities. Terrain reasoning is a specific example of analogical reasoning. Evolutionary reasoning uses models from biology and is implemented via genetic

---

[83] http://plato.stanford.edu/entries/abduction/peirce.html, accessed 02/25/2015.
[84] http://citeseerx.ist.psu.edu/viewdoc/download?doi=10.1.1.434.980&rep=rep1&type=pdf, accessed 02/25/2015.

algorithms. Model-based reasoning varies widely from discrete event, to continuous physics models to producing emergent behavior. Link analysis finds patterns in networks consisting of links and nodes. Textual reasoning leverages natural language processing. Reasoning with physics (Chapter III-12) provides a few elegant algorithms.

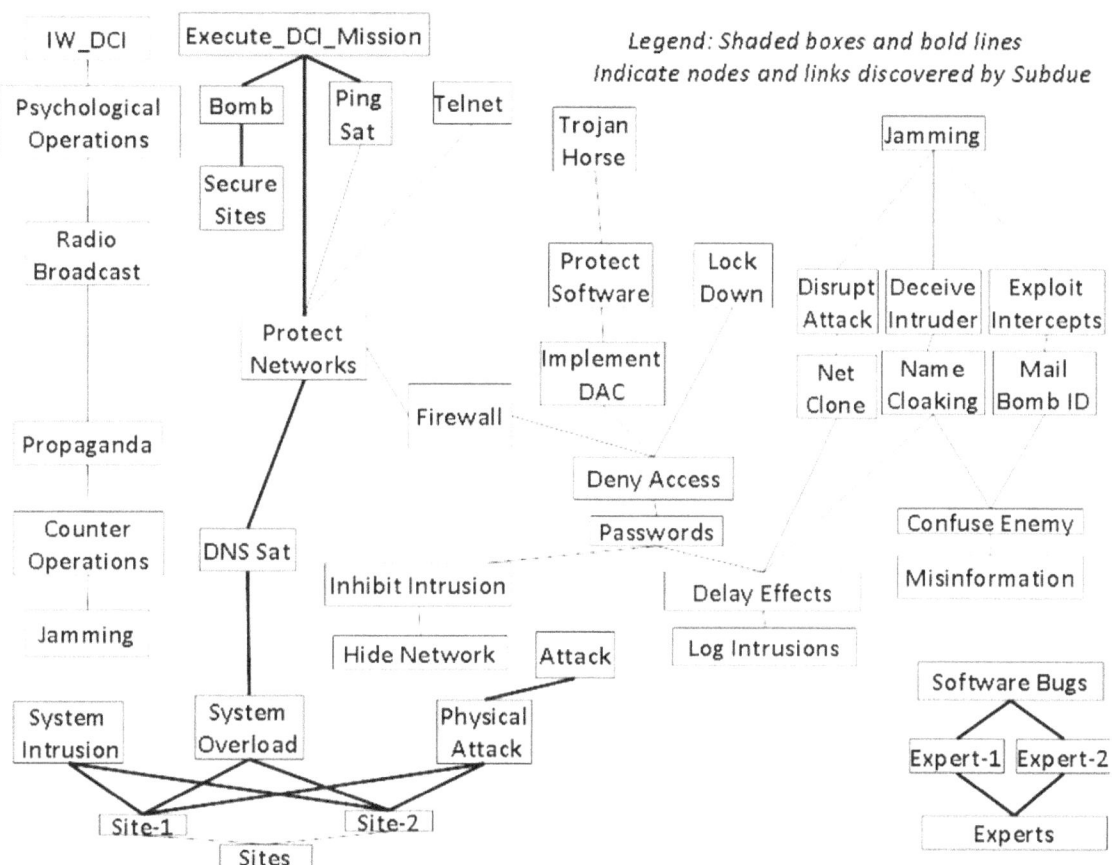

Figure III-11.1: Subdue Hierarchical Clustering – Example

Chapter III-12
Reasoning with Physics: A Few Astrodynamic Algorithms

Space Object Conjunction Prediction

An algorithm is defined to compute the cumulative probability and a confidence interval for space object conjunctions and collisions. A collision is a physical conjunction between two space objects. We assume that space object uncertainties are defined as normally distributed covariance ellipsoids. Monte Carlo integration is used to efficiently solve this seven degree-of-freedom (three positional coordinates for the primary space object, three positional coordinates for the secondary space object, and a time interval about the point of closest approach).

Given the 10 February 2009 collision[85] between two satellites (COSMOS and Iridium) there is a need to efficiently calculate the cumulative probability of space objects (including debris and rocket bodies) collision in orbit. Most current schemes are approximations and only provide a peak probability at a single point (closest approach). Current approximations are based on huge error ellipsoids (5 to 50 kilometers) that erroneously predict 100s of possible conjunctions per day.

Related Work

Most current computations are performed only at closest approach and provide peak, versus cumulative, probabilities of collision. This application is an exception: it approximates covariance ellipsoids as polynomials and computes a per-pass probability of collision. Most calculations determine the separation of two objects at closest approach, with no probability specified. The Center for Space Standards & Innovation offers a Satellite Orbital Conjunction Reports Assessing Threatening Encounters in Space[86] that computes a "true probability" when covariance matrices are available, and a "maximum probability" when no covariance data or when data of insufficient quality is available. The Joint Space Operations Center[87] uses a Computation of Miss Between Orbits algorithm that computes conjunctions based on separation at closest approach A two-dimensional calculation of peak probability based on a ratio of projected ellipsoid axes at closest approach, on linear relative velocity and many other assumptions, is also discussed in the literature. Recent papers[88] discuss calculations based on bundles of parallelepipeds with intersections at compound miters – this also gives only a peak probability of collision. None of the current approaches compute a confidence interval.

Approach

Rather than approximate the solution using geometric constructs (current algorithms) or attempt a formal solution (a seven dimension numerical integration of the covariance volumes versus

---

[85] http://swfound.org/media/6575/swf_iridium_cosmos_collision_fact_sheet_updated_2012.pdf, accessed 03/25/2015.
[86] http://celestrak.com/SOCRATES/, accessed 02/25/2015.
[87] http://www.vandenberg.af.mil/library/factsheets/factsheet.asp?id=12579, accessed 02/25/2015.
[88] http://www.centerforspace.com/downloads/files/pubs/AAS-07-393.pdf, accessed 02/25/2015.

time), we use Monte Carlo integration: a time interval about the point of closest approach is determined empirically, and covariance matrices are propagated to random times in this interval. Six normally distributed numbers within computed position intervals are drawn and assigned to appropriately rotated coordinate axes of the space objects, corresponding to the position coordinates of the two space objects. The distance between the randomly drawn points is calculated. If it is within the specified conjunction or collision threshold, a counter is increased (n = n + 1). The instantaneous probability of collision is computed as P = n/N, where N = total number of trials. These are summed over the close approach time intervals to get the cumulative per pass probability of collision. A confidence interval about this point estimation is also derived, based on standard statistical formulas.

Mission Implications

For the space surveillance mission, highly efficient algorithms are required to produce conjunction analysis results for a large number of primary object such as payloads (4,000) that may be impacted by an even larger number of secondary objects, including other payloads, rocket bodies, and debris (forecast: 100,000). The calculation is required for one week. This is the most computationally intense task that the Joint Space Operations Center faces.

For defensive counter-space, kinetic attack attribution relies on a calculation of the conjunction between two orbiting objects: one the orbital interceptor, and the other the satellite target. Post attack orbital elements are propagated backwards in time to obtain closest approach and associated uncertainty ellipsoids of the weapon and the target. Here, the objective is to calculate the probability of conjunction, which is defined as the intersection of the normal probability distributions of weapon and target. Note that the normal distributions arise not as an assumption, but rather as the output of the batch weighted least squares or extended Kalman Filter that produces a covariance matrix with normally distributed values.

Geometry

The locations of the primary and secondary spacecraft are obtained at closest approach. In addition, the uncertainties are specified as the 1 standard deviation values (Figure III-12.1) from the covariance matrix in the in-track, radial, and cross-track directions.

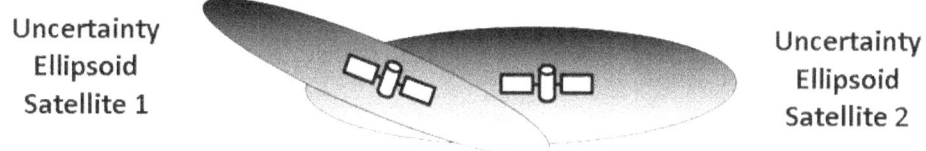

Figure III-12.1: Conjunction Analysis Geometry

The calculation of the common volume at closest approach is the basis for the peak probability of conjunction. However, there is a finite probability of conjunction both before and after closest approach as the uncertainty volumes move through one another. Although one object passes another in about 1-10 seconds, the cumulative probability of conjunction may be an order of magnitude larger than the peak probability

Algorithm Description

Nominal position data, derived from a two-line element sets (TLEs) or special perturbations[89] (SPEPH) calculation, is used to construct coordinate frames centered at the primary and secondary space objects. TLEs, which do not have associated covariance matrices, are currently used in conjunction analysis calculations. However, we intend to compute the covariance matrix, an accessible by-product of the batch weighted least squares filter[90]. Constructed covariance matrices, based on historical data, or estimated, are typically huge (5 to 50 of kilometers in-track). SPEPH vectors and covariance matrices are much more accurate. Even better are precision ephemeris state vectors and covariance matrices from satellite owner operators: this telemetry data is not routinely processed in the Joint Space Operations Center.

Coordinate Conversions

The position vectors that are obtained from the TLEs, SPEPH, or telemetry are in earth-centered inertial coordinates. We need to translate and rotate these to a body-fixed orbital coordinate frame centered on the primary object. Translation from orbital intercept (OI) to Target coordinates (unprimed to primed) is given by:

$$\text{Translation: } x = x' - x_0$$
$$y = y' - y_0$$
$$z = z' - z_0$$

Position coordinates of the OI relative to the target is based on the ECI position and velocity vectors which are readily obtained from the orbital elements. Orbital coordinate axes are obtained as:

Radial Unit Vector: $\mathbf{i}_R = \mathbf{R}/R$, where $\mathbf{R}$ = distance vector from Earth center
Cross-Track Unit Vector: $\mathbf{i}_H$ = unit ($\mathbf{R} \times \mathbf{V}$), where $\mathbf{V}$ = velocity vector
In-Track Unit Vector: $\mathbf{i}_T = |\mathbf{i}_H \times \mathbf{i}_R|$

Using this orthonormal set of right-handed Cartesian axes, vehicle attitudes are specified using three angles and sequential Euler rotations about the cross-track (pitch), radial (yaw), and in-track (roll) axes – in that order. The attitude of the OI, represented by the unprimed axes is specified relative to the target (primed) axes by a rotation of the pitch axis followed by a rotation of the resultant yaw axis, and finally a rotation about the roll axis.

This sequence of Euler rotations is somewhat arbitrary, but is chosen because it corresponds to visually meaningful angles: $\Delta\gamma$ is the angle above or below the local horizontal, ΔA is the out-of-plane angle, and $\Delta\phi$ is the "twist" angle. For prototyping purposes, the three angles are assumed given, or input parametrically, but they can readily be computed from the position and velocity vectors.

---

[89] http://www.afspc.af.mil/library/factsheets/factsheet.asp?id=19466, accessed 02/25/2015.
[90] http://www.centerforspace.com/downloads/files/pubs/AIAA-2008-6770.pdf, accessed 03/01/2015.

Problem Formulation: The probability of conjunction is the ratio of the intersected probability distributions of two volumes divided by the total probability distributions of the two volumes. Note that the "difference" coordinate frame can reduce this to three position coordinates, but our method is faster by drawing six random numbers rather than computing the difference frame.

$$P = \iiint P_2 \left[\iiint P_1 \, dv_1 \right] dv_2$$

      where P = probability of conjunction
           P_1 = probability distribution of the target
           P_2 = probability distribution of the interceptor
           dV_1 = elemental volume of the target
           dV_1 = elemental volume of the interceptor

Monte Carlo Integration: Random sampling is a method of evaluating integrals. Consider an integral is of the form,

$$\int^\infty g(x) \, f(x) \, dx$$

where **f(x)** is a probability density function. For a sample, $x_1, x_2, \ldots x_N$ that has been drawn by chance from the distribution represented by f(x), the quantity $N^{-1} \sum_{i=1}^{N} g(x_i)$ is an estimate of the integral, which becomes more accurate as N gets larger.

Although a technique that deliberately leaves any aspect of a calculation to chance is naturally suspect at first glance, critical examination of the classical process of numerical integration reduces the distinction to insignificance, in view of the uncertainties associated with the alternative solution. To numerically integrate, uncertainties are associated with choice of integration step size, integration formula, finite grid effects, and rounding error accumulation.

The superiority of random sampling becomes evident in the conjunction analysis problem. Systematic sampling for classical numerical integration at closest approach requires N^6 calculations, where N is the grid size. It is not easy to fill six dimensions systematically, but it is easy to give these points equal (or, in our case, normally distributed) representation by permitting them all a normally distributed chance to be drawn in a random sample. Thus, the Monte Carlo method permits use of reasonable numbers of calculated points to provide unbiased estimates of the desired integral.

Confidence Interval

The estimate provided by Monte Carlo integration has a standard deviation given by $\sigma = \sigma_g / N^{1/2}$, where σ_g is the standard deviation of the function g(x) and N = number of samples. For conjunction analysis, the probability formula indicates that g(N) = constant. Thus, the technique exhibits a probability distribution of error proportional to $N^{-1/2}$.

Results

A prototype computer program using Monte Carlo Integration for conjunction analysis is written and tested. An IBM library function (GGNQF[91]) generates two sets of normally distributed random coordinates. The distance between points is compared with the critical radius. If the distance is smaller, the fractional contribution is added to the incremental probability.

This prototype software computes only the point probability of conjunction at closest approach. It is easily modified to choose uniformly distributed random numbers for times near closest approach, perform the same calculations, and from the resultant point probabilities, construct the cumulative probability per pass of a conjunction.

Results (Figure III-12.2) are obtained for $\varepsilon = (6, 2, 2)$, $\delta = (6, 2, 2)$, $d = 18$. Units are in kilometers. Orientations of ellipsoids are such that both objects are in the same plane and have the same radial values. Comparison of classical numerical integration (software is prototyped for this method) versus Monte Carlo Integration indicate that Monte Carlo Integration converges faster, is just as accurate as numerical integration with small step sizes, and executes much faster.

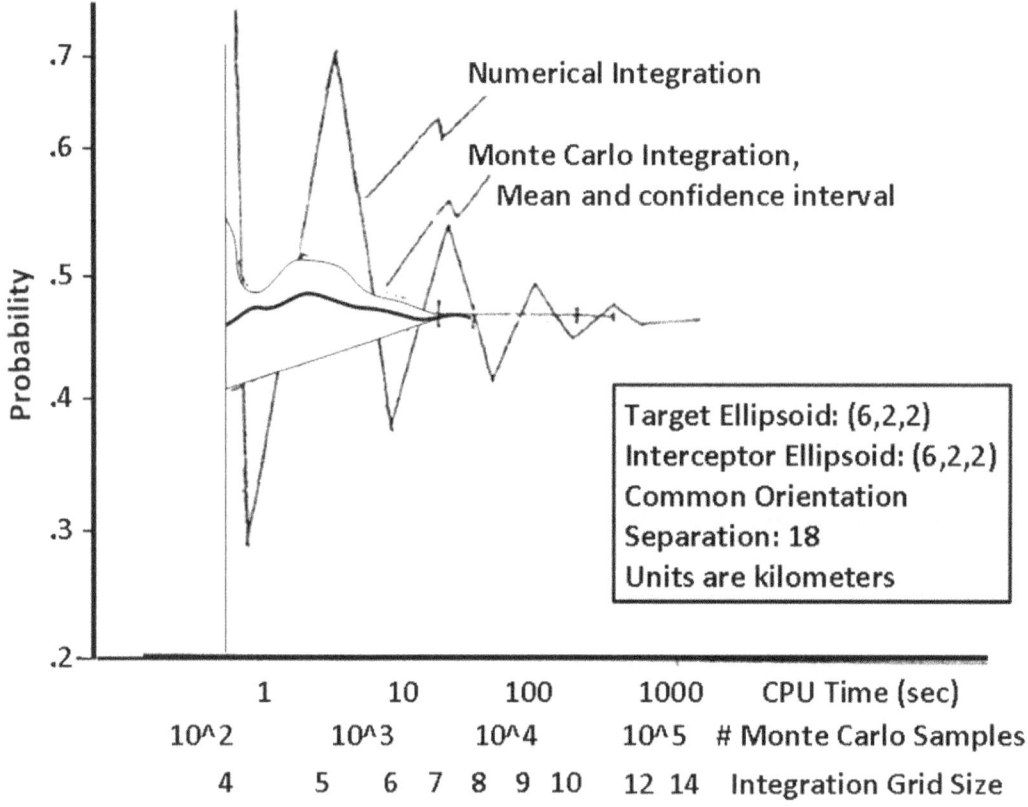

Figure III-12.2: Comparison of Numerical and Monte Carlo Integration

The grid size for integration is given by N, the number of points per kilometer.. The number of

---
[91] https://software.intel.com/en-us/articles/installing-and-using-the-imsl-libraries, accessed 02/25/2015.

calculations required to determine the conjunction probability (P) is N^6. P converges to an answer at N=12, which took about 8.5 minutes of CPU time (in 1983). With N=10, the error in P is ~3% and requires ~ 3 minutes of CPU time. The plot shows of CPU time for numerical integration for various values of N. Monte Carlo Integration results are superimposed: Monte Carlo integration runs in seconds and converges smoothly.

Summary and Conclusions

A Monte-Carlo integration approach for computing the per-pass probability of collision, along with a confidence interval, has been discussed. The approach is fast, provides a "just-in-time" answer, and is easily executed in parallel over multiple computers. This provides an elegant solution to a complex, large-scale computing problem that is of significant current interest.

Space-to-Space Intercept

The guidance problem requires equations to compute the velocity at the interceptor position that is required to intercept the target. Our current Lambert procedure (Figure III-12.3) is for a missile intercept of a ground target. Consequently, the target is assumed to be Earth-fixed.

An algorithm sketch of the Lambert Guidance algorithm is provided:
Given: Interceptor Vector (**r**), Target Vector (**R**), and a time-of-flight (T)
Estimate a value for (tan Γ). Default is tan Γ = -.008
Compute required Interceptor Velocity (**v**)

$\cos \theta = \mathbf{i_r} \cdot \mathbf{i_R}$, the dot product of unit vectors, where θ = angle between interceptor and target
$\sin \theta = SQR (1 - \cos^2 \theta)$, with [0° <= θ <= 180°]

1 $e \cos F = (-R/r + 1 + \sin \theta \tan \Gamma)/ (R/r - \cos \theta - \sin \theta \tan \Gamma)$
   where e = eccentricity, Γ = target flight path angle, F = target true anomaly
$e \sin F = \tan \Gamma (1 + e \cos F)$
$e = SQR[(e \sin F)^2 + (e \cos F)^2]$
$e \sin f = e \sin F \cos \theta - e \cos F \sin \theta$, where f = interceptor true anomaly
$e \cos f = e \cos F \cos \theta - e \sin F \sin \theta$
$\tan \gamma = e \sin f / (1 + e \cos f)$, where γ = flight path angle at the interceptor
$p = r / (1 - e \cos f)$, where p = parameter of the new orbit
$a = p / (1 - e^2)$, where a = semi-major axis of the new orbit
$E_0 = \tan^{-1} [\{SQR(1 - e^2) \sin f \} / 1 - e \cos f)/ \{(1/ e (1 - r/a)\}]$,
   where E_0 = interceptor eccentric anomaly
$E_t = \tan^{-1} [\{SQR(1 - e^2) \sin F\} / (1 - e \cos F) \}/ \{(1/ e (1 - R/a)\}]$
   where E_t = target eccentric anomaly
$t = SQR(a^3/\mu) / (E_t - e \sin E_t - E_0 + e \sin E_0)$, where t = computed time of flight
if (T - t) > .0000001 then $\tan \Gamma = \tan \Gamma_1 + \{[\tan \Gamma_1 - \tan\Gamma_0]/ \{T_1 - T_0\}\} *(T-t)$ and go to 1
$v = SQR(\mu/p) (1 + 2 e \cos f + e^2)$, where v = velocity needed by interceptor
$\mathbf{v} = v (\sin \gamma \ \mathbf{i_r} + \cos \gamma \ \mathbf{i_R})$, where v = velocity vector needed by interceptor

The reason we include this algorithm in enough detail to code is that it is a very robust way to calculate a maneuver from a current location to any other point at any other time. This algorithm is unique to this book. It is published nowhere else.

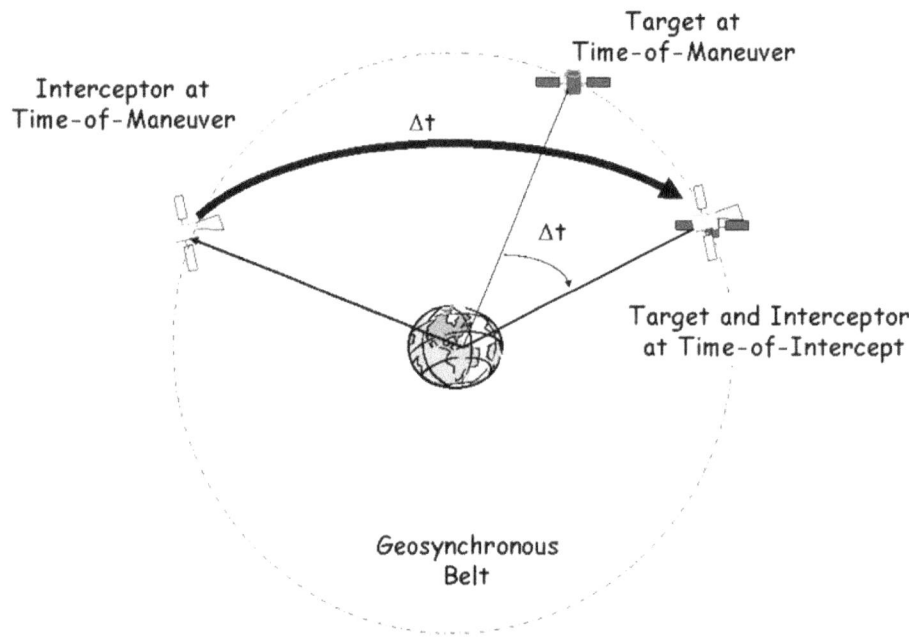

Figure III-12.3: Space-to-Space Intercept

For the space-to-space problem, the target is in orbit. As with the missile intercept problem, both the interceptor and the target are moving. For the space-to-space problem, three changes are required:
1) the target is specified using orbital elements
1) the target is propagated to time-of-intercept, which produces a state vector
2) the position components of the target state vector at intercept are used to compute the range angle parameter.

We compare our Lambert routine with the Lambert routine found in the Java Astrodynamics Toolkit[92] (JAT). An experiment computing delta velocity for Lambert maneuvers among random places near the geosynchronous belt shows that our Lambert routine performed in about 1/50th the time of the routine found in JAT. The delta velocities output from the method ported from BASIC and the method in JAT are also compared and found to be the same within a centimeter per second. We ran 100,000 iterations and obtained the following performance metrics: LAM takes 891 milliseconds, VAT takes 49150 milliseconds => Ratio: VAT/LAM = 55.2

---

[92] http://www.academia.edu/5317237/Java_Astrodynamics_Toolkit, accessed 02/25/2015.

Part IV
Multi-Strategy Reasoning and Learning

Chapter IV-1: Uncertainty Management
Chapter IV-2: Combining Algorithms, Graphs and Simulation
Chapter IV-4: Processing Synergy and Feedback Loops
Chapter IV-5: Automated Discovery of Unknown Unknowns
Chapter IV-6: Network Metrics Define Centers-of-Gravity

Part IV
Introduction

In previous chapters, we introduce the various types of automated reasoning and the algorithms used to implement them. Here, we examine multi-strategy reasoning: combining one or more algorithms to produce a meaningful result to a user needing to make a decision.

Chapter IV-1
Uncertainty Management

This chapter describes the problem of uncertainty management (UM) in decision support systems, the engineering process for designing UM into them, and why effective presentation of uncertainties in support of human decision makers is critical to system success. Differences between uncertainty management and information management are cited. Types of uncertainty are defined. Uncertainty is related to engineering domains. An architecture that supports uncertainty management is defined. Uncertainty measures are formulated. Applications that demonstrate the utility of uncertainty management are explored. Technology insertion projects and metrics are discussed.

Introduction

Real world data usually exhibit a wide range of errors, are often incomplete, and rarely exhibit structure or organization suitable for precise analysis. However, most traditional *information management* (IM) systems implicitly assume that all relevant data are available, accurate, properly organized, current, without conflict, and complete. This incongruity forces engineers to make assumptions that can compromise the validity of inferences made and the information presented to decision makers on the basis of the available data. *Uncertainty management* (UM) is an emerging system engineering discipline[93] that identifies the uncertainties in data entering a system and offers design approaches for storing and manipulating data without filtering, thresholding, or making system design decisions that may compromise the integrity of the input data and of the system's results.

UM implementation opportunities generally result from the desire to enhance the functionality of an existing IM system, although opportunities for designing UM systems from inception occasionally occur. UM enhancements to IM systems are usually due to system evolution or the need to manage uncertainty per the "voice of the customer" (that is, requirements). UM requirements are recognized by a number of "tip-off" phrases; these phrases also indicate the class of uncertainty – error, incompleteness, etc. – at issue. Regardless of the motivation for UM requirements, there are differences between the engineering methodologies needed for enhancing a legacy IM system with UM capabilities and those needed for designing a system with UM from inception.

Propagating and presenting uncertainty effectively for human decision makers is critical to system success. The value of UM is quantified in terms of how much more often the decision maker makes the right decision, their confidence in the decision, and the time necessary to arrive at the correct decision. At the heart of quantifying the value of UM are two questions: 1) do humans make better decisions when presented with the uncertainty associated with the situation, and 2) what additional UM support makes humans more capable of recognizing and processing uncertainty correctly? This work presents experimental evidence 1) for identifying situations in which humans need to appreciate uncertainty to avoid bad decisions, 2) for how best to handle the uncertainty in these situations and 3) for a variety of visualization techniques that cue the

---

[93] http://www.springer.com/us/book/9780792398035, accessed 03/28/2015.

human's attention to key uncertainties associated with pertinent information.

Uncertainty Management versus Information Management

The current trend in Information Technology systems is to provide a cloud-based, n-tier architecture using a Commercial Off the Shelf (COTS) framework. Specific services are added to meet the unique requirements of the user. These COTS products often are developed using several underlying assumptions consistent with an IM system. Unfortunately, these underlying assumptions may not apply in specific systems of interest to planners, businesses, cyberspace operations, and the intelligence community. The result is a system that is missing significant features needed by the users. As the complexity of the problem increases, these missing features place an impossible burden for which the user must compensate. This paper focuses on specific applications in military planning systems and shows why the principles of Uncertainty Management provide a better systems engineering foundation than the typical Information Management approach.

Most traditional IM systems implicitly assume that all relevant data are available, accurate, properly organized, and complete. This incongruence forces engineers to make assumptions in IM systems that can compromise the validity of inferences made and the information presented to decision makers on the basis of real world data. UM is an emerging system engineering discipline that identifies the uncertainties in data entering a system and offers design approaches for storing and manipulating data without filtering, thresholding, or making system design decisions that may compromise the integrity of the input data and of the system's results.

Definition of 12 Types of Uncertainty

A hypothesis is an assertion about a concept. Evidence is information that supports a hypothesis. Hypotheses and evidence are subject to many different kinds of uncertainty. These sources of uncertainty are defined and examples are provided.

- **Understanding.** Ideally, hypotheses and evidence should be intelligible. Uncertainty in the human understanding of a hypothesis or evidence may result in a fundamental cognitive problem. For example, "I can't understand what you're saying?"

- **Random.** A hypothesis or evidence may be dependent on random variables; that is, measurable quantities that randomly vary. A defining characteristic of a random variable, in classical statistics, is that it has a well-defined mean and standard deviation, based on an assumed underlying distribution. An example of a random variable is the result of a coin toss.

- **Measurement.** This is also known as systematic error. Examples are modeling errors, bias error, and confidence intervals due to lack of a sufficient number of measurements. An example is polling error, often expressed as a confidence interval or margin of error: Candidate has a 78% acceptance, +- 3% margin of error (based on an assumed normal distribution, a sample size of 1000, and other characteristics of the population).

- **False**. This category of uncertainty captures the idea that we may not be sure that the hypothesis is valid. A bi-variate distribution may be used to define a degree of belief in the credibility of a hypothesis. The non-parametric case is also of interest. For example, a 10% disbelief in a hypothesis doesn't mean that the degree of belief is 90%, it means that our degree of ignorance is 90%.

- **Conflicting**. This uncertainty is defined for evidence that contradicts a hypothesis. The lack of conflicting evidence is called plausibility (Pl). In an evidential interval [0, 1] the degree of conflict or disbelief is 1 – Pl.

- **Missing Known**. Hypotheses have characteristics, including quantified degrees of belief or disbelief, and answers to questions such as who, what, where, when, why, and how. A missing known is an empty slot indicating missing evidence about a characteristic. For example, if the hypothesis is that we will capture a fugitive, then the slot for "where is he" may be empty.

- **Missing Unknown.** This category of uncertainty is defined as missing hypotheses or links between hypotheses; that is, we don't know what we don't know. An example is an unintended consequence of an action.

- **Ambiguous**. This type of uncertainty reflects the fact that either evidence or a hypothesis may be understood or interpreted in more than one way. For example, "All is well" does not have sufficient context (with whom, with what, where, when,..) to be interpreted unambiguously.

- **Obsolete**. This uncertainty stems from the evidence not being up to date. Evidence ages over time. This induces temporal uncertainty that must be accounted for when fusing evidence arriving at different times. For example, last year's weather forecast is obsolete, and likely has no value, while some assertions may always be certain.

- **Vague**. This uncertainty arises from spoken language. It denotes a lack of crispness, or fuzziness in interpretation. Words like probably, tall, and soon are vague.

- **Undecidable**. This uncertainty applies to a collection of evidence, hypotheses, and links that is ill-posed. The test for undecidability is whether an answer can be generated in a finite number of steps (inspired by Godel's work[94]). For example, if we are attempting to achieve a set of effects, but the environment is not rigorously specified, it may not be possible to identify unintended effects. Another class of undecidable hypotheses arises from computational and evolutionary game theory where players follow an optimal strategy that is probabilistic.

- **Chaotic**. The indicator that this uncertainty is present is sensitive dependence to initial conditions. One measure of chaotic behavior is the Lyapunov exponent. Examples are the "butterfly effect": a butterfly flapping its wings in Aspen produces a snowstorm in Denver. Many systems in nature, like the weather, exhibit this behavior.

---

[94] http://www.scientificamerican.com/article/what-is-godels-theorem/, accessed 04/09/2015.

These types of uncertainty are organized (Figure IV-1.1) as occurring in data or in processes. Uncertainties in data are distinguished according to whether they represent inaccurate or incomplete data. Uncertainties in processes are grouped according to whether they are conceptual or dynamic.

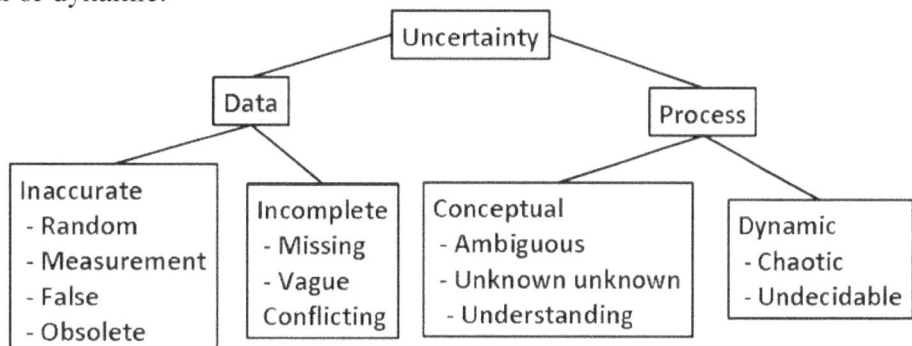

Figure IV-1.1: Taxonomy of Types of Uncertainty

Tipoffs

How do we spot uncertainty? In typical statistical settings, a variable is sampled and the sample mean, standard deviation, and sample size may be specified. This explicitly provides the parameters that define random and measurement uncertainties. Likewise, obsolete data can be identified by an associated time tag. In all other cases, the underlying uncertainty is implicit. We must derive it from indicator words. We experimented with passages extracted from the Internet that postulated terrorist activity. We "tagged" words, phrases, and other indicators of the 12 types of uncertainty and generalized the way in which they occurred. As expected, missing unknowns (by definition), chaotic behavior, and undecidable propositions are most difficult to discern.

Type of Uncertainty Indicators

- **Understanding**   Content is not clearly stated
- **Random**   Data, mean, variance, (random) variable, curve-fit
- **Measurement**   Confidence interval, numerical sampling, models and simulations
- **False**   Deceive, camouflage, deny, mimic, lie, bluff
- **Conflicting**   Differing, opinions, disbelief
- **Missing Known**   Missing data, incomplete, lack of evidence
- **Missing Unknown**   Intent, new concept, surprise, discovered
- **Ambiguous**   Concepts or evidence not clearly defined, multiple meanings
- **Obsolete**   Time sensitive, old information, stale
- **Vague**   Use of hedge words, qualified opinion not concrete
- **Undecidable**   Untestable, key concepts or connections missing
- **Chaotic**   Sensitive to initial conditions, fractal, nature processes (weather)

Unifying Framework for Computing Total Uncertainty

We derived an architectural foundation, consisting of an organized and sequenced set of applications (Figure IV-1.2), as a foundation for uncertainty management. Information is extracted from raw data, including hedge words defining vague content and ambiguous concepts. The extracted evidence fills frames in the knowledge base, which may add false and missing known data. Errors due to random deviation, measurement error, conflicting information, and obsolete data are derived from the frames in the knowledge base. Hypotheses are subject to tests for understanding. Data mining tools may provide missing unknown hypotheses or links between hypotheses. A fog-of-war module perturbs solutions to determine chaotic behavior of the belief network (for example, large variations in results from small perturbations). Finally, human analyst may run analytical tests to see if the belief network is decidable (based on best possible evidence, are high level goal nodes deterministic and achievable?).

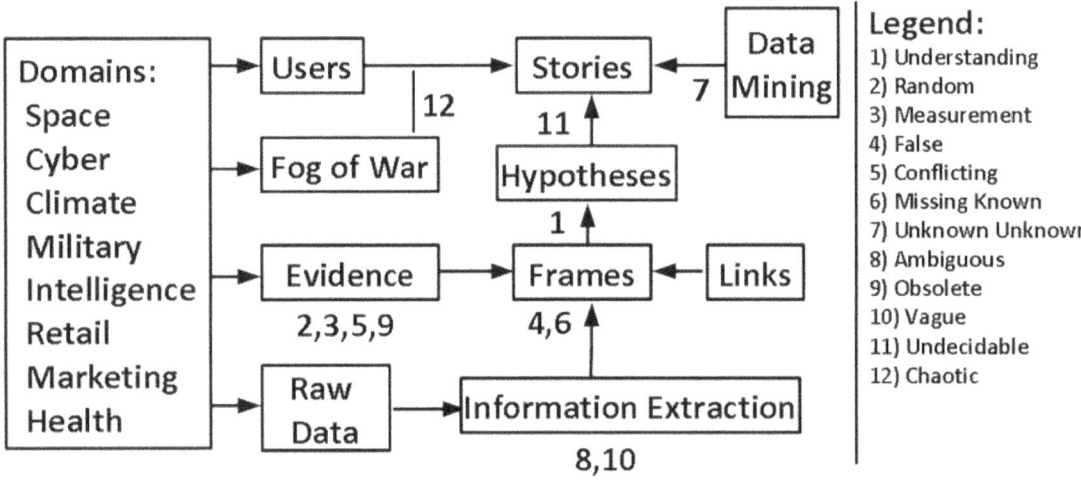

Figure IV-1.2: Uncertainty Management Applications Architecture

Total uncertainty is calculated as the root-sum-square of point estimates, upper estimates and lower estimates for the 12 types of uncertainty. Monte Carlo sampling of belief networks is used to determine the point, upper, and lower estimates for each of the types of uncertainty, based on and evidence combination rule, such as Bayes Rule or the Dempster-Shafer Combination Rule. Three tricky uncertainties to compute are unknown unknowns, understanding, and undecidable. Unknown unknowns are computed using data fusion tools and data mining tools to discover new hypotheses and links between hypotheses[95]. Understanding (of a hypothesis) is assessed using the hypothesis and it's characteristics in a Web search to find hypotheses that could be confused with the stated hypothesis. Undecidable is determined by maximizing belief in all evidence nodes in the belief network to see if the belief in the top node is above threshold, and hence decidable. We also envision a game-theoretic approach, using a freeware application like Gambit[96], to determine whether a hypothesis is decidable, based on an active environment; for example, a responsive adversary.

---

[95] http://www.google.com/patents/US8078559, accessed 02/26/2015.
[96] http://gambit.sourceforge.net/, accessed 04/09/2015.

Prototype Code

Based on our domain analysis, we defined and implemented software prototypes of six uncertainty management tools:

1. Kiviat diagram showing 12 components of uncertainty, driven by a belief network,
2. Geographic uncertainty management, based on explicit rendering of sensor "holes"
3. Probable threat locations computed from a trafficability algorithm
4. Logical uncertainty portrayed by the belief network editor linked to the Kiviat
5. Temporal uncertainty computed based on rendezvous schedule
6. Dynamic plan update based on hybrid belief networks driven by information extraction

Figure IV-1.3 shows a Kiviat diagram with the 12 components of uncertainty. These are normalized relative to the uncertainty threshold, show as the circle. At a glance, the decision maker can easily ascertain that the out-of-limit uncertainties are Vague, Missing Known, Understanding, Obsolete, and Random.

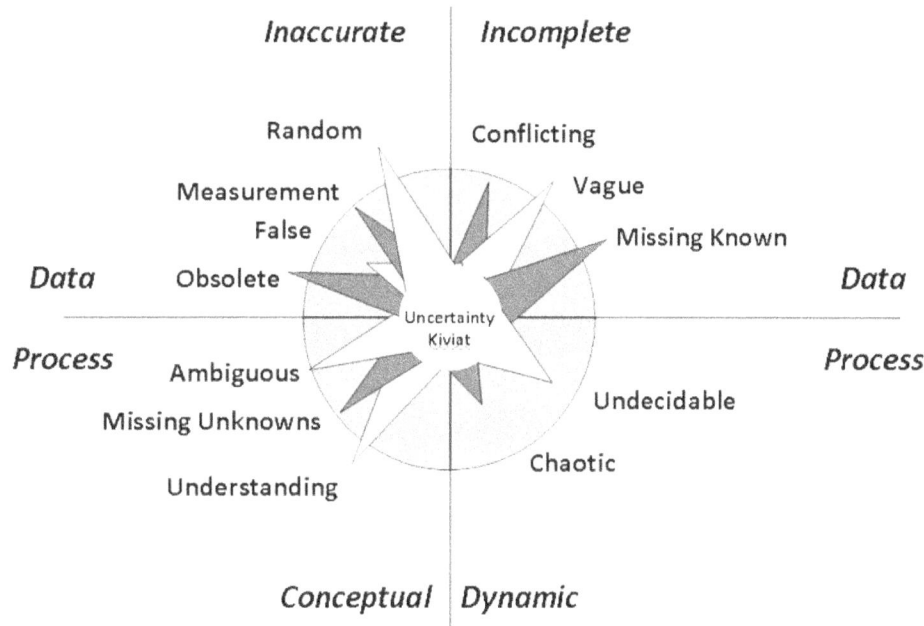

Figure IV-1.3: Kiviat Diagram Identifies Problematic Uncertainties

The basis for automating the management of uncertainty is our ability to extract hedge words that denote vagueness and other types of uncertainty from unstructured text. A text message is parsed and the resulting input into a knowledge base frame that contains hypotheses and their associated characteristics.

We constructed a Java version of an High Interest Event as described earlier. The message and icons, with shape of the icon indicating uncertainty (square icon means little uncertainty, and round icon means moderate uncertainty) are posted automatically, subject to analyst review of information extracted from text.

A new application is constructed to demonstrate the ability to combine Bayesian and Dempster Shafer belief networks to leverage the best features of both. The Bayes net is used to fuse voluminous, probabilistic sensor data related to pipeline outages in the scenario. The best next observation[97] identifies the best evidence nodes to update, as reflected in the uncertainty Kiviat. Dempster-Shafer Belief Networks accept this information and use it to reason about enemy intents, subject to vague, conflicting, and obsolete evidence. By clicking on the suggested evidence node, the dynamic plan update uses case-based repair to identify options that worked well in the past to update the plan.

Summary and Conclusions

We identify the need for uncertainty management, rather than information management, define the underlying uncertainties and related tip-offs, and characterize the problem space for managing uncertainties. Armed with this background, we formulate an application-oriented architecture for representing uncertainties and calculating total uncertainty and its components.

---

[97] http://www.researchgate.net/publication/228950168_Inference_to_the_Best_Next_Observation_in_Probabilistic_Expert_Systems, accessed 02/26/2015.

Chapter IV-2
Combining Algorithms, Graphs and Simulation

Introduction

A key technical objective of multi-strategy reasoning is to combine the functionality of planning, assessment, and simulation tools. We derive an approach for orchestrating these tools: a framework with a MySQL[98] database. This framework approach provides loose coupling among the applications and is successfully used earlier – to combine inference algorithms - as a foundation for multi-strategy reasoning. The remaining question is: how do we effectively combine planning tools and graphs with simulation?

This chapter explores the similarities and differences between the underlying structure of planning tools, graphs, and simulations. Based on a similarity in the goal; that is, influencing social networks, for these disparate computational techniques, we identify common ground and provide a deeper connection between planning tools, graph-based representations and simulations to provide substantive interoperability.

Need

We have a collection of analyst tools: 1) Strategy Development Tool (SDT) for planning, 2) Belief Network Editor (BNE) for assessment, 3) the System Effectiveness Analysis Simulation (SEAS) for "exploratory analysis of transformational, information-driven warfare"[99], and 4) the MySQL database – which is also part of the framework. Orchestration of these tools (Figure IV-2.1), from a software development perspective, is accomplished by having the tools interface loosely with one another through the framework. Users access the tools through a "dashboard".

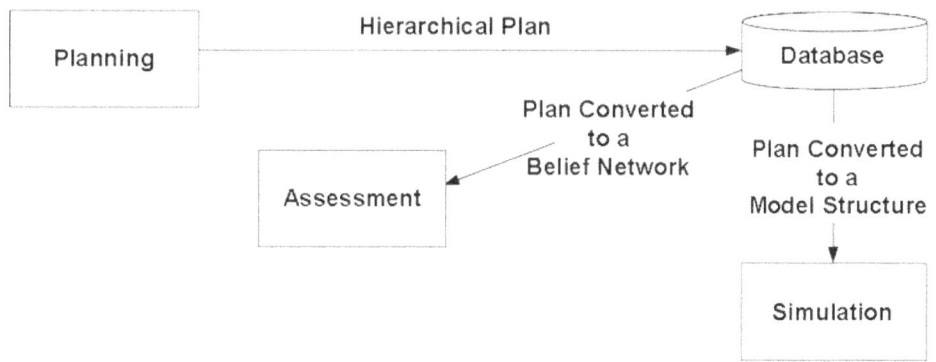

Figure IV-2.1: Combining Graphs and Simulations - Processing Flow

A typical processing flow is for the analyst to edit or create a plan using a template-based planning tool. This produces an .XML file that is sent to the database and is accessed by the

---

[98] https://www.mysql.com/, accessed 03/28/2015.
[99] https://www.teamseas.com/, accessed 02/26/2015.

belief network to automatically create a belief network graph. The data base contains evidence and graph characteristics for the belief network and operational parameters from the plan. How could this database content drive SEAS?

We expect that a simulation expert, with knowledge extracted from subject matter experts (SMEs), creates or edits a SEAS file based on an understanding of the plan and the BNE assessment of incoming evidence. It seems like significant analyst judgment is required, and that it is difficult or impossible to validate that the SEAS agent-based simulation actually represents the plan from the SDT and assessment from BNE.

With this discussion as background, the need is to determine how to connect the SDT plan and BNE graphs with the SEAS simulation? This is challenging because the input, processing and output of SDT and BNE look nothing like that of the SEAS simulation.

Related Work

As discussed in Chapter II-2, we face a dilemma trying to integrate multiple decision algorithms. We have three dozen decision support tools that need to interface with one another for input and output. Our goal is to produce a long processing thread with drill-down to explanations. The problem is that we spend most of our time interfacing algorithms with one another. We encounter quadratic scaling $(2n)^2$ where "n" is the number of applications. While this means plenty of work for our software engineers, it isn't productive. The solution[100] is to have all algorithms interface with a knowledge base, structured as a collection of hierarchical, characteristics-based semantic networks, and to use a single defined interface for both import and export. This results in linear scaling of interface development with the number of applications.

This knowledge-based solution is derived from insights from the machine learning community[101]: The largely unsupported proposition is that five types of inference, all with very different formulations and representations, are expressible as semantic networks. The five types of inference are: rule-based systems, belief networks, case-based reasoning, genetic algorithms, and artificial neural networks. In our work, we prove that underlying reasoning strategies important to us (abductive, analogical, deductive, inductive, and probabilistic/evidential) are elegantly expressible as semantic networks, and can be tightly integrated with a knowledge base that is, itself, a semantic network.

We note that other approaches to a unified framework for multi-strategy reasoning include intelligent software agents (which SEAS uses), introspection, and simulation. We did not, in the earlier chapter, attempt to incorporate intelligent software agents or simulation into the semantic network framework. That is the purpose of this chapter: to incorporate simulation, based on intelligent software agents, into a common framework.

An impediment to combining plans and graphs, which arise in real-world environments, and

---

[100] P. Talbot, "Semantic Networks: A Unifying Concept for Multi-strategy Reasoning", Northrop Grumman Technology Review Journal, Spring/Summer 2003, Volume 11, Number 1, Pages 59-76.
[101] P. Langley, "Elements of Machine Learning", Morgan Kaufmann, San Francisco, 1995

with virtual simulations, that are typically used in an off-line or background mode, is that the underlying applications are used by different communities of interest. Decision support tools that automate real-world missions are embedded in the natural world; however, simulations do not get nature "for free". Simulations must model all relevant aspects of nature. Schemes for interoperability vary: applications may use a Service Oriented Architecture[102], while simulations may use High Level Architecture[103] for distributed interactive simulation.

Here is an instantiation of a "Guess and Check" idea. A planner, based on a directive from higher authority, postulates a set of adversary actions and corresponding friendly Courses of Action (COAs) using a tool like the Strategy Development Tool[104] (SDT) that produces graphical views for mission assessment (BNE). Using the same directive, another planner, with simulation expertise, constructs a SEAS model, with Monte Carlo numerical experiments simulating BNE confidence intervals, and executes it to provide an independent "check" on the adversary COA and to also perform what-if analysis to identify emergent behavior, perhaps in the form of unintended effects. We use the "Guess and Check" idea to assure that all of our tools are explicitly modeling the same scenario.

Objectives

We want to achieve "substantive interoperability". This phrase, borrowed from the simulation community, really specifies what we need: tools that truly do work together to provide synergy, without the need for an analyst to mentally "bridge the gaps"! We achieve substantive interoperability by defining a few sub-objectives:

1) define requirements – satisfying these constitutes success,
2) explore the similarities and differences between graphs and simulations – two tools are graph-based (SDT and BNE), and one tool is simulation-based (SEAS),
3) find common ground between the two representations – common goals, common input, processing, and output parameters,
4) produce a design that leverages the similarities, and resolves the relevant differences,
5) translate the design to impacts on graph and simulation design,
6) determine database schema impacts,
7) implement the design,
8) integrate and test the design.

Approach

The sub-objectives defined above are now converted to specific tasks to implement a design with substantive interoperability.

Task 1: Requirements. The information provided here is suitably augmented to assure that it

---

[102] http://en.wikipedia.org/wiki/Service-oriented_architecture, accessed 02/26/2015.
[103] http://www.informs-sim.org/wsc97papers/0142.PDF, accessed 02/26/2015.
[104] http://www.amazon.com/Effects-Based-Operations-using-Strategy-Development/dp/B00NYGVRU4, accessed 01/27/2015.

clearly specifies our intent: to make graphs and simulations substantively interoperable.
- **Strategy Development Tool Requirements**. The Center-of-Gravity (COG) Articulator module of SDT shall produce a simplified network (graph with nodes and links). The graph shall provide a planner's view of the scenario. The graph shall be posted to the database. Incoming evidence and user interaction related to the graph shall be posted to the Framework and available to the graph. The Strategy Development Tool graph shall be a hierarchical characteristics-based ontology (knowledge structure). Results shall be displayed to the analyst.
- **BNE Requirements.** The BNE network topology (graph with nodes and links) shall be obtained automatically from SDT. The belief network shall process uncertain information. Incoming evidence and analyst interaction related to a graph shall be mathematically fused and propagated through the belief network graph. The belief network graph shall be a hierarchical characteristics-based ontology. Results shall be displayed to the analyst.
- **SEAS Requirements**: The SEAS network topology (graph with nodes and links) shall be shall be manually constructed to mirror the graph in the database. The SEAS network topology shall be obtained from SDT. The graph shall provide a simulation view of the scenario. Incoming evidence and user interaction related to the simulation shall be propagated through the simulation. The SEAS graph shall be a hierarchical characteristics-based ontology. Confidence intervals established by BNE shall be simulated in SEAS. Results shall be displayed to the analyst.

Task 2: Similarities and Differences

From the requirements section, we see that the three tools have much in common:

Similarities:
- All provide an abstraction of reality, based on a scenario
- All can be formulated as hierarchical, characteristics-based graphs.
- All rely on a common graph from the database
- All rely on incoming evidence and user interaction
- All propagate evidence through the network as a function of time
- All provide an analyst-oriented display
- All are complementary; same scenario, measures of effectiveness, and commensurate results.

On the other hand, it is obvious that there are significant differences among the three tools. In addition, there are subtle differences that also require discussion.
- All have different primary purposes: planning (SDT), assessment (BNE), and simulation (SEAS).
- SEAS has more specificity than the other tools. While SEAS can simulate the graphical layers between strategic intent and evidence sources, it may also require lower level models; for example, terror groups, cliques, and individual persons.
- Although the three tools may have nodes and links in common, the characteristics of

these nodes and links may be different: SDT may include mission, purpose, end state, and risk; BNE may include degree of belief and disbelief, along with textual evidence from which it is derived; and SEAS may include traits of individuals, groups, and social situations.
- Computing algorithms are different: SDT is a template-filling tool with most computation limited to preparing data for display, BNE does uncertain reasoning, and SEAS is an agent-oriented simulation.

Task 3: Common Ground

This section provides a synthesis of previous discussion. The overarching goal that these tools share is to reason about networks, to postulate adversary actions, plan friendly actions, execute a mission based on a realistic scenario, update the tools with evidence, dynamically update adversary and friendly actions, and assess mission success.

The challenge these tools have in common is that the tools are built for kinetic warfare rather than influence operations; consequently, effects, targets, assets, and timelines are less tangible. Assessment is also more difficult in non-kinetic versus kinetic battle spaces.

The tools also have a common structure; namely, a hierarchical, characteristics-based knowledge structure. The visualization of this structure is a graph consisting of nodes and links. Nodes represent objects (hypotheses, people, places, things) and the links represent the influence one node has on another.

Quantitative results are produced by all of the tools. Specifically, nodes in the graph can represent hypotheses about objects and have values of belief, unknown, and disbelief. Links have values that quantify the supporting or detracting influence that one node has on another.

Each of the tools produces a common, well-defined topology that allows many metrics that are relevant to social networks to be computed. Among these are betweenness, degree, dynamical importance, clustering coefficient, and link efficiency. These computable metrics provide operational measures of effectiveness to assess whether desired effects are being achieved.

Task 4: Design

This task is directed toward configuring a SEAS simulation. SEAS is a COTS product for which we do not have the source code; however, the product does allow new models to be configured within it. The design is geared to provide commonality between the SEAS and the other three graph-based tools. As earlier indicated, lower levels of granularity are likely required to get agent-oriented behavior. The hierarchical structure of SEAS allows inheritance of characteristics. The products that constitute a SEAS design are: User Interface Requirements, User Display mockup, Software Interface Requirements: a High-Level View (Figure IV-2.2), Use Case Diagram, Top Level Design, Sequence Diagram, and a Class Diagram.

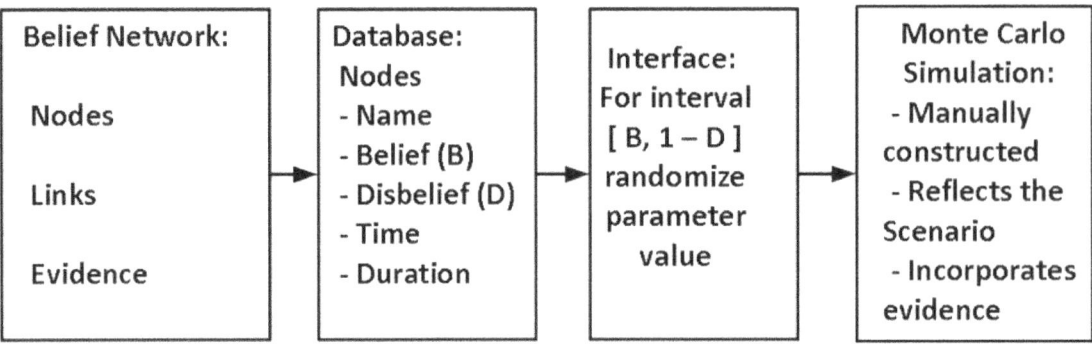

Figure IV-2.2: Belief Network Interface with Simulation Tools

Task 5: Impacts

The most significant impact that substantive interoperability has on the configuration of SEAS is the following constraint: the hierarchy of the simulation must mirror the hierarchy in the database to some extent. One or more layers of the graph produced by SDT or BNE must be evident in the simulation hierarchy. Typical SEAS simulations are about damage by kinetic weapons on geographic targets.

Organizations and assets provide the basis for hierarchical decomposition. For Blue forces, the top rung of the hierarchy is the National Command Authority (the President), and the bottom rung is an individual asset. It appears possible to associate this seven-level hierarchy with Effects-Based Operations, with the following correspondence:

Level	Effects
National Command Authority	Strategic Intents
Blue Force Commander	Operational Objectives
Joint Force Local Command Center	Tactical Effects
Mechanized BCT (Brigade)	Tasks
Blue Reconnaissance Battalion	Threat Prosecution
Avenger Pedestal Mounted Missile	Engage
SINCGAARS radio	Assessment Evidence Source

The Tactical Programming Language, which looks a lot like BASIC, allows significant flexibility. Limitations may occur in Monte Carlo sampling, data aging, and the actions that objects can take – especially those that are non-kinetic.

A slight change in the nature of the SEAS hierarchy enhances compatibility with SDT and BNE graphs. Rather that building a hierarchy of things; for example, organizations, platforms, sensors, the hierarchy is recast to be about hypotheses – statements about things. For example, sensor becomes sensor effectiveness, weather becomes weather impact on laser-guided munitions, Radar becomes radar report credibility. This allows us to assign degrees of belief and disbelief to nodes in the hierarchy. For a simulation, these are obtained directly from the evidence, from rules, or from simulation modules that inject evidence. We calculate the state of a situation based on simulation modules, and convert the output to degree of belief and disbelief based on rules.

The SEAS schema, once converted to hypothesis nodes, requires links between nodes. Links indicate the influence that one node has on another. For SEAS, this translates to the degree to which a node inherits the characteristics of another. Characteristics for both nodes and links are specified for SEAS in a way that is consistent with SDT and BNE.

Task 6: Implementation

A data schema that is consistent with representations for planning tools and graphs is best formulated by experimentation. A simple problem with a direct analogy between influence operations and kinetic warfare is proposed. Time Critical Targeting (TCT) is a challenging problem. The task is to execute the "kill chain" - find, fix, track, target, engage, assess (F2T2EA) within about 10 minutes. We construct a simple three-layer graph that allows us to post uncertain, incomplete, and sometimes conflicting evidence to achieve semi-automated data fusion.

The TCT belief network consists of a lower level that shows evidence sources. Note that no evidence source suggest that this is a kinetic engagement. Replacing the evidence sources with those relevant to our influence operations scenario makes this graph a simple first implementation.

The evidence nodes are modified to match a simplified version of the Irhabi007 scenario (Figure IV-2.3). " For almost two years, intelligence services around the world tried to uncover the identity of an Internet hacker who had become a key conduit for al-Qaeda. The savvy, English-speaking, presumably young webmaster taunted his pursuers, calling himself Irhabi -- Terrorist -- 007. He hacked into American university computers, propagandized for the Iraq insurgents led by Abu Musab al-Zarqawi and taught other online jihadists how to wield their computers for the cause."[105]

Lower level nodes are added hierarchically, although for the SEAS simulation many more individuals and cliques are included. This provides an implementation that tracks with the TCT plan and graphs for the BNE representations, but also allows SEAS to interact low-level entities in an agent-based simulation. The hypotheses that go with these lower-level nodes are the degree of belief or disbelief that these nodes are members of the nodes above them. Link strength is the degree to which they contribute to the nodes above them.

Summary and Conclusions

A hierarchical, characteristics-based data structure is a unifying framework for multi-strategy reasoning. It has been used in the past for a wide variety of inference algorithms, but never for simulations. In this paper, we bridge the gap between the graph-based formulations for planning (SDT), data fusion (BNE), and simulation (SEAS). The result is empirical evidence in support

---

[105] http://www.washingtonpost.com/wp-dyn/content/article/2006/03/25/AR2006032500020.html, accessed 02/26/2015.

of the following proposition: a simulation can be represented as a hierarchical characteristics-based semantic network if nodes are defined as hypotheses and properties of links and nodes are calculated from computer algorithms.

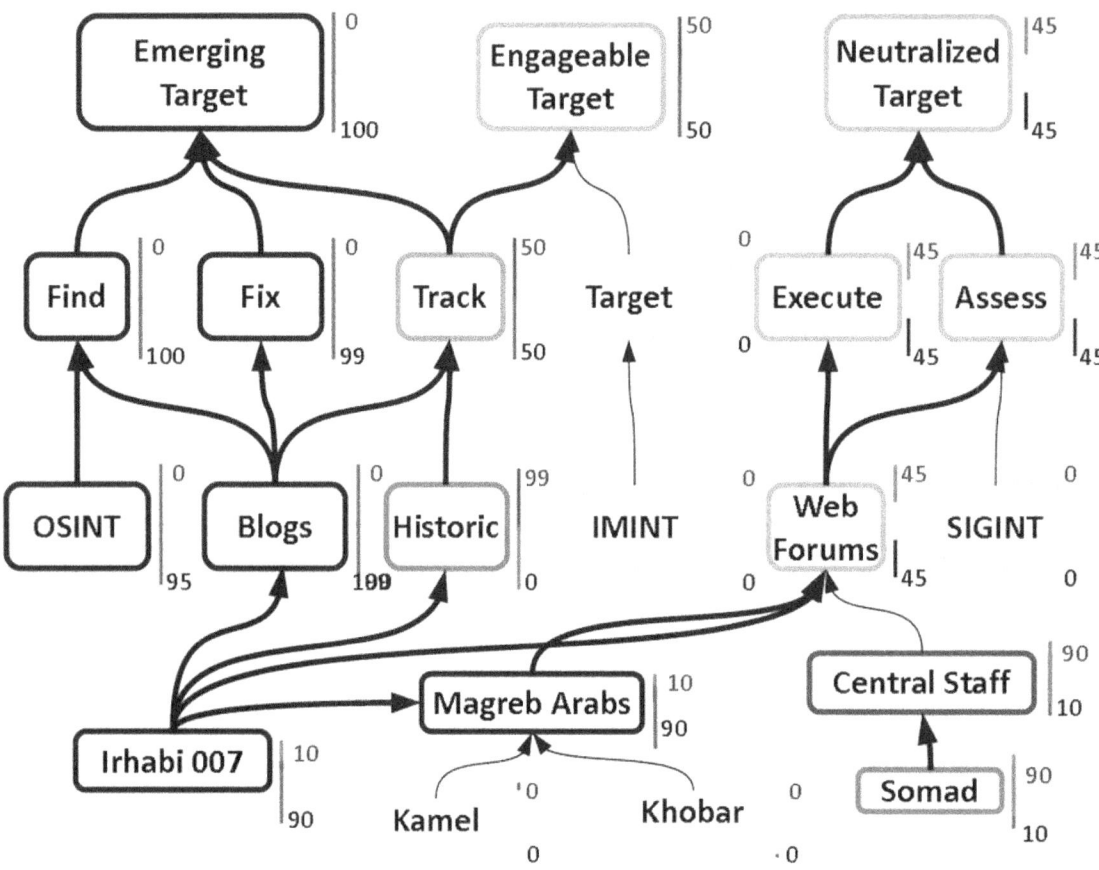

Figure IV-2.3: Augmented Time Critical Targeting Graph

Chapter IV-3
Processing Synergy and Feedback Loops

Earlier discussion centered on the concept of long mission threads. We monitor (M) the situation, assess (A) the state of ongoing missions and adversary activity, plan (P) missions to achieve specific effects, and execute (E) approved plans. These MAPE functions are performed under routine and quick reaction tempos. In fact, these activities require significant processing synergy and rely on feedback loops rather than processing threads. The result is a double MAPE loop: an outer loop is traversed during routine operations. It forms the foundation, with predictive analysis and on-the-shelf plans, for execution of a quick-reaction MAPE loop.

Introduction

In this chapter, we address processing synergy and feedback loops as an instantiation of multi-strategy reasoning and learning technology. For many missions, including response to bad weather, computer network defense, space control, military operations, and intelligence missions, the double MAPE loop (Figure IV-3.1) is continually traversed. Space control is shown here for concreteness.

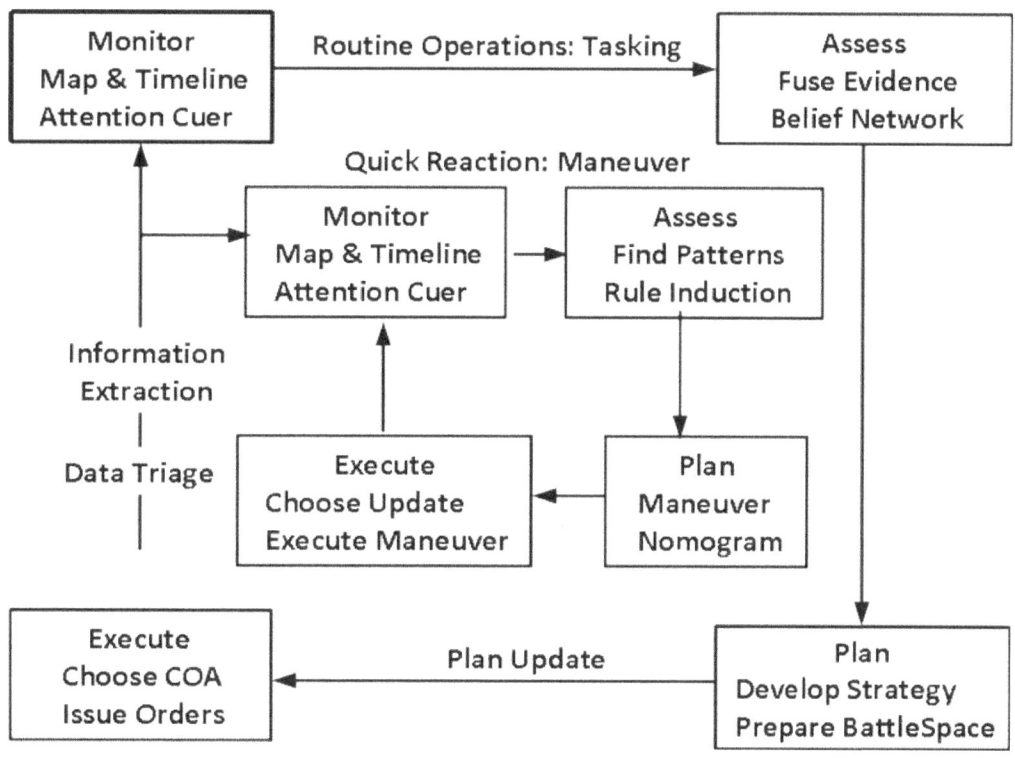

Figure IV-3.1: Double Monitor – Assess – Plan – Execute (MAPE) Loop – Example

During routine operations, situation awareness is provided by a monitoring function (upper left) that features mission summaries, and linked map, timeline, and logical views. The link from routine monitoring to assessment is tasking. Routine assessment (upper right) is evidence fusion with a belief network for threat prediction. Routine planning, also called deliberate planning (lower right), receives the belief network automatically from the Strategy Development Tool (SDT) as an input to planning. SDT is used for both developing strategy and preparing the battle space. A single click on the Plan produces recommended actions (lower left) which may be a broad Course of Action (COA) or dynamic plan updates. Here, the action is to choose a COA and issue orders. Based on sensor responses to tasking orders, additional information is ingested through data triage and information extraction. This feeds both routine and quick-reaction monitoring and the loops continue.

Quick reaction operations are initiated via automated data triage and information extraction (middle left). Quick reaction monitoring is facilitated by a real time map with icons indicating status. Quick reaction assessment is accomplished with support from data mining tools (inner upper right). Not shown, but also used for synergy with data mining tools are data fusion tools (Chapter IV-4: Automated Discovery of Unknown Unknowns). Quick reaction planning, for example using a nomogram[106] (inner lower right), is actuated directly from the belief network. The nomogram (shown later in Figure V-14.3) helps the user identify a plan update, in this case a maneuver to avoid intercept by a hostile satellite.

Objective

The purpose of this chapter is to describe how displays, algorithms, and knowledge are orchestrated to incorporate processing synergy (the whole is more useful than the collection of the parts) and feedback loops. The goal is to provide a holistic view of mission processing that is controlled by users and supported with automated decision support tools.

Approach

The methodology for describing synergy and feedback focuses on interfaces between routine and quick reaction activities that form the double MAPE loops. An interface is provided for displays, algorithms, and knowledge. As discussed earlier, knowledge drives algorithms that update displays that help the user make decisions.

Results

N-squared interface diagrams are constructed for displays, algorithms, and knowledge. All applications import information from the knowledge base and export processed information to the knowledge base. Knowledge interfaces refer to the automated transfer of information to and from the knowledge base. The double MAPE loop provides a unifying functional view of the components of a long mission thread. MAPE functions also map to the major categories of user decisions: what's going on (Monitor), how well are we doing (Assess), what to do (Plan), and do it (Execute).

---

[106] http://en.wikipedia.org/wiki/Nomogram, accessed 02/26/2015.

Chapter IV-4
Automated Discovery of Unknown Unknowns

Decision-makers entrusted with difficult situations are on full alert – and with good reason. We live in a dangerous world. The toughest problem for those who assess the ever-evolving threat, whether it be climate change, obesity, weapons of mass destruction, or terrorism, is how to cope with unknown unknowns. Analysts with lots of time, well-structured data, and broad domain expertise appear to manage this task reasonably well. Students discover, assimilate, and learn to reason about entirely new and previously unexpected domains of knowledge. For computers, this automated reasoning capability is more difficult. In fact, it is an open research problem[107].

Our objective is to automate the discovery of unknown unknowns. We use automated reasoning and learning algorithms separately and in combination to process the content of a knowledge base. Our solution leverages extremely rich knowledge structures. We use this knowledge representation to specify what is known to a knowledge base. Differences between what is specified as known, and what new evidence suggests, are computed. These differences, dubbed unknown unknowns, are of three types: new hypotheses, new links, and new story fragments. The result of our research is to prototype six methods for the automated discovery of unknown unknowns. We focus these methods in three domains: strategic planning and analysis, information operations, and evidence marshalling to combat terrorism.

Introduction

Military commanders, intelligence analysts, and homeland security personnel are among many decision makers that are operating at full alert status – and with good reason. Our enemies wish us harm and are "out to get us". The toughest problem for our personnel who assess the ever-evolving threat is to solve the following problem: how to cope with unknown unknowns such as "pop-up" threats and the unintended effects that invariably result from executing a battle plan. Stated another way, how to proceed when you don't know what you don't know? These unknown unknowns are of significant concern because unknown threats such as new terrorist organizations, new threat strategies, and changing alignments of third party nations, significantly impact our national security. Analysts often focus narrowly on known unknowns, such as missing data, to acquire and assimilate information. During the course of their analysis, they may stumble upon unanticipated new facts that reshape their views, or cause them to revise existing theories.

More broadly, people with lots of time, well-structured data, and good common sense appear to manage this task reasonably well. In our narrow domains, we are all adept at this. We discover new hypotheses, "connect the dots" by linking previously unassociated hypotheses, and can sometimes make a "leap of faith" by advancing a revised or radically new theory based on empirical evidence and innate human creativity. Students discover, assimilate, and learn to reason about entirely new and previously unexpected domains of knowledge. However, this is a highly manual, human intensive effort that benefits greatly from automation. For computers, this automated reasoning is difficult. To date, computers deal only with expected data, static

---
[107] In 2004, when we submitted a patent application, "Automated Discovery of Unknown Unknowns" produced no Google results. As of 04/09/2015, the same query (in quotes) produces 493 hits.

hypotheses, and prearranged links between hypotheses. In fact, this open research problem – to automate the discovery of unknown unknowns - motivates our research.

Related Work

Different kinds of unknowns are presented in the literature, both pictorially and verbally. The spectrum of knowledge (Figure IV-4.1) depicts unknown unknowns as serendipitous and vast in extent. Moving up the knowledge uncertainty pyramid, we see that known unknowns translate to missing content, known unknowns reflect unusable data, and known knowns indicate collectively shared facts in a domain.

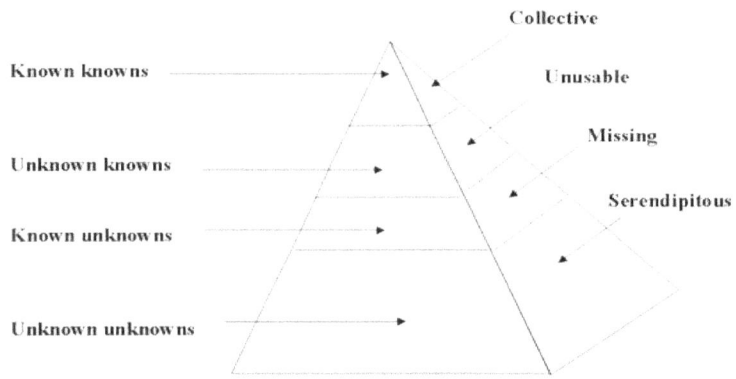

Figure IV-4.1: Uncertainty Pyramid

Donald Rumsfeld, Secretary of State in the George W. Bush administration, put it this way: "Reports that say that something hasn't happened are always interesting to me, because as we know, there are known knowns; there are things we know we know. We also know there are known unknowns; that is to say we know there are some things we do not know. But there are also unknown unknowns -- the ones we don't know we don't know. And if one looks throughout the history of our country and other free countries, it is the latter category that tends to be the difficult ones." The sidebar provides a poetic view.

The Unknown   (Sidebar)
As we know,
There are known knowns.
There are things we know we know.
We also know
There are known unknowns.
That is to say
We know there are some things
We do not know.
But there are also unknown unknowns,
The ones we don't know
We don't know.
Donald Rumsfeld — Feb. 12, 2002, Department of Defense news briefing

---

The Defense Advanced Projects Agency (DARPA) addressed a similar problem set in the Terrorist Awareness Initiative[108] with emphasis on unanticipated links between known hypotheses. The bioinformatics and proteomics domains are also interested in discovering unknown functionality in proteins[109]. The Evidence Extraction and Link Discovery (EELD) program combined information extraction from text with link analysis[110]. The automated creativity community seeks novel computer-generated solutions to problems[111]. Quantification of unknowns, with reference to unknown unknowns, is also a challenge for the modeling and simulation community[112]. Finally, InfoTame addressed the explication of tacit information[113].

Objective

We want to automate the discovery of unknown unknowns (ADUU). A contributing goal is to define an extremely rich formulation of knowledge. We use this knowledge representation to specify what we know to a computer knowledge base. As time passes and new content is input to the knowledge base, differences between what is specified as known, and what the new evidence suggests, are computed.

Our state of knowledge increases in two ways: new content may fill in previously missing data (known unknowns), and new content may provide new hypotheses, new links between hypotheses, or new story fragments (combinations of hypotheses and links). We define this latter use of new content as discovering unknown unknowns because it is unanticipated rather than merely missing. Our approach is to use automated reasoning and learning algorithms separately and in combination to process the content of a knowledge base. The results of our research are prototypes of many methods for the automated discovery of unknown unknowns.

Approach

To discover unknown unknowns, our technical approach (Figure IV-4.2) consists of seven steps. We begin with a definition, based on literature research. A user interface is then established and driven by military and anti-terrorism scenarios to establish context. We define belief networks with associated characteristics of hypothesis nodes and links, along with evidence about hypotheses as a story, and we construct a knowledge base from such stories. We use multi-strategy data mining algorithms to find hypotheses and links different from those in the knowledge base, integrate the algorithms and user interface into a prototype test bed, and demonstrate the results to our customer set.

---

[108] http://www.jbholston.com/periodical.php 3/17/04
[109] http://pir.georgetown.edu/pirwww/search/textsearch.html, accessed 02/26/2015.
[110] https://w2.eff.org/Privacy/TIA/eeld.php, accessed 02/26/2015.
[111] http://www.doc.ic.ac.uk/~sgc/research/gc.html, accessed 02/26/2015.
[112] http://www.pnl.gov/scales/abstracts.stm, accessed 02/26/2015.
[113] http://www.infotame.com/Support/faq.shtml, accessed 02/26/2015.

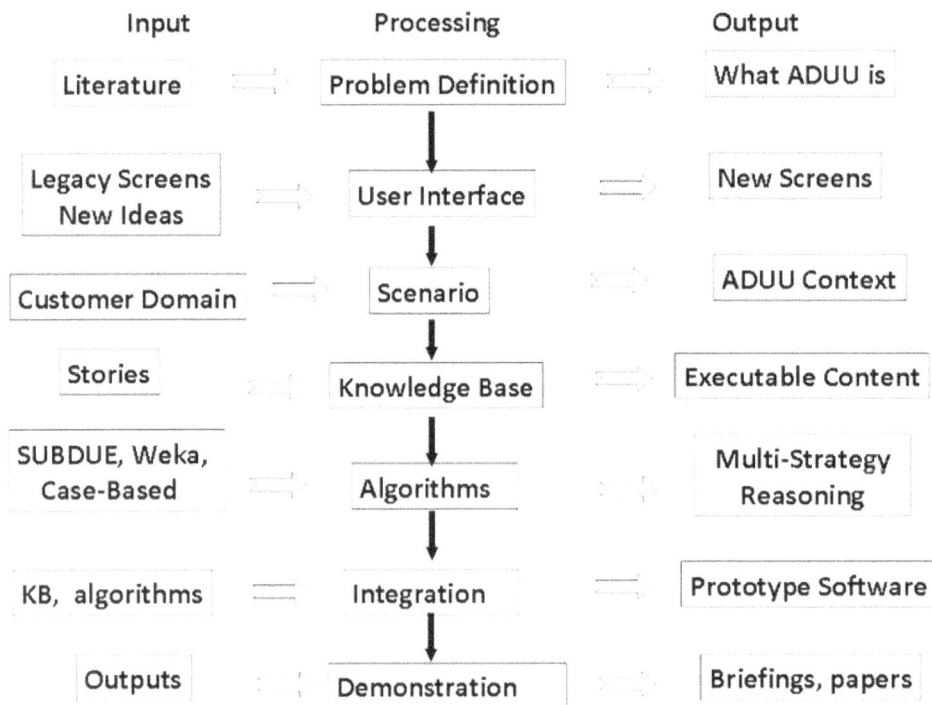

Figure IV-4.2: Approach to Automated Discovery of Unknown Unknowns (ADUU)

Results

Our formulation and solution to the ADUU problem is guided by a wide range of problems. To provide decision tools, we pursue the following problems areas:

- **Knowledge Representation and Processing:** A challenge is to bring new information from the "open world" that is currently not in a system into the "closed" world within the system.
- **Dynamic Plan Update**: for a variety of planning segments, users need to adaptively and efficiently "patch" a plan to optimize effectiveness based on unintended consequences of execution.
- **Strategic Planning and Analysis:** for next-generation planning and analysis, military and business planners need to track unintended effects of implemented Courses of Action.
- **Effects Based Operations:** we analyze plans and plan updates for continuous assessment based on combat assessment – as the feedback loop. Understanding of unintended and higher-order effects must be assessable.
- **Predictive Analysis:** users need to fuse snippets of information to produce predictive battlespace awareness (PBA) and to revise our determination of enemy intent when new evidence indicates new threats, changes in enemy tactics, or previously unknown enemy intentions.

- **Coping with Information Overload:** the "big data analytics community" needs to identify and track emerging information based on the semi-automated extraction of information into a knowledge base from unstructured text. There is currently no capability for automated augmentation of knowledge bases with new hypotheses, new links, or new story fragments.
- **Improved Data Marshalling:** for future crisis action centers, a user needs rapid and accurate fusion of battle damage assessment, based on evidence that in uncertain, incomplete, and conflicting, is required. Combat assessment must be extended to assess unintended consequences, 2nd order effects, and enemy reactions. There is currently no automated means of assessing unintended effects.

The ADUU project is focused on integrating and tailoring innovative technologies to assist analysts in performing their military command and control mission. Analysts are inundated with too much data to read, understand, and apply. As volume increases and the type of data analysts must absorb becomes complex, they must rely on multiple automated reasoning algorithms to reduce the processing load. To efficiently interface these different algorithms in a substantively integrated system, we extend our unified framework for multi-strategy reasoning. Evidence fusion and knowledge discovery technologies, although varied in maturity, are quickly evolving from barely capable products; that is, link analysis, to decision-aid components for new functionality, such as evidence marshalling.

Story fragments are computed as differences between what is specified in a story and new story fragments computed by automated reasoning algorithms. These knowledge-processing algorithms are used to put content into the knowledge base and to process the content that resides in the knowledge base. A subset of the latter – processing knowledge base content produces new story fragments. However, pre-processing algorithms are discussed to explain what they deliver to the knowledge base. Both single-strategy and multi-strategy reasoning produce new story fragments: of these, multi-strategy reasoning provides automated analyst support and produces the most powerful results.

Pre-processing Algorithms

Unstructured text is the assumed input. Semi-structured text such as E-Mail or structured data from a database is easier. The steps required are to segment and filter the text, associate a hypothesis with each text segment, parse the content into template, and register the content of each template into the knowledge base, and fuse the new evidence with all stories that contain the hypothesis.

Importantly, the result of these pre-processing steps is that characteristics are extracted from text, associated with a hypothesis, and fused with existing evidence.

Single-strategy Reasoning

Automated reasoning algorithms that generalize from specific information produce new story fragments. Five types of automated reasoning that provide generalization are: induction, abduction, analogical, probabilistic, and connectionist. New story fragments arise by comparing existing semantic networks (graphs composed of nodes and links) with those produced with newer information. In order to automate the discovery of unknown unknowns, the computer-processing environment is provided with a structured specification of what is known (for example, quantified degrees of belief and conflict in linked hypotheses), and what is unknown (for example, missing data). New information is compared and contrasted with the current understanding to identify differences between what we think is occurring and what evidence extracted from observations tells us may be occurring. These differences are then related to the current understanding to provide new hypotheses and links that revise the story to accommodate the new information. Visualization is required to convey these unknown unknowns to the user in a meaningful way. Based on this discussion, single strategy reasoning is a necessary, but not sufficient in ADUU.

Unified Multi-Strategy Reasoning

To automate knowledge processing, we have a variety of algorithms that perform abductive, analogical, deductive, inductive, probabilistic, and evidential reasoning. Each algorithm type has a unique formulation (tree, lattice, tabular). However, we discovered that the underlying structure of these algorithms is a hierarchical, frame-based semantic network[114]. We structured our knowledge base to reflect this structure so that all algorithms have a single interface with the knowledge base for both import and export. Real-world applications, such as predictive battlespace awareness and space situation awareness analysis, often require a combination of substantively integrated algorithms. We integrate multi-strategy reasoning algorithms with a knowledge base in many domains, including strategic forces, missile defense, air operations, information operations, and homeland security.

Knowledge Base

The knowledge base abstracts the domain structure to provide a central knowledge repository. We began with Protégé 2000, but recently switched to a MySQL-based solution we call the Visual Knowledge Base. The knowledge base is not a monolithic hierarchy of concepts (for example, CYC[115]). Instead, it consists of stories. A story, or belief network, is a hierarchical set of hypotheses with frames that provide metadata about the hypotheses. Hypotheses are linked to indicate the influence of one hypothesis on another. Stories can be nested, combined (union and intersection), overlaid, and contain feedback loops. Knowledge visualization provides the analyst with an understanding of these stories using displays that capture geographic, logical, and temporal knowledge. Our algorithms interface via the knowledge base. This permits us to develop algorithms in parallel, based on a single interface, rather than creating a custom interface to interact with other algorithms. Although the knowledge base is an algorithm itself, we identify it as the hub of the processing architecture as it relates concepts to one another. The knowledge base concept benefits missions by allowing analysts to absorb information more quickly and powerfully. It also allows new tools to be integrated more easily and cost effectively.

---

[114] Talbot, P., Semantic Networks, Northrop Grumman Technology Review Journal, Spring/Summer 2003, Pages 59-76.
[115] http://www.cyc.com/, accessed 01/18/2015.

Data Mining

Each of the six data mining algorithms (unsupervised clustering, supervised rule induction, abductive reasoning, analogical reasoning, link analysis, and Bayesian belief networks) works in a different but complementary way. A few of these are discussed.

Unsupervised clustering shows the density of evidence in support of various hypotheses, new clusters, and changes in clusters. New clusters may be an indicator of unknown unknowns in the form of newly discovered hypotheses. Supervised rule induction computes rules (IF, AND, AND, THEN) and changes in rules. The change is based on rule induction using the old evidence supporting the old theory versus rule induction based on exceptions to the old rules. That is, new rules are computed from the aggregate of old plus new evidence and may produce new hypotheses.

The abductive reasoner (we use Subdue from the University of Texas at Arlington) reasons to the best explanation using positive examples of the old theory. Abductive reasoning output is provided in a hierarchically structured way with different degrees of granularity computed that best compresses the graph. We prototype (Figure IV-4.3) the automated interaction between a belief network and the abductive reasoner. We input hypotheses from a belief network into Subdue, and obtain hierarchical differences between these hypotheses versus new evidence. A "new node" indicator is then posted to the belief network that allows an analyst to add the node, name it, and have it automatically inserted into the belief network. The belief network then automatically reconciles itself, based on a back-propagation algorithm.

The analogical reasoner reasons by similarity. It finds cases in the knowledge base that are most similar to the old theory and suggests differences in hypotheses based on historical cases. To date, we have used analogical, or case-based reasoning, in a number of applications, including automated terrain reasoning, optimum location of transportable resources, case-base planning, and case-based plan repair. We have not experimented with it for discovering unknown unknowns. However, this technology has significant promise – others are using it for computer creativity; here, we want to use it to create new hypotheses.

These data mining tools provide information to support predicting enemy intent. This is an end goal of our research. Our technique would allow military command and control centers, as well as intelligence agencies, to identify surprising patterns earlier than usual. In addition, the analyst would no longer have to query a database for information. This issue of "you can't ask what you do not know to ask" is eliminated.

Data Fusion

A story, or belief network, is a model for representing uncertain knowledge and the primary building block of our knowledge base. Nodes represent hypotheses and links represent influence. The Belief Network Editor (Figure IV-4.3) is our tool for qualitative and quantitative fusion of data and exploration of a story. BNE is used to propose stories "waiting to happen," and prove hypotheses based on fusion of uncertain evidence. We use the Dempster-Shafer Combination rule to fuse evidence and a back-propagation algorithm to reconcile belief networks after an analyst overrides a belief or disbelief value. We find that Dempster-Shafer evidential reasoning is far superior to Bayesian Belief Networks because we can easily distinguish between disbelief and conflict regarding a hypothesis. Prior probability distributions are not required; conflict resolution is straightforward and works well with sparse evidence.

1) The ADUU proof-of-principle shows nodes in the belief network (unshaded).
2) New evidence is then made available from the knowledge base to both the belief network and SUBDUE. The new evidence is not recognized by the belief network because the evidence does not have associated hypotheses being tracked in the belief network. SUBDUE, however, produces the modified graph by adding new nodes (shaded).
3) The new nodes and links change the situation. What is apparent is that new treats (Bomb, Ping Saturation, System Intrusion, System Overload, Physical attack , and Software Bugs) have emerged, based on new evidence.
4) The belief network editor is then used to "wire in" the new hypothesis with it's new links. The belief network with the new structure is automatically reconciled and saved as a new belief network.

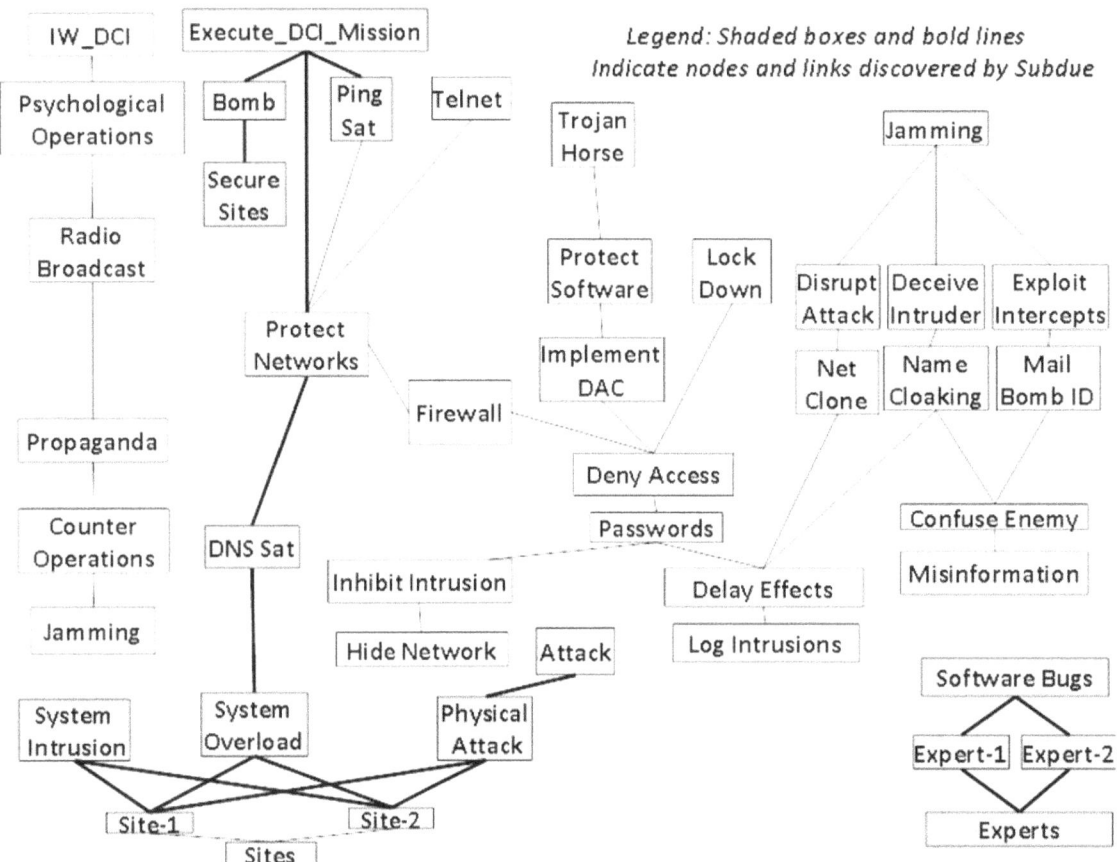

Figure IV-4.3 Implementation of Automated Discovery of Unknown Unknowns (ADUU)

BNE supports structured argumentation, showing nested hypotheses and supporting evidence. It allows non-programmers to create executable belief networks. In addition, the BNE automatically propagates belief values and shows quantitative information graphically and supports interactive sensitivity analysis.

Researchers, and investigators often must take imprecise and unstructured information and form hypotheses and structured arguments around it. However, existing tools do not allow researchers to work with precise, numeric estimates of confidence and influence in an intuitive way. BNE allows users to structure arguments and work with the quantitative values in a qualitative, graphical, and intuitive manner. BNE allows a user multiple dimensions of exploration of the quantitative values in a complex mathematical model using an easy-to-use qualitative interface. It does this by drawing a tree of nodes, giving each node and line visual properties corresponding to their numeric weights. The user adjusts these numeric values and views the impact of the numeric changes. In concert with the Subdue data mining algorithm, which accepts hypotheses from the BNE and identifies new concepts based on new evidence, hidden hypotheses are made explicit and are automatically added to the BNE.

Integration

We demonstrate ADUU by integrating three components: a knowledge base, data fusion algorithms and data mining algorithms. New links and nodes are discovered. A common integration strategy suffices for six multi-strategy reasoning techniques. All exhibit the following similarity: they are data mining schemes that contrast discovered story segments with story segments in the knowledge base.

Data Mining Technique	Inference Type	Product	Unknown Unknown
Hierarchical Clustering	Abductive	Subdue	links and nodes
Unsupervised Clustering	Inductive	Weka	nodes
Rule Induction	Inductive	Weka	nodes
Link Analysis	Analogical	Saffron	links
Case-Based Reasoning	Analogical	MyCBR	fragments
Bayesian Classifier	Probabilistic	AutoClass	nodes
Bayesian Belief Nets	Inductive/Deductive	JavaBayes	nodes

Conclusions

A general class of solutions to the problem of providing automated discovery of unknown unknowns is described. A proof-of-principle for the most promising – abductive reasoning – is provided. The foundation is a rich knowledge representation, accomplished using executable stories. Data mining algorithms that are driven by new evidence produce new hypotheses, new links, and new story fragments consisting of nodes connected by links are then discovered and added to existing stories. The components of this technology are integrated and demonstrated in software.

Chapter IV-5
Network Metrics Define Centers-Of-Gravity

Need

We analyze Internet Forums based on a set of network metrics. These metrics are chosen based on an analysis (Chapter III-5) of a large set of metrics organized in a taxonomy. These metrics were discussed with our Subject Matter Expert, who identified this meaningful subset for implementation. We need a process for defining how to use these for determining which node or nodes are critical from a centers of gravity (COG) perspective. We relate roles of people in forums to their effects on the network. The metrics are: Betweenness, Degree, Clustering Coefficient, Path Believability, Importance under Uncertainty, Dynamical Importance, and K-core.

Objective

A COG is defined as "the source of power that provides the moral or physical strength, freedom of action, or will to act"[116]. The goal of this paper is to define how to use these network metrics to identify an adversary COG from a network depiction of adversary objectives, effects, tasks, activities, and events.

Approach

Each of the identified network metrics tells us something different. Four of the metrics (betweenness, degree, clustering, and importance under uncertainty) provide information about individual nodes. The remaining three metrics (path believability, dynamical importance, and k-core organization) provide information about collections of nodes and links in the graph. Should we compute node metrics or sub-graph metrics first? As will be discussed, this is a difficult question: a strategy which computes sub-graph metrics first provides a general-to-specific view, but because there are a nearly infinite number of possible sub-graphs for any fairly large network (hundreds to thousands of nodes), it is not feasible. On the other hand, computing node metrics first leads to unproductive analysis if, for example, a small collection of nodes is locally significant but isolated from the mainstream activity of the network.

Another consideration is that some metrics are computationally more expensive than others (Shown earlier in Chapter III-5). This also provides a choice: in what order should we compute network metrics so that the nodes and sub-graphs of interest are filtered before computationally intensive operations are performed.

A final consideration is that network metrics must be suitable for the mission. For example, a mission to "decapitate" the command and control of a nation-state might lead to a COG consisting of the Dictator and his inner circle as determined by the importance/rank metric. On the other hand, if the mission is to interdict the flow of nuclear material to a terrorist organization, the COG might consist of key individuals in the supply chain as determined by the betweenness metric. This is an analyst decision.

---

[116] http://www.dtic.mil/doctrine/jel/new_pubs/jp1_02.pdf, accessed 03/01/2015.

Our methodology is to identify a processing sequence, rules-of-thumb, and supporting rationale for our Internet scenario. The ranking of computational complexities suggests a sequential strategy:

- Calculate **degree** to identify (at low computational cost) nodes with the most connections, which in our scenario suggests the most highly trafficked nodes. Alternatively, or in conjunction with degree calculations, we may also elect to compute **betweenness**.
- Compute **k-cores** based on the largest k-core that contains nodes that are potentially of interest (analyst decision). This effectively filters the graph.
- Determine the **importance and rank** the high-degree nodes in the k-core partition. This ranks high-degree nodes based on link weights and belief values.
- Calculate the **clustering coefficient** for the most highly-ranked nodes to determine tightly-connected neighbors who may be part of the COG. Analyst decides whether to keep any or all of the nodes with large clustering coefficients and whether to keep their neighbors.
- Choose nodes with high importance, high clustering coefficients, and a subset of clustered neighbors as a candidate COG. Calculate the **dynamical importance** of this candidate COG; that is, determine the % eigenvalue changes in the k-core with & without the COG.
- Calculate activity threads that include best candidate COGs and determine **path believability** to justify prosecution of these activity threads based on link values and node confidence [Belief, Unknown, Disbelief].

Betweenness is probably not a good metric for this particular scenario because the adversary network is "dark", meaning at least partially unknown, and we are unlikely to determine and locate individuals with "courier" roles because they are very likely to maintain a low profile. The extent of a courier's true links would not be known. On the other hand, forums are visible and may exhibit high betweenness. If the neutralization of a node with high betweenness is desirable, this metric can substitute for, or augment, the degree calculation.

Clustering coefficient provides a global view of how tightly connected the network, or a k-core of the network, is. This metric could be calculated after Step 2 to provide supplemental information on a k-core. Each of these network metrics is computed as the ratio of the metric value to the maximum value. This normalized the value so that it lies on the [0, 1] interval.

Results

We used the info-structure layer of an Internet Forum model to try out the methodology described above. A graphical view (Figure 4-5.1) of this network, for a single forum, clearly indicates the presence of six discussion threads which are analytically found to be the COGs.

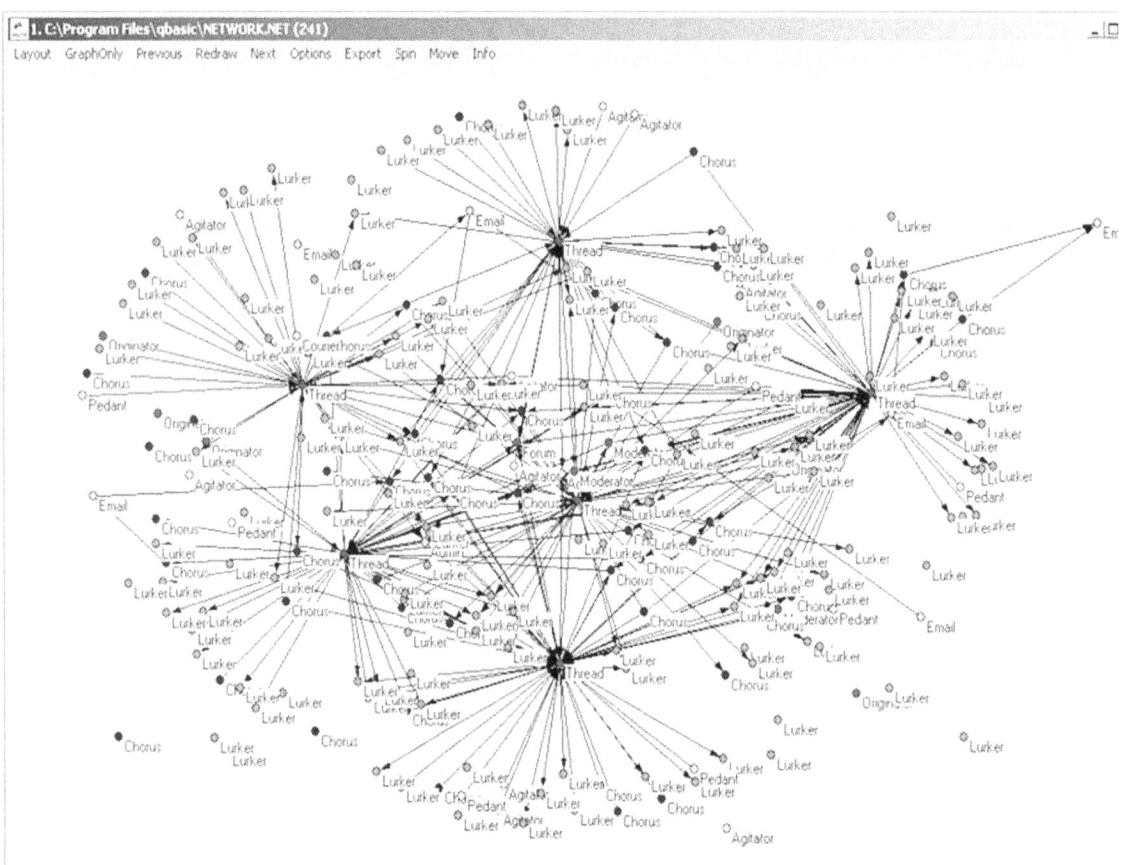

Figure IV-5.1: Info-structure Network, 1 Forum, 6 Threads

A simpler, but less intuitive description is provided by the adjacency matrix. Entries in the matrix indicate the presence of links from one type of node to another.

	From										
To	Forum	Thread	Email	Admin	Moderator	Originator	Courier	Agitator	Pedant	Chorus	Lurker
Forum		1					1				
Thread			1	1	1			1	1	1	
Email											1
Admin	1										
Moderator	1										
Originator						1					
Courier											
Agitator	1										
Pedant		1									
Chorus		1	1								
Lurker	1	1									

Discussion

The strategy outlined above is based on our Internet Forum scenario. Other influence operations scenarios are, however, also dark. Although networks depicting adversary physical nodes, activities, effects and intents are based on evidence to the extent possible, expert opinion to identify hidden nodes and links is required for a meaningful representation – along with a quantitative representation of the associated uncertainty.

Conclusions

Sequential filtering and refinement of a graph, based on the results of network metrics, appears to be helpful in identifying candidate centers-of-gravity.

Chapter IV-6
Emergent Behavior

Introduction

Emergence refers to the way complex systems and patterns, such as those that form a hurricane, arise out of a multiplicity of relatively simple interactions[117]. Perhaps the most elaborate recent definition of emergence is provided by Jeffrey Goldstein[118] in the inaugural issue of Emergence. To Goldstein, emergence refers to "the arising of novel and coherent structures, patterns and properties during the process of self-organization in complex systems." The common characteristics are: (1) radical novelty (features not previously observed in systems); (2) coherence or correlation (meaning integrated wholes that maintain themselves over some period of time); (3) A global or macro "level" ; that is, there is some property of "wholeness"); (4) it is the product of a dynamical process (it evolves); and (5) it is "ostensive" - it can be perceived.

There are (at least) four ways to identify emergent behavior: by definition, visualization, analogy, and metrics. This paper analyzes an executable model of a graph that is implemented in a belief network and a simulation. Agent roles and behaviors are defined. Anticipated results, which include emergent behaviors analogous to a market, predator/prey, and flocking are discussed. Metrics to quantify infrastructure layer behaviors are defined. These are linked to produce a metrics hierarchy.

Emergent Properties: An emergent[119] behavior or emergent property appears when a number of simple entities (agents) operate in an environment, forming more complex behaviors as a collective. If emergence happens over disparate size scales, then the reason is usually a causal relation across different scales. In other words there is often a form of top-down feedback in systems with emergent properties. These are two of the major reasons why emergent behavior occurs: intricate causal relations across different scales and feedback. The property itself is often unpredictable and unprecedented, and may represent a new level of the system's evolution. The complex behaviors or properties are not of any single such entity, nor can they easily be predicted or deduced from behavior in the lower-level entities: they are irreducible. No physical property of an individual molecule of air would lead one to think that a large collection of them will transmit sound. The shape and behavior of a flock of birds or shoal of fish are also good examples.

One reason why emergent behavior is hard to predict is that the number of interactions between components of a system increases exponentially with the number of components, thus potentially allowing for many new and subtle types of behavior to emerge. For example, the possible interactions between groups of molecules grows enormously with the number of molecules such that it is impossible for a computer to even count the number of arrangements for a system as small as 20 molecules.

---

[117] http://en.wikipedia.org/wiki/Emergence, accessed 03/01/2015.
[118] https://www.questia.com/library/journal/1P3-3044051551/complexity-and-philosophy-re-imagining-emergence, accessed 01/18/2015.
[119] http://en.wikipedia.org/wiki/Emergence, accessed 03/01/2015.

On the other hand, merely having a large number of interactions is not enough by itself to guarantee emergent behavior; many of the interactions may be negligible or irrelevant, or may cancel each other out. In some cases, a large number of interactions can in fact work against the emergence of interesting behavior, by creating a lot of "noise" to drown out any emerging "signal"; the emergent behavior may need to be temporarily isolated from other interactions before it reaches enough critical mass to be self-supporting. Thus it is not just the sheer number of connections between components which encourages emergence; it is also how these connections are organized. A hierarchical organization is one example that can generate emergent behavior (a bureaucracy may behave in a way quite different to that of the individual humans in that bureaucracy); but perhaps more interestingly, emergent behavior can also arise from more decentralized organizational structures, such as a marketplace. In some cases, the system has to reach a combined threshold of diversity, organization, and connectivity before emergent behavior appears.

Unintended consequences and side effects are closely related to emergent properties. Luc Steels[120] writes: "A component has a particular functionality but this is not recognizable as a sub-function of the global functionality. Instead a component implements a behavior whose side effect contributes to the global functionality. Each behavior has a side effect and the sum of the side effects gives the desired functionality". In other words, the global or macroscopic functionality of a system with "emergent functionality" is the sum of all "side effects", of all emergent properties and functionalities. Systems with emergent properties or emergent structures may appear to defy entropic principles and the second law of thermodynamics, because they form and increase order despite the lack of command and central control. This is possible because open systems can extract information and order out of the environment."

Emergent Behavior in Internet Forums: Our characterization of terrorist forums on the Internet has many characteristics described above. In these forums, members are driven by self-interest. Sageman[121] states that " All the factors assumed to have high relevance in predicting terrorism… do not apply to the global Salafi Jihad, which is characterized by decentralization, a fluid horizontal structure, and a surprising absence of periodic purges of leadership".

Agents: A wide variety of agents participate in these forums, including explosives suppliers, web hosts, forum moderators, webmasters, content distributors, active members (who post content and lead discussions), the "choir", terrorist leaders, and cross-pollinators from other forums. ***The interactions of these agents, as stated by our expert, a self-dubbed internet activist, appear to form a market.***

The 'story" is that a supply chain consisting of originators, distributors and webmasters provide content to Website forums. Moderators may receive complaints from leaders and members related to the content – often gruesome videos – and remove such content. Active members are promoted by moderators, receive praise from leaders, other active members, and the membership at large for postings. Active members may radicalize forum members who may commit terrorist acts.

---

[120] http://en.wikipedia.org/wiki/Luc_Steels, accessed 03/01/2015.

[121] Sageman, M, *Understanding Terror Networks*, University of Pennsylvania Press, 2004, page 90.

Objective

Our goal is to produce an executable version of this formulation (Figure IV-6.1), along with metrics to help quantify and understand the results. This allows us to understand the circumstances under which this collection of agents, each motivated primarily by self-interest, produces emergent behavior as described above.

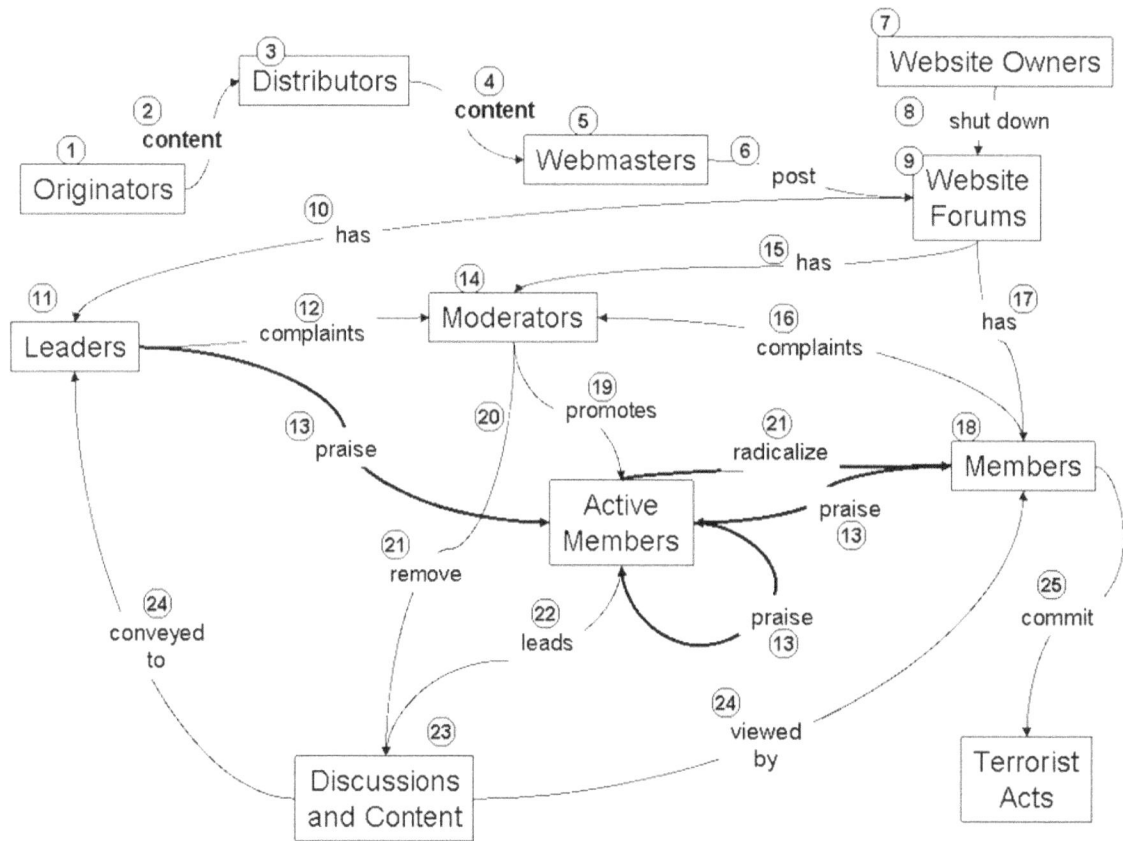

Figure IV-6.1: Agent Interactions In Internet Forums

Specifically, our premise is that a market emerges that efficiently organizes the flow of ideas, content, and motivation among its agents. A second premise is that this market increasingly radicalizes angry young men, causing a small number to commit terrorist acts.

An equally important goal is to introduce actions into this market to show how these forums can be rendered inefficient. A list of 25 such tactics, the basis for Measures of Effectiveness, are discussed.

Technical Approach

To affect any or all of these tactics, we derive more detailed characteristics of the agents involved. Agent roles and behaviors are created that provide the characteristics required for detailed simulation of their activities. Behavior rates and rules are estimates from earlier

Subject Matter Expert (SME) meetings. These behavior rates and initial conditions are added to the earlier models. An additional feedback loop is added to indicate that terrorist acts provide new content to originators.

Techniques for Impacting the market flow of goods and services:
1. Neutralize originators (e.g., arrest)
2. Disrupt flow of goods from originators to distributors
3. Neutralize distributors (e.g., arrest)
4. Disrupt flow of goods from distributors to webmasters
5. Neutralize webmasters (e.g., arrest)
6. Undermine the ability of webmasters to post compelling content
7. Inform website owners of terrorist activity on their sites
8. Request that website owners shut down terrorist websites
9. Disrupt website forum (e.g., distributed denial of service attack)
10. Monitor and dissuade leaders from visiting web forums
11. Capture leaders
12. Modify web forum content to induce leadership complaints
13. Reduce leadership and member praise of active members (e.g., introduce "issues")
14. Neutralize moderators (e.g., discredit)
15. Discourage active members from being moderators (e.g., harass)
16. Increase member complaints (e.g., plant moderate or commercial content)
17. Reduce membership (e.g., shut down site, slow down site access, plant virus)
18. Monitor, track, harass, and neutralize members (e.g., harass, arrest)
19. Disrupt moderator tasks to limit promoting of active members
20. Influence moderator to remove discussion threads and content (e.g., complain)
21. Decrease active member's ability to radicalize other members (e.g., discredit)
22. Discourage active members from leading discussions and posting content (e.g., harass)
23. Delete discussion threads and content (e.g., disrupt specific pages)
24. Limit viewing and forum discussion by members (e.g., introduce latencies, deface pages)
25. Minimize member ability to commit terrorist acts (e.g., crack down on explosives dealers)

Anticipated Simulation Results

Nine types of agent composites are defined. Of these, the numbers of originators, distributors, webmasters, website owner, leader, and moderators are fixed. The numbers of forums, members, actives, discussions/content, and terrorist acts vary. Each of these is discussed in more detail.

Forums

The number of forums on a website is initialized at some reasonably small number. The number of forums is decreased by website owners in response to complaints. It is increased by moderators by a similar rate. Depending on simulation rules; for example, lag time to replace a closed forum, cyclic behavior in the number of forums versus time may be observable. The metric is the number of forums at a given time, following some deliberate action to reduce them.

Members: The number of members is initialized at some middle-sized number, based on a snapshot of forum membership. Members have many behaviors: they complain to moderators and website owners. They praise active members. They access web pages. They are radicalized and commit terrorist acts. The number of members on a forum is based on discussion and forum dialogs lead by active members: the number of members is empirically related to the goodness of the content, based on two rates. Primary metrics are the number of members, and the number of terrorist acts. Secondary metrics are the number of complaints, and the number of praises. The number radicalized appears not to be measurable, unless it is assumed to be proportional to the number of terrorist acts tied to the website.

Actives: The number of active members is initialized at 20% of the number of members. Actives lead discussions and are "helpful" in furthering the purposes of the forums with the result, perhaps not explicitly intended, that they radicalize members. They praise other actives and, in turn, receive praise from leaders, moderators, members, and other actives. Actives may complain to the moderator. Primary metrics are the number of actives and their stature, based on observables. Secondary metrics are the number of praises they receive, and the number of complaints they tender.

Discussions & Content: The number of active discussion threads and active postings is initialized at some typical number per week. This is decreased by the amount removed by moderators in response to complaints (or content quality) and increased by postings from actives. The metric is the quality of content, and is based on the number of page hits per week

Terrorist Acts: The number of terrorist attacks attributable to internet forums is initialized at some rate averaged over the last decade and tempered with judgment from our SMEs. Terrorist acts may be attributed to radicalized forum members (with access to explosives) and may increase and decrease with forum membership. The metric is the number of attacks attributable to terrorists.

Evidence of Emergent Behavior: There are (at least) four ways to identify emergent behavior: by definition, visualization, analogy, and metrics. Identifying emergent behavior based on its definition has already been discussed in earlier sections – its not very convincing. The second and perhaps the most compelling way to identify emergence is visualization; for example, the flocking behavior in BOIDS[122] is so obviously a behavior that emerges that no further proof is necessary. A third way to identify emergence is to draw the analogy that internet terrorist forums portray a behavior that is similar to a well-known emergent behavior. We are already arguing that the behavior has characteristics of a market. We could identify predator versus prey cycles. We could even argue that it forums have flocking behavior[123] similar to BOIDS by producing analogous effects to the BOIDS rules. A preliminary System Effectiveness and Analysis Simulation (SEAS) simulation of the info-structure layer does, in fact, show members flocking from one forum to another.

---

[122] http://www.red3d.com/cwr/boids/, accessed 03/01/2015.
[123] Modeling Opinion Flow in Humans using BOIDS Algorithm and Social Network Analysis, http://gamasutra.com/features/20060928/cole_01.shtml, accessed 03/01/2015.

Force	Boids Effect	Effect on Thought
Alignment	Birds steer in the general direction of the group.	People tend to give credence to an idea shared by many people. They build their picture of what the 'group' thinks based on information from their news sources.
Cohesion	Birds move toward their near by flock mates.	People tend toward the ideas of the people they respect.
Separation	Birds maintain a safe distance from their flock mates.	People maintain a distance from ideas held by people for whom they have contempt.

The fourth technique for identifying emergent behavior is metrics. While not the most intuitive, this is the most powerful technique. The anticipated simulation results provide a glimpse of agent behavior. A plot of these metrics versus time may show cycles. The question remains: where is the emergent behavior, and how do we measure it? Based on the definitions given earlier, evidence for emergent behavior is "integrated wholes that can maintain themselves over a period of time". Plots of membership versus time, number of forums versus time, number of active members versus time, and number of terrorist acts versus time are metrics that suggest emergent behavior. In this vein, the number of gigabytes of content on an internet website is also an indicator. Some or all of these five metrics may show cyclic behavior. For a particular scenario excursion, cycles may be correlated (a classic example is the anti-correlated predator versus prey cycle where cyclic increases of predators result in corresponding cycles of decreasing prey). From the definition of emergence given earlier, "coherence or correlation" has a temporal interpretation that signifies emergence. We may find that a temporal cycle of increased complaints is anti-correlated with gigabytes of content.

The relative insensitivity of the model to perturbations is another metric indicating emergence, based on the definition of emergence as "coherent structures". For example, suppose the number of instances for agents is used in a SEAS as the basis for Monte Carlo sampling. If we simultaneously perturb multiple agent behaviors, but obtain a fairly tight bound on resultant behaviors (for example, average membership and confidence interval for membership do not vary much), we have demonstrated a "coherent structure".

An example (Figure IV-6.2) from a related project is pertinent[124]. We loop on JavaBayes (a freeware tool that computes Bayesian Belief networks) with a Monte Carlo sampling procedure. The simulation is a terrorist-oriented belief network to predict the likelihood that a terrorist attack occurs at Lloyds of London (the Use Citrus node). We uniformly perturb the Threat, Have Bomb, Alert, and Defense Success nodes and compute the distribution for a "successful attack" (right hand side): the 95% confidence interval is normally distributed and an order of magnitude smaller than the input perturbations. This simple model gives us confidence that our threat model constituted a coherent structure that is relatively insensitive to perturbations.

---

[124] Anderson, M., JavaBayes Sensitivity Analysis Documentation, September 2005, unpublished

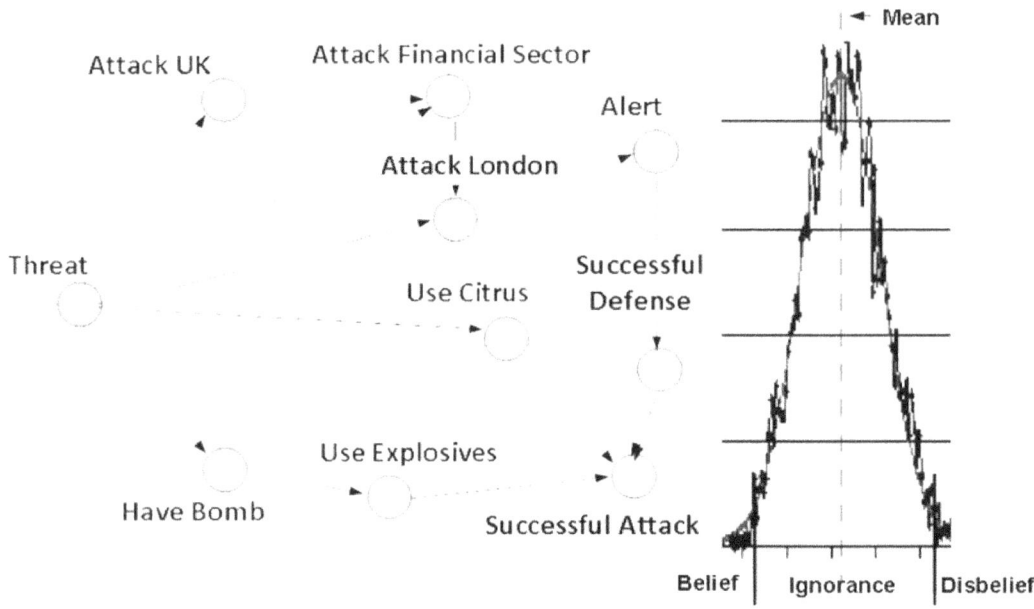

Figure IV-6.2: Sensitivity Analysis

Higher-Level Metrics: So far, we have discussed metrics related to individual nodes in the graph. These metrics represent summary data at low granularity. These must be "bubbled up" to the indicators and outcomes layers to provide the "so what". A hierarchical view of metrics is provided (later in Figure V-2.6) for "Offense-Defense Integration" and this is used as a point of departure.

The higher-level metric is the probability of mission success, which we tailor as the probability of COA success. The force multiplier metric is relevant because we seek to show the advantage of synergistically combining offense and defense. For our project, we tailor the force multiplier metric to tout the synergistic combination of Scotland Yard and the USAF in reducing internet forum effectiveness.

Ratio of Effects Achieved is the effect-based operations equivalent of the "exchange ratio" and accounts for how well friendly versus adversarial forces fare during all mission phases. Since we are calculating effects achieved for Blue Planning and Blue Perspective on Red, this ratio is a relevant metric. For our project, the effects metrics include strategic effects, operational effects, and tactical effects which are available from belief networks.

Complexity Theory Metrics: Metrics that identify emergent behavior for complex adaptive systems (of which an Internet Forum is certainly an example) are of interest. The fractal dimension is a measure of volatility and the Lyapunov exponent is a measure of stability. A fractal is a geometric pattern that is repeated at ever smaller scales to produce irregular shapes and surfaces that cannot be represented by classical geometry. Fractal Dimension allows us to measure the degree of complexity by evaluating how fast our measurements increase or decrease as our scale becomes larger or smaller. The fractal dimension is defined as DB = log (# of self-similar pieces) / log (magnification factor). Sub-networks can cluster based on fractal dimension (Figure III-5.8)

Another use of fractal dimension is for analyzing the volatility of a time series. The fractal dimension measures dispersion, but provides more information than the variance in classical statistics: there is no distinction between smooth and rapid changes in the measure of variance. For example, the stock index volatility[125] may be likened to the change in page hits on a particular internet forum versus time. Activity may be fairly stable or very volatile with wide swings in short periods of time. As shown the fractal dimension increases with increasing volatility.The Lyapunov exponent is another measure of interest in dynamic systems as discussed in Chapter III-5. See Figure III-5.7 for dynamic stability based on eigenvalues related to the Lyapunov coefficient[126].

Summary

This paper defines emergent behavior, defines four ways to determine whether emergent behavior is obtained, produces an executable model of the info-structure layer of the Internet Forum scenario, and identifies metrics that provide the most compelling measures of emergence.

---

[125] http://papers.ssrn.com/sol3/papers.cfm?abstract_id=425300, accessed 02/26/2015.
[126] http://flneerh.home.xs4all.nl/publications/Neerhoff_59.pdf, accessed 10/28/2014.

Part V
Applications

Chapter V-1: Simulated Commander for Missile Defense War Games
Chapter V-2: Strategic Offense/Defense Integration
Chapter V-3: Fog-of-War
Chapter V-4: Systems of Systems Analysis
Chapter V-5: Defensive Space Control
Chapter V-6: Sensor Fusion
Chapter V-7: Computer Network Defense
Chapter V-8: Uncertainty Management – Army Scenario
Chapter V-9: Pattern-preserving Extrapolation of the Space Catalog
Chapter V-10 Starship Cybernetics
Chapter V-11 Longevity Prediction
Chapter V-12 Fusing Data to Estimate Total Uncertainty
Chapter V-13 Monte Carlo Sampling for Benchmarking in the Cloud
Chapter V-14 Incorporating Uncertainty in a First Responder Simulation
Chapter V-15 Retail Problem Solver

Part V
Introduction

Although previous chapters gave examples of practical implementations, this section provides detailed scenarios and processing threads for applications ranging from longevity prediction, to a retail problem, to an artificial intelligence module for a starship.

Chapter V-1
Simulated Commander for Missile Defense War Games

Missile defense war-games are cutting-edge command and control simulations developed at the Joint National Test Facility in the United States. This paper describes the development of a Simulated Commander for such a war-game. A new requirement for its Theater Missile Defense (TMD) mission is to simulate decision-makers. Although the Synthetic Theater of War project[127] successfully advanced technology in the field of simulated human behavior by creating computer-generated forces, these players are low-echelon forces, rather than high-level decision-makers. The objective is to evolve existing research to meet war-game requirements. Especially challenging requirements include modeling "fog-of-war" effects on decision-makers, and dealing with incomplete, uncertain, and possibly conflicting data.

The focus of our work is to prototype a Patriot Battalion Commander's behavior in the execution-monitoring phase of a Theater conflict. The architecture consists of a parallel discrete event simulation framework called SPEEDES with applications consisting of a missile defense model, a battle planner, the Fuzzy CLIPS[128] expert system, and decision algorithms drawn from earlier research. Results of the conceptual prototype are presented to show how this analysis is used to influence the design of the war-game. We discuss the prototype software, lessons learned, and future challenges.

Introduction

We prototype Patriot Battalion Commander behavior in the execution-monitoring phase of a Theater conflict. A unique feature of our design methodology is that it is decision-centered: decisions are derived from mission requirements and are supported by displays and algorithms. The architecture consists of a parallel discrete event simulation framework called the Synchronous Parallel Environment for Emulation and Discrete Event Simulation (SPEEDES). Applications include a missile defense model, a battle planner, the Fuzzy CLIPS expert system, and decision algorithms drawn from our earlier research. The parallel simulation framework and the formulation of the simulated commander as a swarm of intelligent agents are vital for high performance on a supercomputer. Performance requirements that stress the resulting architecture are the need to employ up to 50 simulated commanders for system test and to execute the war-game at up to 100 times faster than real time.

Requirements

Functional requirements for the war-game Simulated Commander are summarized in the Table. The need for mid echelon decision-making: for example, a Patriot Missile Battalion Commander, and the need for vertical interactions with higher-echelon and lower-echelon players (real or simulated) is apparent. The ability to switch between a simulated and a real commander also needs to be provided. The basis for the initial delivery is a rule-based, but derived requirements point to the need for a variety of algorithms.

---

[127] http://en.wikipedia.org/wiki/Dynamic_Terrain, accessed 01/20/2015.
[128] CLIPS Reference Manual, Volume I, Basic Programming Guide, Version 6.05, November 1st 1997, accessed 01/20/2015.

Four requirements in the table specify what the Simulated Commander must do. Target prioritization by threat type and launch location exemplifies a decision that is amenable to rule-based algorithms. Perceived data is inherently uncertain and incomplete – this led us to consider Fuzzy CLIPS and belief networks for data fusion. Cognitive functions, such as the ability to recognize and react appropriately (as a human would) in self-defense situations, require rapid decision-making, especially with missile launch scenarios where timelines are short. The simulated commander is required to interact with either an automated or a manual battle manager: in the Theater Missile Defense case, default rules and common sense reasoning may provide solutions.

The realistic environments requirement is the most challenging. Although this is a far-term requirement, the initial formulation must provide extensibility. Decision-making based on attributes such as level of training, morale, fatigue, national resolve, political or religious influences have not yet been demonstrated in command and control simulations. We favor a society of intelligent agents[129] software paradigm for providing this functionality because it is inherently extensible, matches the Object-Oriented nature of the war-game, and should allow fine-grain partitioning for high-performance computing.

- **Command Level:** Initially, the Simulated Commander (SC) shall be capable of mid-level command functions; for example, to generate orders
- **Multi-Eschelon Interaction:** Each SC shall interact with other SCs.
- **Toggled Control:** the war-game shall provide the means to switch a player position between human control and computer control.
- **Rule-Based:** SC decisions shall initially be based on predefined rule sets.
- **Target Prioritization:** SC shall be capable of target prioritization based on target type and launch location
- **Perceived Data:** SC shall act only on perceived data that is available to the live counterpart
- **Cognitive Functions:** SC shall be capable of recognizing and reacting appropriately in self-defense situations.
- **Capabilities:** SC usually acts on data provided by automated command, control, and battle management systems, but in some low level theater air and missile defense instantiations, the SC shall perform "manual" target prioritization, track fusion, engagement planning.
- **Realistic Environments:** SC shall be capable of making decisions based on factors such as level of training, morale, fatigue, national resolve, political and religious influences.

Conceptual Design

The overarching philosophy that guides our effort is the concept of a decision-centered methodology discussed in Part I. The mission requirements are translated into a scenario to provide a concrete foundation for understanding and visualizing the TMD domain. In addition to the functional requirements, quantitative performance requirements (QPRs) are derived – the driving QPR is to perform a TMD mission with 50 simulated commanders at 100 times real

---

[129] Minsky, M. The Society of Mind, First Touchstone Edition, 1988, accessed 01/18/2015.

time. Requirements are then allocated to processes: we choose the Joint Operational Planning and Execution System[130] because it is widely used by the United States military. The focus of our effort is the TMD Patriot Missile Battalion Commander – this processing thread is rich enough to provide interesting rules and algorithms for the simulated commander to execute.

Operator decisions (human or simulated) are the most important part of the methodology. Decisions are derived from the processes allocated to a human Patriot Battalion. This keeps the interface between the simulated commander and the battle manager clear! The simulated commander does only what a human would. Displays to the simulated commander exist only to help with decisions; however, we also use displays to characterize the global situation, show inputs to the Patriot Battalion Commander, and to summarize the results of decisions. Displays showing the decision process within the Simulated Commander's "head" - and interactions with other simulated or human players - are more challenging.

Algorithms are derived from the flow down of requirements, processes, decisions and displays. In our previous research, this top-down approach to deriving algorithm needs resulted in a clear mapping between process types and algorithm types; for example, case-based reasoning is excellent for planning processes, expert systems are preferred for mission execution, and belief networks provide data fusion for engagement assessment. Importantly, a fancy algorithm is not a solution in search of a problem! The decision paradigm employed is that a decision is a state transition from a current state to a future state according to a plan. Current states, plans, and future states may be uncertain. Three advantages of this view of decisions are:

- The state transition diagram is a staple of object-oriented design.
- Types of decisions map cleanly to types of state transitions - and suggest useful algorithms.
- Uncertainty is represented explicitly in the decision-making process.

Typical Limitations of Direct Manipulation Interfaces	Advantages of Agent-Oriented Approach
– Large problem spaces scale poorly	- Scalability
– Actions rely on user interaction	- Scheduled or event-driven actions
– No composition	- Abstraction and delegation
– Rigidity	- Flexibility and opportunism
– Functional Orientation	- Task orientation
– No improvement in behavior	- Adaptive functionality

Agent-oriented design[131] is a generalization of object-oriented design that overcomes many limitations of direct manipulation interfaces. The biggest advantage of an agent-oriented approach for the simulated commander is scalability: we estimate that up to 50 players at up to

---

[130] Joint Operational Planning and Execution System (JOPES) Planning Policies and Procedure, Volume 1, 4 August 1993, accessed 01/20/2015.
[131] Knapik, M. and Johnson, J., Developing Intelligent Agents for Distributed Systems, McGraw-Hill, 1998, accessed 01/18/2105.

100 times real-time may require WG2K execution on 32 or more processors. Additionally, the simulated commander is a substitute for a human in a discrete-event simulation, and event-driven actions rather than user interaction is essential. Finally, future functionality will require machine learning, which is a feature of the agent-oriented approach.

Prototype Development Architecture

The prototype test bed for developing the Simulated Commander is shown below. The hardware, operating system, and framework (Silicon Graphics shared memory computer running the IRIX operating system and the SPEEDES framework) are the same as the war-game development environment. SPEEDES provides the event synchronization and data distribution management utilities to run discrete event simulations in parallel. The SPEEDES environment contains a display interface for situation awareness that we upgraded to OpenGL for portability.

- **Applications: Decision Algorithms, commercial and government off-the-shelf products**
- **Graphics: Open GL**
- **Framework: Synchronous Parallel Environment for Emulation and Parallel Discrete Event Simulation (SPEEDES).**
- **Communications: TCP/IP**
- **Hardware: SGI/IRIX**

SPEEDES also contains low-fidelity models of threats, weapons, sensors and a battle planner developed by Los Alamos National Laboratory. Fuzzy CLIPS is obtained from the National Aeronautics and Space Administration. These Government-Off-The-Shelf applications are used to produce input stimuli to the simulated commander and an expert system for executing rules. Fuzzy CLIPS is also used to call decision algorithms.

The object classes and associations that we model are shown in Figure V-1.1. The TMD problem that we portray is defensive: a ballistic missile launch is detected by radar and space-based infrared systems. Threat detections are reported to a Battle Manager. Data is displayed to a simulated commander who makes decisions about deploying Patriot Missile Interceptors. The Simulated Commander code only interfaces with the Battle Planner code within the Battle Manager composite. Based on these decisions, and in-flight updates on threat trajectories, ground-based kinetic interceptors are launched to intercept enemy ballistic missiles.

Although not explicitly shown, the simulated commander is involved in all phases of the mission: situation awareness, mission planning, execution monitoring, replanning engagements, assessment, and reconstitution. Our focus is on the execution-monitoring phase.

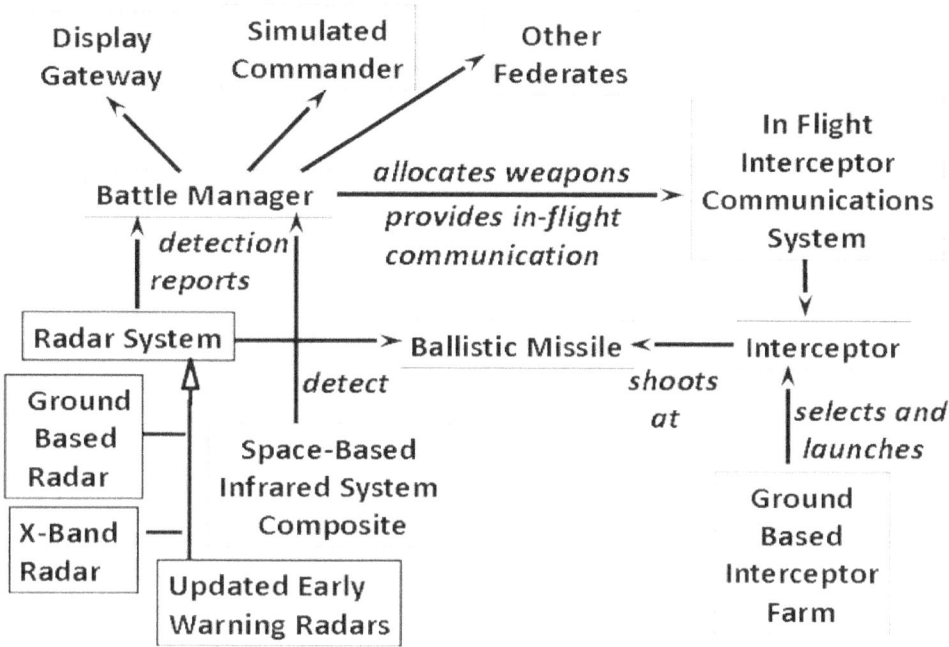

Figure V-1.1: Object-Oriented Design

Fuzzy CLIPS is a NASA software product that provides a comprehensive environment for developing and executing an expert system rule base. Rules are developed to allow a Simulated Commander to make decisions in the execution-monitoring phase of a TMD scenario. These include: Threat Condition Updates, Rules of Engagement Updates, Mission Objective Updates, and Battle Plan Updates.

The functional model provides a SPEEDES-based TMD simulation with realistic (but low-fidelity) physics models that execute in parallel. Simulation time, track data, asset inventories and engagement plans are provided to the battle planner and displayed to a human or simulated commander. Interactions between the Simulated Commander and the Battle Planner are conducted via side-by-side display windows with dialog boxes moving back and forth to "embody" the Simulated Commander decisions and the Battle Planner implementation of the decisions.

Decision Algorithm Execution

We are using Fuzzy CLIPS to execute other decision algorithms; for example, IF decision_type = data_fusion THEN CALL belief_network. This hierarchical call structure is convenient and readily extensible. Two examples of promising decision algorithms to augment Fuzzy CLIPS are discussed.

Long-term planning, based on broad goals, is a difficult problem for rule-based reasoners. The range of possible solutions is an exponential function of the number of decision criteria, which can number 12 or more. A case-based solution strategy (Figure III-6.1) is favored.

The problem of optimally locating transportable resources has been discussed (Chapter III-6) in a related domain using case-based reasoning (CBR). It allows an operator (here, a Simulated Commander) to readily identify locations for resources (here, a Patriot Battery) based on weapon effectiveness calculations. CBR is easily described as a generalization of an engineering trade study: rank pre-stored options according to weighted selection criteria and choose the option, or combination of options, with the highest score

Data fusion for engagement assessment (did the Patriot missile intercept the target?) is another problem that is not conveniently solved using expert systems. Perceived data is uncertain, incomplete, and possibly conflicting – and the appropriate rule may not "fire" at the appropriate time! In risky decision-making, three factors are important: belief, ignorance, and disbelief. To account for these two parameters in an intuitively satisfying way, we built a Dempster-Shafer Belief Network[132]. For the TMD problem, the context for data fusion (Figure V-1.2) is data arriving asynchronously.

To update the Rules of Engagement, the Simulated Commander must fuse incoming data with the currently perceived situation and issue a new directive as appropriate. Raw data (Space-Based Infrared, Early-Warning Satellite, Strategic Radar, and Intelligence Community) updates indicators in the middle layer (Ballistic, Threat, Valid) of the belief network, and forms outcomes (Launch Assessment) at the upper level. These are thresholded based on rules about the necessary degree of belief to produce a decision (Employ Ballistic Missile).

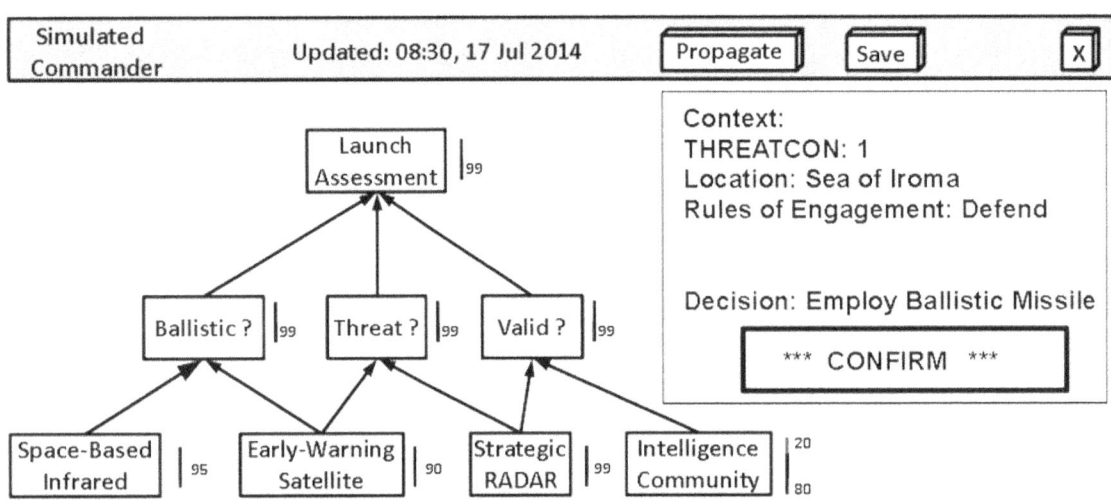

Figure V-1.2: Belief Network for Data Fusion

Summary

Based on war-game requirements, a conceptual design for a Simulated Commander is complete. The agent-oriented design relies on fuzzy rules initially and is extensible to a variety of algorithms. The methodology is decision-centered and defines a decision as a state transition for a current state to a desired state, based on a plan. Use of SPEEDES allows parallel processing

---

[132] Pearl, J., Probabilistic Reasoning in Intelligent Systems: Networks of Plausible Inference, Morgan Kaufmann, 1988., accessed 01/20/2015.

(scalable), provides extensibility, and tackles a realistic TMD scenario. Decision algorithms are called from the Fuzzy CLIPS expert system. We favor case-based reasoning for planning and belief networks for data fusion. We have integrated the collection of hardware and software to produce a prototype test bed with a simulated commander making decisions based on a TMD scenario driven by realistic wargaming models.

Many other artificial intelligence methods have been analyzed, prototyped, and assessed for applicability to our task. For example, case-based reasoning is a good for planning, because we already have cases for human commanders and previous war-games. On the other hand, operations research may be good for scheduling, but is too inflexible for modeling human decision-making. We experimented with neural networks, but find them unsuitable, in their basic form, because they do not readily provide an explanation.

Chapter V-2
Offense/Defense Integration Performance Metrics

This chapter summarizes performance metrics for integration of offensive and defensive military forces. The objective of this effort is to develop a hierarchy of performance metrics to quantify the military utility of combining offensive and defensive missions. The focus is the potential benefit of integrating the U.S. Triad, which consists of Intercontinental Ballistic Missiles, Submarine Launched Ballistic Missiles and Strategic Bombers, with the emerging Missile Defense System.

We formulate a fundamental performance metric – the Ratio of Damage Expectancies - for quantifying the military utility of integrated offense and defense. The advantages of this metric are that it is: familiar to strategic offense analysts, simply related to metrics such as "leakage" used in strategic defense, intuitively understood, and easily tailored to tactical missions. A decision tool kit, which incorporates this metric as an objective function for planning integrated offensive and defensive missions, is prototyped. Knowledge acquired from our studies suggests that the Ratio of Damage Expectancies is a useful intermediate metric in a hierarchically structured set of performance metrics for analyzing the potential benefit of integrating offensive and defensive missions.

Introduction

The strategic and tactical environment within the U.S. Department of Defense is changing. The advent of missile defense signals a shift from an "offense only" policy for deterrence to a weapons arsenal containing strategic offensive and defensive systems. How will these offensive and defensive systems be employed? What is the military utility of coordinating their use? The answers are unknown. Wide-ranging mission planning and execution challenges for strategic and theater commanders are apparent. The need is therefore to identify, and quantify to the extent possible, the military utility associated with integrating strategic and tactical offensive and defensive systems in future command and control centers.

The objective is to investigate the implications of offense/defense integration (ODI) on the command and control of the battlefield. A spectrum of operations concepts is defined. A performance metric, the Ratio of Damage Expectancies (rDE) is formulated and implemented as an objective function for battle planning. Simulations of future command and control systems for integrated offense and defense are developed and executed. Simulated results indicate the potential utility of the rDE performance metric. Future work entails hierarchically expanding the metric to compute force multipliers for rogue nation scenarios and the probability of integrated mission success for multi-mission integration. The rDE performance metric appears tailorable to tactical missions and related mission areas such as computer network warfare.

Approach

Our top-down methodology is performance-centered. Performance metrics required for ODI are derived from Strategic Deterrent Forces (SDF); that is, Nuclear Triad[133] and Missile Defense System[134] (MDS) mission areas. The methodology begins with an understanding of the mission domain, the concept of operations, and mission-related requirements. We leverage the Joint Operations Planning and Execution System[135] (JOPES) to provide a common foundation for analyzing military processes. Performance metrics are particularly relevant to the Execution Planning phase of JOPES because they provide objective functions for rating detailed plans. Requirements are levied on candidate performance metrics: familiar to military analysts, intuitive and easy to understand, computable from readily available information on targets, weapons, and constraints, and tailorable to other mission areas.

A flow diagram of the approach (Figure V-2.1) indicates that a description of the mission context forms the basis for a use case, which defines mission objectives, threat, assets, and constraints. Broad Courses of Action (COAs) then provide a foundation for a detailed scenario composed of an event timeline along with required activities and operator decisions. Based on the detailed scenario, performance metrics are postulated from a review of current measures of effectiveness (MOEs), commonality among missions, and applicability to the scenario. Feedback is solicited to refine the process. Chapter III-7 describes how the finding the optimal COA using the rDE metric is automated.

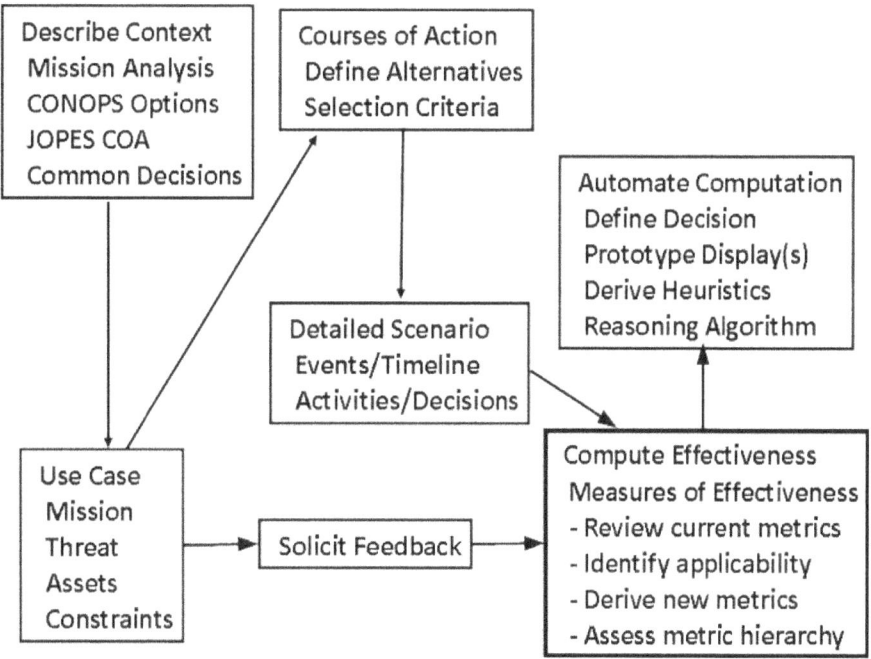

Figure V-2.1: Approach to Deriving Performance Metrics

Derivation of rDE

Performance metrics provide a natural way of computing the relative worth of candidate plans.

---

[133] http://en.wikipedia.org/wiki/Nuclear_triad, accessed 02/27/2015.
[134] http://www.mda.mil/system/system.html, accessed 02/26/2015.
[135] http://www.dtic.mil/doctrine/jfs/jdtc/jdtc_jopes_training.pdf, accessed 02/26/2015.

Based on training in strategic planning that we received at the STRATCOM[136], the dominant planning metric for strategic deterrent forces such as intercontinental ballistic missiles is Damage Expectancy (DE). This performance metric for offensive forces is a probability of inflicting damage. Although the factors comprising it vary among the various analysis groups, a typical representation is that damage expectancy has three conditional probability factors: probability of pre-launch survivability, probability of arrival and probability of damage. Sophisticated physics codes are used to compute these factors and they are continuously updated.

On the other hand, a key MDS performance metric is leakage (L). In defending the U.S. against a limited missile threat from a rogue nation, leakage is the number or percentage of threat missiles that impact in a defended area. The insight that produces a connection between offense and defense is that pre-launch survivability of U.S. forces is a function of leakage; that is, if they are not intercepted, incoming missiles targeted at U.S. strategic weapons reduce their pre-launch survivability. Further, U.S. preemptive strikes on rogue nation missile facilities may reduce enemy pre-launch survivability. Therefore, it is reasonable to postulate a linkage between offensive Damage Expectancy and defensive Leakage.

The linkage between offensive and defensive performance objectives is formulated based on an offensive "Blue-on-Red" damage expectancy (DE b-r) and a defensive "Red-on-Blue" damage expectancy (DE r-b). Throughout this discussion, "Red" refers to enemy, "Blue" connotes friendly. The goal of offense is to maximize DE b-r while the goal of defense is to minimize DE r-b. Since offense and defense are interdependent, based on the effects of leakage and preemptive strike mentioned above, a useful goal of ODI is achieve to a desired ratio of damage expectancies (Figure V-2.2). Since rDE increases without bound, it is not reasonable to maximize rDE. This strategy leads to infinite cost and complexity!

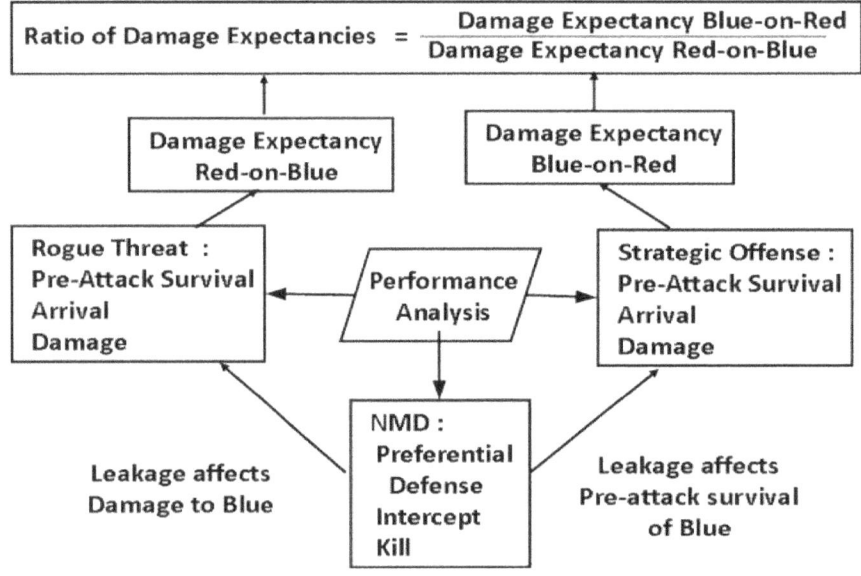

Figure V-2.2: Ratio of Damage Expectancies

---

[136] http://www.stratcom.mil/, accessed 02/26/2015.

To show the expected behavior of rDE as a function of leakage, assume that rogue missiles are targeting U.S. strategic weapons. In such a scenario, DE_{b-r} is directly proportional to pre-launch survivability, which is proportional to threat missiles intercepted ($1 - L$)

$$DE_{b-r} \propto (1 - L)$$

Given that defensive forces are targeting incoming rogue missiles, the enemy perspective is that DE_{r-b}, which is directly proportional to the probability of arrival of the rogue missiles, is therefore proportional to leakage, which is the percentage of missiles arriving

$$DE_{r-b} \propto L$$

The ratio of damage expectancies for this simple example is given as

$$rDE = \propto (1 - L) / L$$

Behavior of this family of curves (Figure V-2.3) is given for various values of a "lumped" efficiency parameter (e). For small values of leakage, rDE increases to infinity. An rDE of unity may connote parity: friendly and enemy forces perceive the same damage expectancy based on what their respective weapons are targeting. This simple illustration is geared to explaining potential synergy between offensive and defensive forces; clearly, significant analysis using sophisticated simulation of the scenario is required to obtain a reasonable understanding and validate these hypotheses.

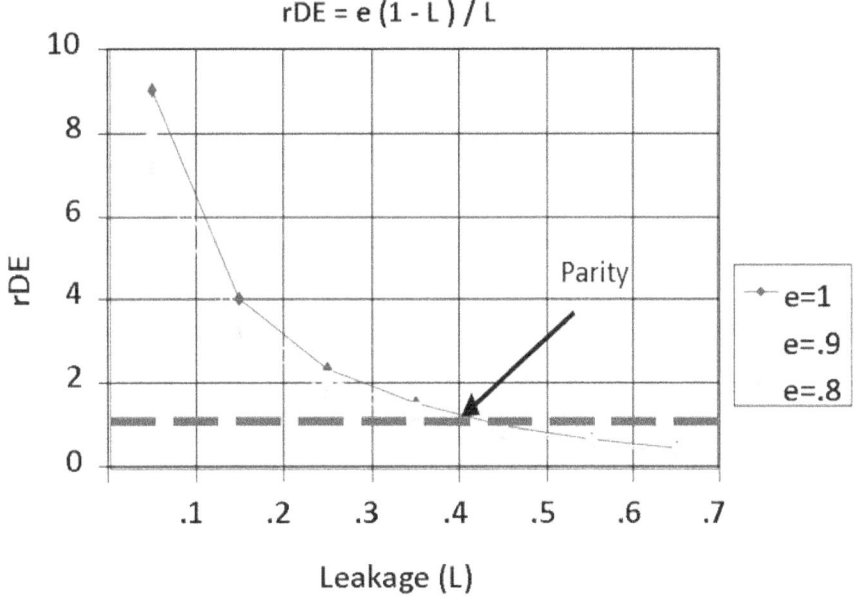

Figure V-2.3: Ratio of Damage Expectancy versus Leakage

Combination Rule

The phrase "based on what their respective weapons are targeting" leads to a significant algorithmic effort to define and update the basis for damage expectancies; that is, the targeted assets for Red and Blue. Probabilities are necessarily computed based on an underlying population; for example, the probability of drawing a certain number depends on the characteristics of the sample (sample size, how many times that number occurs) from which it is drawn. For ODI, it is straightforward to identify Red and Blue objects involved in a single action; for example, rogue nation missile launches toward West Coast submarine port. However, combining an interleaved set on offensive and defensive Blue actions with offensive (and possibly defensive) Red actions is more challenging.

Detailed planning is based on attaining a specified rDE. Offensive and defensive response options are defined. The parameters that define a response option are: name, weapon munitions (conventional or nuclear), arena (strategic of theater), type and timing of action (preempt, defend, destroy, deny, retaliate), six probability factors, and a default rDE.

The conditional probability factors are estimated for Survival (S), Arrival (A), and Damage (D) based on the following rationale:

- S_{b-r}: independent on arena and munitions, highly dependent on timing; for example, $S_{b-r} = 1$ for preempt, $s_{b-r} = .5$ for retaliate

- A_{b-r}: very high (.9) overall, less high (.8) for long-range trajectories (strategic) during (destroy) and after (retaliate) first wave

- D_{b-r}: poor for long range conventional (.3) and deny (.4, .5), otherwise good to very good (.7 - .9)

- S_{r-b}: no influence (1.0) on defend and retaliate. Low (.1) for nuclear preempt (.1, .2). Variable otherwise.

- A_{r-b}: Unlikely (.1) for defend options, otherwise arrival is very likely (.9)

- D_{r-b}: Moderate damage (.7), except for tactical defend (failed attempt) allowing much damage (.9)

These rules-of-thumb are incrementally refined using a data mining technique (rule induction tree) from a freeware package. This technique allows the heuristics to be revised for consistency and completeness.

Targets are also defined (Figure V-2.4). The parameters that define a target are: name, locale, location coordinates, type (airbase, submarine port, chemical facility, missile base, etc.), value or importance, hardness, and mobility. Response Option / Target pairs are defined as actions. A Course of Action (COA) is defined as a collection of actions and has an overall rDE that is computed based on the combination rule with rDE perturbed by few simple rules based on target parameters.

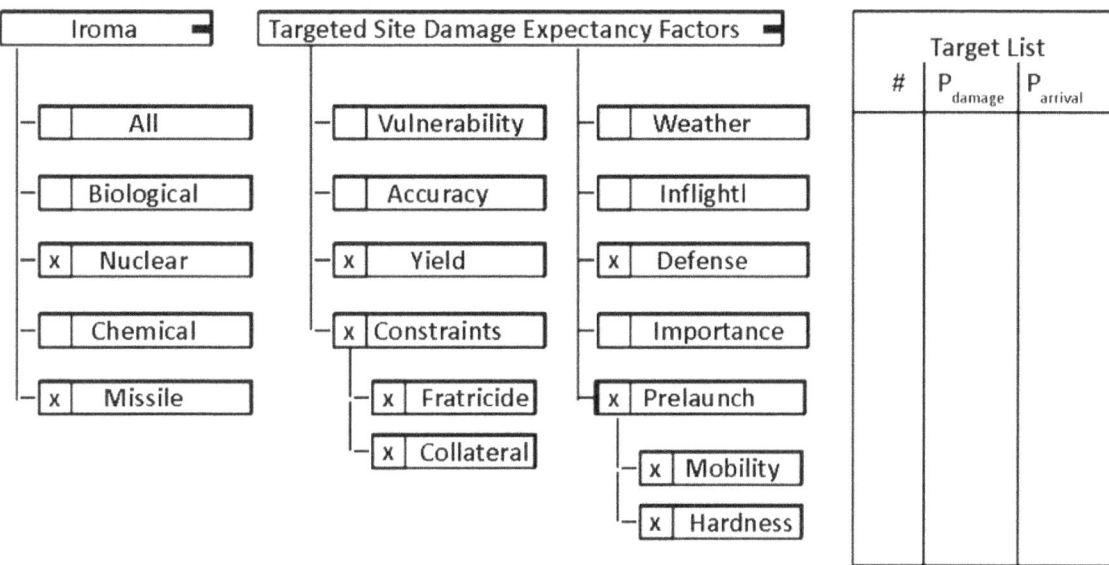

Figure V-2.4: Target Definitions

The combination rule is an algorithm that provides a prescription for combining the six probability factors that are used to calculate rDE:

```
For each factor (for example, SVbr)
    For all actions on the same target class
        if different actions, use AND rule (Q = q₁ * q₂*,...)
        else use OR rule ( P = { 1 - [1-p₁] [1- p₂],...})
    For all actions on different target classes weight worth of
            each target class relative to others
        normalize so sum of weights is unity
        apportion factors according to weights
```

As an example (Figure V-2.5) of synergy that integrated offense and defense can produce, consider two actions against a rogue nation missile base. A Strategic Defense action, taken alone, produces an rDE of 7.2 while a tactical nuclear preemptive action, taken alone, produces an rDE of 12.9. However, if these are combined, the probability of Red pre-launch survival is low (.1) due to the preemptive action, while the probability of Red missile arrival is also low (.1) due to the defensive action. The result is a combined rDE of 72. The "force multiplier", defined as the ratio of these results with and without integration is 3.6!

Ratio of Damage Expectancies

Action:	SVbr	ARbr	DMbr	SVrb	ARrb	DMrb	rDE
Strategic Defend (C)	.8	.9	.7	1	.1	.7	7.2
Tactical Preempt (N)	.6	.9	.9	1	.9	.7	12.9
Combined:	.7	.9	.8	.1	.1	.7	72.0

Blue pre-launch survivability is based on defend (.8) and Preempt (.6). If the worth of each is equal (.7), the combined Probability is for the union of Blue assets:
P = (.8*.7 + .6*.7)/(.7 + .7) = .7

Strategic defend has no impact on Red pre-launch survivability, but Tactical Preempt does: same Population with OR rule:
P = 1 * .1 = .1

Offense/Defense Integration Force Multiplier:
• Ratio of Damage Expectancies with/without integration
• Force Multiplier : 72.0 / (7.2 + 12.9) = 3.6

Figure V-2.5: Example of ODI as a Force Multiplier

The next step is to postulate a realistic rogue threat scenario and use physics-based simulations to compute more realistic rDE factors. We expand the rDE metric to a hierarchy of metrics (Figure V-2.6), and compute performance metrics such as force multiplier, and probability of integrated mission success.

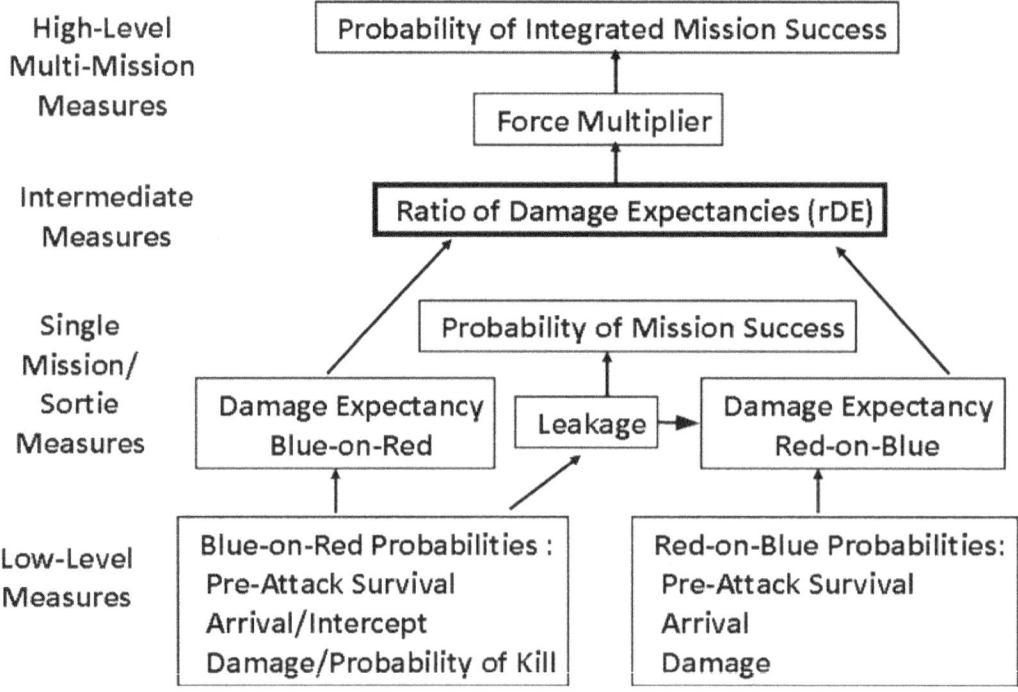

Figure V-2.6: Hierarchy of Performance Metrics

Summary

We postulate rDE factors for 14 response options. We use a data-mining algorithm (rule induction tree) to refine these for consistency and completeness. Target parameters are defined based on unclassified data obtained from the Internet. A few rules are implemented to perturb the rDEs for various target classes. A combination rule integrates actions to produce COAs based on quantitative performance metric called the ratio of damage expectancies.

Conclusions

The rDE performance metric has desirable attributes:

- Familiar to analysts who have worked with the Damage Expectancy and Leakage metrics,
- Intuitively suggests maximizing damage to threats while minimizing damage to defended areas
- Tailorable to tactical and information operations missions
- Provides an explicit objective function for plan optimization (we used a genetic algorithm, see Chapter III-7)

Chapter V-3
Fog-of-War

A next generation command and control simulation for analyzing future ballistic missile and air defense systems is under development. An important component of this simulation is the representation of a realistic, automated decision-making process, the "Simulated Commander." This paper describes our work toward coping with a particularly difficult requirement: simulate the effects of "Fog-of-War" on decision-making. This is defined as the influence of real-world uncertainties, conflicts in information, and attention deficits that degrade human decision-making.

Our approach to representing and reasoning in the presence of fog-of-war is to apply perturbations to the input, processing, and output of various reasoning algorithms to appropriately insert a variety of its effects. These include lost inputs, latencies, modified priorities, degraded confidence, operator confusion, and misplaced outputs. The module is formulated to allow selectable "fog settings". A test bed prototypes a "thinker" module, a missile launch scenario, and related decision-making. A graphic user interface is implemented and connected to a Dempster-Shafer Belief Network for data fusion. Results to date are intuitively satisfying; for example, it does appear that perturbations to a probabilistic belief network credibly mimic fog-of-war effects on decision-making.

Introduction

In the conduct of a war-game, a considerable number of players are required, and many travel to the wargaming facility from distant locations. Allowing the "players" to participate from their home locations by using remote terminals and established communication nets substitutes human travel for purchased hardware and communications costs. In some situations, this is cost effective; however, many people are still required to devote considerable time in every war-game that we conduct. This is especially true if the war-game is focused on some particular role, then the other "players" are needed to keep the test realistic but benefit little from participating. Selectively replacing human participants with Simulated Commanders is an approach that is intended to minimize monetary and human costs while still providing realistic "players" for a wide range of war-game functions.

The war-game has requirements for fully automated command decision-makers that realistically represent human commanders at any level in the command hierarchy of a joint services Ballistic Missile Defense war-game. An important component of this simulation is a realistic automated command decision-making process, or a "Simulated Commander." This paper provides a conceptual design for coping with a particularly difficult requirement: simulate the effects of "Fog-of-War" on decision-making. This is defined as the influence of real-world uncertainties, conflicts in information, and attention deficits that degrade human decision-making. Our approach to representing and reasoning in the presence of fog-of-war is to apply perturbations to the input, processing, and output of various reasoning algorithms to appropriately insert a variety of its effects. These include lost inputs, latencies, modified priorities, degraded confidence, mis-attribution of evidence, and misplaced outputs. The module has been formulated to allow selectable "fog settings". A test bed prototypes this "thinker" module, using a missile launch scenario, and probabilistic decision-making for concreteness. Results to date indicate that perturbations to a probabilistic belief network credibly mimics fog-of-war and its effects on decision-making.

Approach

The overarching approach to demonstrating fog-of-war in simulated commander decision-making (Figure V-3.1) consists of defining causes and using operators to map them to effects on input, processing, and output. Causes are identified by a requirements analysis, identifying tools (decision algorithms) for testing the design, organizing sources of "fog", and factoring in uncertainty, unreliable of information, and human errors that make fog-of-war difficult to simulate. Causes are identified as pertaining either to the data, the human, or the computer processing. Effects are allocated to inputs, processing, and outputs of reasoning algorithms to foster universal applicability. The resulting "cause and effect matrix" is then applied using "operators" that quantifies the effects. Significantly, causes are defined according to how often a perturbation occurs (frequency) and how intense the perturbation is.

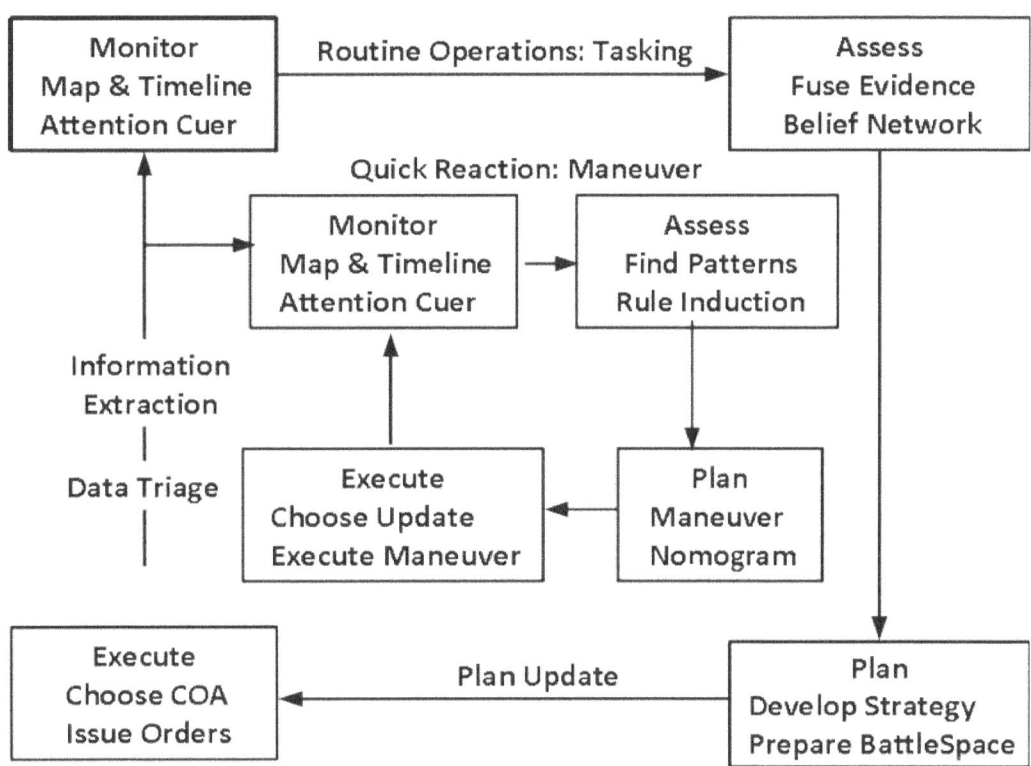

Figure V-3.1: Fog of War Approach

We derive requirements for simulating the fog-of-war from requirements in a war-game specification. The gleaned phrases are: ambiguity, disruption, timeliness, overload, indecision, confusion, stress, fear, stroke, fatigue, and unanticipated threats. These causes of "fog" are supplemented based on discussions with war-game developers to produce a more complete set of causes: data-related, human, and processing-related.

The war-game shall be capable of supporting the simulation of the experience of fog-of-war which includes:

- Ambiguity in attack assessment
- Disruption of critical data links and nodes connecting sensors to the battle management, command and control system
- Exceptionally heavy weather effects such as floods, blizzards, volcano, and heavy fog that reduce performance of sensors and battle management
- Inability of key sensors and nodes to perform their mission during enemy attack
- Inability to gain prosecution decisions from higher authority in a timely manner
- Insufficient bandwidth or extensive communications delay in intelligence traffic, including excess information causing overload
- Indecision or confusion caused by battle-related stress such as fear, stroke, or fatigue
- Unanticipated threats or countermeasures that degrade sensors sources, and methods resolution in confusion in Battle Management command, and control.

Because we anticipate many forms of reasoning to be performed by the Simulated Commander (heuristic, probabilistic, case-based, and possibly neural) causes of "fog" are defined based on their impact on inputs, processing, and outputs of reasoning algorithms. This has the added advantage of applying to other war-game objects subject to "fog-of-war"; namely, humans, weather, sensors, weapons.

Formulation

Causes of fog-of-war that impact human decision-making are derived from requirements and discussions with domain experts. These are mapped to effects on reasoning algorithms; specifically the Dempster-Shafer Belief Network, using simple "operators". A combination of operators is defined to convert causes into effects. Default values for these operators are provides and correspond to "low", "medium" and "high".

Each of these operators is defined for the first reasoning algorithm for which we apply this fog-of-war formulation - a Dempster-Shafer Belief Network for data fusion (shown earlier in Figure III-8.4). We define, via operators, data-related and human causes of perturbation to evidence (input), weights (processing), and outcomes (output) of a belief network. Typically, causes are related to effects by two operators; for example, random (quantifies frequency of occurrence) and severity (quantifies strength of the effect).

The final step is to provide a graphic user interface and integrate this fog-of-war formulation into the existing simulated commander. Results are reported in the next section.

Algorithm Integration

A major hurtle is to show how already uncertain inputs to a reasoning algorithm (Dempster-Shafer Belief Network) can be perturbed by fog-of-war (Figure V-3.2). Resources, such as a battle manager or intelligence terminal, provide evidence related to various hypotheses (for example, is the foreign space launch hostile?). The information source that specifies the degree of belief in the hypothesis, and the time of the input are collected. Here, we find that fog-of-war operators directly perturb one or more of these three parameters. The random operator is first employed to determine whether to apply a perturbation. If so, the source may be changed, the belief may be decreased, or the time of evidence receipt may change, depending on the operator. After applying fog-of-war to the input, the fusion algorithm[137] proceeds in the regular way: a Cartesian product fuses the new evidence with previous evidence, normalization drops unsupported beliefs, belief and plausibility are computed, and results are plotted on a timeline.

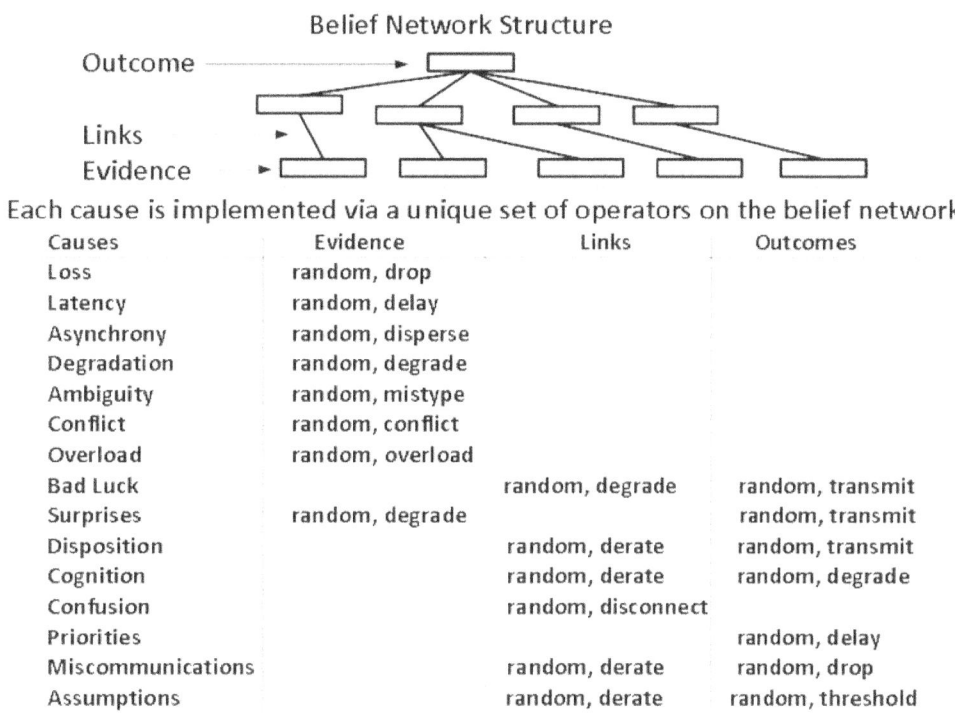

Figure V-3.2: Belief Network Perturbations

The belief network has three layers: evidence, indicators, and outcomes - that successively refine data. These are connected by links called "weights". We find that processing is easily perturbed by modifying the weights in the belief network. They may be modified in value, scrambled, or deleted - depending on the operator. Similarly, outputs may be delayed, degraded in quality, or not sent.

Processing

A more detailed view of the code that implements fog-of-war in a belief network reveals the processing steps:

1. Determine if fog-of-war is enabled,
2. Get "fog factors"
3. Determine whether to apply

---

[137] Stein,R., The Dempster-Shafer Theory of Evidential Reasoning, AI Expert, August 1998., accessed 01/20/2015.

4. Apply severity factor

In the example, the confidence in evidence is decreased due to a "loss".

Displays

A user interface is constructed using a commercially available display builder. The "Fog of War Control" allows users to load, edit, and save settings. Defaults are modified by selecting data-related or human causes, or by selecting effects evidence, weights, or outcomes. Fog is turned "on" or "off" globally in one click. related or human causes, or by selecting effects evidence, weights, or outcomes.

Architecture

Having integrated fog-of-war with the belief network algorithm, the ensemble is interfaced with our "Automated Decision Support System". This provides proof-of-concept for simulated commander functionality and for semi-automated decision support to human decision-makers (Figure V-3.3).

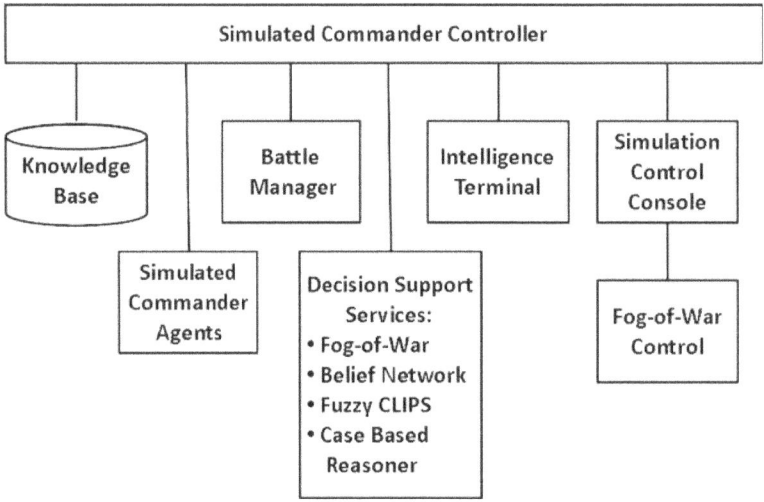

Figure V-3.3: Automated Decision Support System

The Simulated Commander Controller contains the fog-of-war control, a knowledge base, the Simulated Commander, Agents, and decision support services. A "white" control, or simulation conductor, manages fog-of-war unknown to the simulated commander who only receives perceived data inputs from battle managers and an intelligence terminal. We have integrated these code modules and they work together.

The Simulation Control Console is the summary display used by the Simulated Commander Controller - a member of the simulation conductor team. The console provides visibility into behavior, tracks algorithm and fog-of-war settings, shows evidence and history traces, and scores decision quality.

Fog-of-War Example

The task of Foreign Launch Assessment (shown earlier in Figure V-1.2) is chosen as a scenario to show how decision quality and timeliness are impacted by fog-of-war. Simply stated, the simulated commander must decide whether a particular missile launch is a deliberate, foreign, Intercontinental Ballistic Missile threat. Situation indicators and a map are shown. New data, which is uncertain, incomplete, and sometimes conflicting, is received. A timely Rule of Engagement directive must be issued based on the new data. The decision process is based on a Dempster-Shafer Belief Network which fuses new data, forms indicators. (ballistic, threat, valid), and produces as assessment based on this causal network.

Applying fog-of-war factors (Figure V-3.4) has significant effects:

- Without fog, the decision threshold is reached in time: confidence in the assessment (upper right) that the threat is a hostile ICBM is above the threshold (horizontal dotted line) before the time deadline (vertical dotted line). The decision is successful.
- With fog, confidence in new data is lower and sufficient confidence in the assessment (lower right) does not occur in time. The decision is too late.

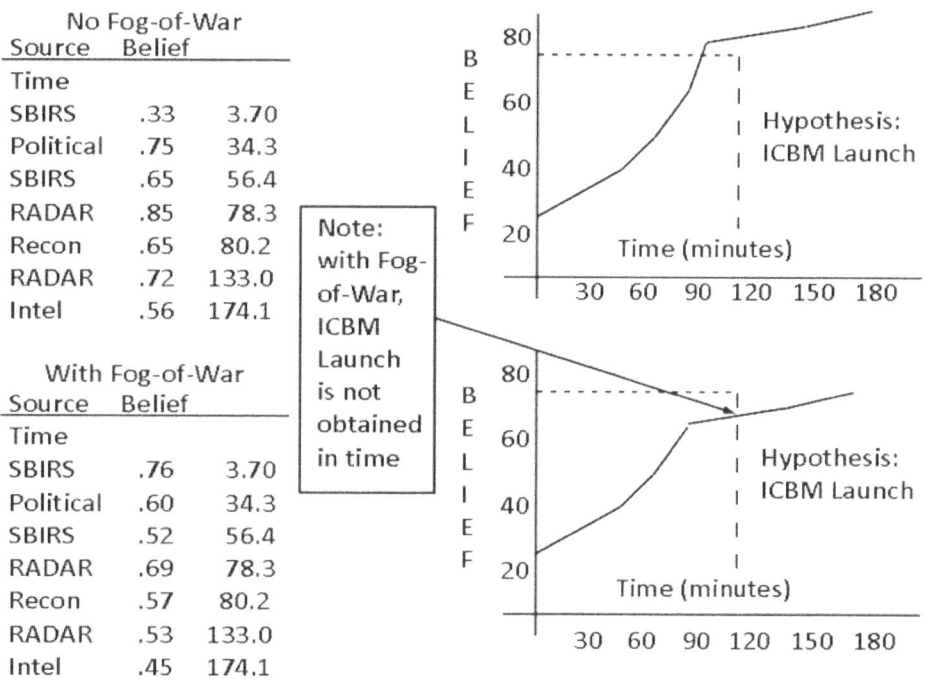

Figure V-3.4: Fog-of-War Example Results

Discussion

We show a proof-of-concept for adding fog-of-war to decision support. We define requirements, provide a scenario context, formulate a fog-of-war algorithm, produce a detailed technical approach, and apply the solution to a belief network.

The "operator-based" solution is applied to rule-based reasoning - we use Fuzzy CLIPS - for a Patriot Battalion Commander decision making process. It is extended to a case-based reasoner for the Patriot Battalion Commander planning process. Based on the observation that a little "fog" produces serious degradation in decision performance, metrics are collected and default values for operators are modified to enhance realism.

Chapter V-4
Systems-of-Systems Analysis

Introduction

Analysis of complex systems requires an understanding of many perspectives of the operating environment. Systems-of-systems analysis (SoSA), defined as the holistic investigation of multiple interacting system perspectives, is a key challenge. We need to provide a system representation that accounts for multiple layers of networks; for example, the Political, Military, Economic, Social, Infrastructure, Informational (PMESII) and other perspectives (adversary, friends, unaligned) of the environment. From the literature, "System of Systems is a relatively new term that is being applied primarily to government projects for addressing large-scale interdisciplinary problems involving multiple heterogeneous, distributed systems that are embedded in networks at multiple levels and in multiple domains"[138]. Another definition of systems-of-systems is "systems that are formed of substantially independently operated elements. That is, they are formed of systems that do not solely contribute to an overall purpose or set of functions, but rather individually fulfill useful purposes"[139].

Guidance

Military publications require "visualizing the operational environment beyond the traditional military battle space as an interconnected system-of-systems comprised of friends, adversaries, and the unaligned"[140]. This systems perspective (Figure V-4.1) shows that SoSA is an interconnected collection of networks that define the operational environment.

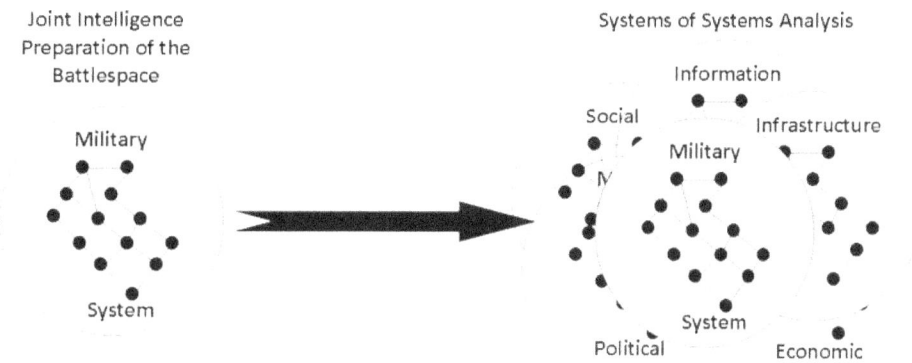

Figure V-4.1: Systems Perspective of the Operational Environment[141]

Systems-of-Systems Analysis (SoSA) is a technique to perform a holistic assessment of the operational environment. The process allows staff to gain a baseline appreciation of the

---

[138] http://en.wikipedia.org/wiki/System_of_systems, accessed 02/26/2015..
[139] http://sebokwiki.org/wiki/Systems_of_Systems_(SoS), accessed 02/26/2015.
[140] Commander's Handbook for an Effects-Based Approach to Joint Operations, Joint Warfighting Center, Joint Concept development and Experimentation Directorate, 24 February 2006. Unclassified. page vii
[141] ibid, page II-2

environment and to organize information in a form useful to the Commander[142]. A SoSA approach helps define Centers-of-Gravity and also provides a way to help identify where to look for changes in behavior or capability that signal adversary activity in multiple systems, potentially resulting in a more accurate and complete assessment of the outcomes of joint force actions[143]. Joint Intelligence of the Battle Space uses a SoSA technique to portray the operational environment as an interactive system composed of nodes (relevant people and things) and their links (interrelationships for friendly, adversary, and unaligned systems.[144]

In planning guidance, systems-of-systems analysis is defined as an analytical process that holistically examines a potential adversary and/or operational environment as a complex, adaptive system, including its structures, behavior, and capabilities in order to identify and assess critical factors and system interrelationships.

Problem

Joint publications provide planning guidance for Effects-Based Operations. This alternative to attrition-based warfare fits within the Joint Operations Planning and Execution system (JOPES). The guidance found in these documents[145] explains what SoSA might look like, but provides no further details on its implementation as a process.

With this discussion of the military planning context complete, the problem is that this "analytical process" is not defined. The literature gives no useful approach for implementing this SoSA process. Visualization of these interconnected networks can be easily drawn (imagine multiple planes, each with a network diagram), but not analytically defined: this is the challenge! Currently, a link drawn from on layer to another simply means that there may be a connection that an intelligence analyst or planner should be aware of and possibly consider when attempting to understand adversary networks or when planning a Course of Action.

Objective

We define processes based on analytical, automated techniques to implement SoSA. The techniques implement current guidance, leverage the systems perspective that uses networks consisting of nodes and links to characterize multiple layers (for example, PMESII) from multiple perspectives (friends, adversaries, and the unaligned), and provide multiple hierarchical levels (strategic, operational, tactical, Wing). The techniques we define, working together or separately, provide a range of options from purely manual to significantly automated processes.

Sub-objectives are to accomplish a set of tasks that provide a range on implementation options. Tasks are defined as follows:

- **Literature Survey**: state-of-the-art in SoSA for techniques currently in practice.

---

[142] ibid, page II-2
[143] ibid, page II-12
[144] ibid, page III-13
[145] Commander's Handbook for an Effects-Based Approach to Joint Operations, Joint Warfighting Center, Joint Concept development and Experimentation Directorate, 24 February 2006.

- **Requirements Analysis**: SoSA capabilities required for our cyber domain.
- **Detailed Design**: SoSA formulation, implementation, test, and demonstration.
- **Data Schema**: efficient data storage mechanism for multiple networks.
- **Multi-Network Formulation:** algorithms that implement multi-layer graphs
- **Cross-Link Computation:** method(s) to quantify links across networks.
- **Implementation:** code a prototype for a computer network defense scenario
- **Test and Refine:** assess operational effectiveness and limitations.
- **Demonstrate and Document:** defined a scenario and documented results.

Guidelines and Assumptions

- Emphasis is on a spectrum of semi-automated techniques.
- All network layers are considered co-equal: grafting a few nodes from a secondary network to a primary network may be necessary but is not sufficient.
- A meta-model subsuming and generalizing network layers is not useful because no single collaborating group has a detailed understanding of all layers.
- The two most important implementation challenges are:
    - Mathematical formulation of the multi-layer, multi-sided, networks
    - Cross-layer link functions (for simplicity, called "cross-links") that properly connect a node in one layer (or one side) with a node in another.
- Visualization mockups are provided for adding cross-links, but little emphasis is expended on new graph visualization techniques – there are already plenty.
- The mathematical network and cross-link formulations are not to impact data schemas or SoSA visualizations.

Approach

This is a summary discussion of our methodology. Based on an early literature survey and understanding of requirements, we move quickly to the detailed design and implementation.

Literature survey

In addition to military needs discussed earlier, Maier[146] cites challenges such as the social-technical equilibria that perhaps use agent-based simulations, upper-layer model description methods which apply "above the transport layer in a network stack" (influence rather than physical links), and upper-layer analysis based on quality attributes (response time of applications and the reliability of data synchronization) that subsume technical and social metrics. Bounava[147] identifies state space and nonlinear dynamics models for SoSA and proposes a state-space like augmented model base on node refinement, link refinement, rules of dynamics (for example, for constructing cross-link functions), physical simulation, and iterative refinement.

---

[146] http://ieeexplore.ieee.org/iel5/10498/33257/01571630.pdf?arnumber=1571630, accessed 02/26/2015.
[147] http://knowledgetoday.org/wiki/index.php/ICCS06/227, accessed 02/26/2015.

Requirements Analysis

A formal statement of requirements is readily gleaned from Joint Publications[148]. These are tailored for application to the computer network defense domain. Important points are: systems perspective, account for multiple layers, multiple sides (friendly, adversary, unaligned), and multiple levels (strategic, operational, tactical, and task), use a network or graph representation, and allow a distributed collaborative interactions among experts in the various layers. Requirements for data schema, processing functionality and new visualization (for example, cross-link functions) are included.

Detailed Design

As earlier indicated, we understand the need for SoSA, the network layers, and the war fighter utility. We need an analytical, automated process for specifying how to implement the capability. The process begins with choosing, filtering, and registering the data sets. Vectors defining nodes and matrices specifying links provide the mathematical formulation, cross-links are identified and defined with cross-link functions. Output depends on the selected visualization, metrics, and algorithm chosen. Here are the steps:

1. Choose data sets
2. Filter data sets
3. Register data sets
4. Initial Processing
5. Identify cross links
6. Define cross link functions
7. Select visualization
8. Choose metrics
9. Iterate results

Data Schema

To specify a network we define nodes and links. We used Pajek[149]. Optionally, node labels, node sizes, node shapes, node colors, link weights, link line types, and link colors, along with many other attributes are specified for better description and visualization. However, we see that a minimal amount of information is needed to mathematically specify a network: a vector of nodes (vertices) and a matrix of links (arcs). A single integer number specifies each node, and two integers (from node, to node) specify each directed link.

The node name; for example, "Originator", is followed by three (x,y,z) position coordinates; for example, 0, 1, 0. If we have six layers, we'd put these in a single table by specifying the number of nodes as the sum of the nodes in all the networks. The nodes are numbered sequentially, and positioned at different z-coordinates depending on their layer. For example, the nodes constituting the political layer have a z-coordinate of "0", the military layer nodes

---

[148] Commander's Handbook for an Effects-Based Approach to Joint Operations, Joint Warfighting Center, Joint Concept development and Experimentation Directorate, 24 February 2006. Unclassified.
[149] pajek.imfm.si/doku.php?id=download, accessed 11/24/2014.

have a z-coordinate of "1", and so on. Nodes corresponding to adversaries, friends, and unaligned could be on the same layer, but have different colors: blue for friends, red for adversaries, yellow for unaligned. This could be drawn three-dimensionally to look like a stack with multiple layers of networks.

Links specify connections between nodes regardless of layer (z-coordinate). What is important is that all nodes and links, regardless of layer and side are specified with a single nodes vector and a single links two-dimensional matrix. As we'll see, this means the mathematical formulation of the multi-layer problem is two-dimensional. This does not compromise our ability to provide a high-dimensional data storage or visualization.

Multi-Network Formulation

At first glance the multiple networks that comprise a system-of-systems appear to constitute a high-dimensional space. Each of six network layers for each of three sides produce as many as 18 networks. Each network evolves over time. Each network consists of multiple nodes and links, each of which has associated characteristics. Nodes may have evidence and supporting explanations. How many distinct dimensions (parameters or measurements required to define the characteristics of a system) is that? More importantly, what is the minimum number of mathematical dimensions (the parameters required to describe the position and relevant characteristics of any object within a conceptual space)?

The innovative insight – dimension minimization - is that a large number of distinct dimensions are collapsed into a two-dimensional mathematical space for computation without sacrificing the large number of dimensions that are useful for measurement and visualization. This is similar to Carley's meta-matrix.[150]

Fortunately, the visual representation of complex multi-layered networks need not dictate the mathematical representation of the problem. Networks and fields are most often formulated mathematically as tensors. In computer science, they are called arrays. More commonly:

- **scalar**: tensor of rank 0, a constant.
- **vector**: tensor of rank 1, a one-dimensional array.
- **matrix**: tensor of rank 2, a two-dimensional array.

Our strategy is to reduce the dimensionality of the multi-layered network problem (Figure V-4.2) and formulate it as a solved problem. Importantly, nodes and links in different layers may have common characteristics and even the same labels, but they are different nodes and treated that way.

Matrix Equations

To compute the characteristics of a network, a matrix equation is constructed and easily solved by iteration. The simplest form is: $\mathbf{x}_{n+1} = \mathbf{A}\mathbf{x}_n$, where $\mathbf{x}_{n+1}$ is the vector of node values at the

---

[150] http://www.casos.cs.cmu.edu/publications/papers/2012DNAandORA.pdf, accessed 02/26/2015.

n+1 iteration, **A** is the link (also called adjacency) matrix, and $\mathbf{x}_n$ is the vector of node values at the nth iteration. The matrix equation in Figure III-5.5 is solved in this way.

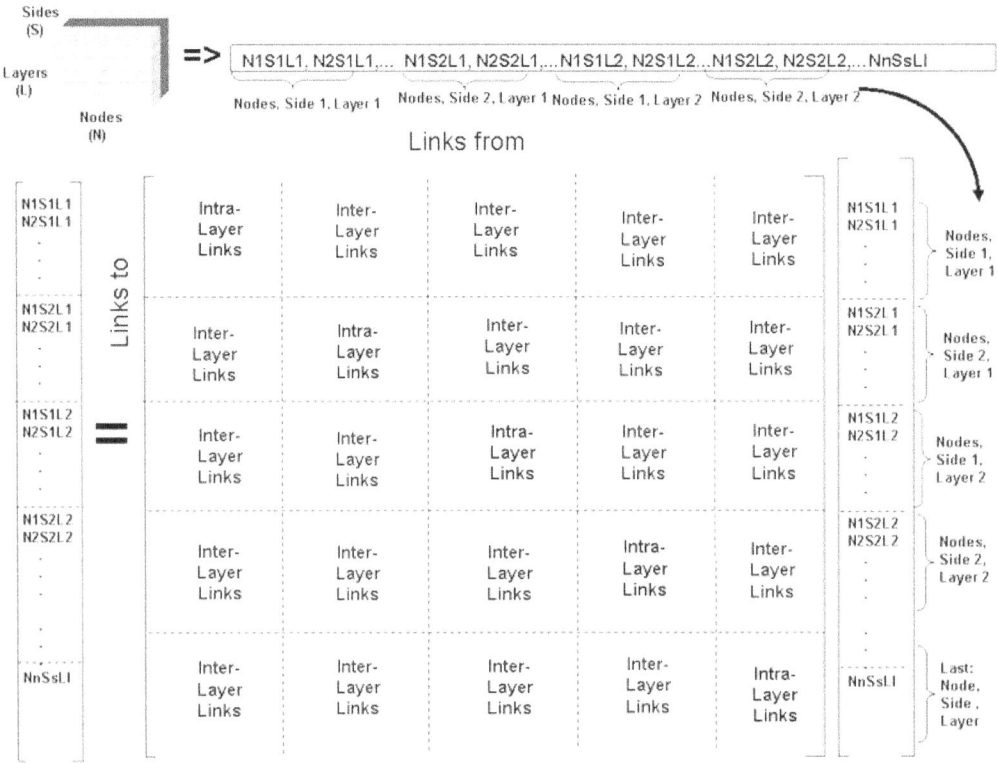

Figure V-4.2: Dimension Minimization

The size of the node vector is the number of nodes. The size of link matrix is the square of number of nodes. The nodes vector is formed putting the nodes of the first network into a column and appending the nodes of the additional networks to the column. The link matrix is formed by starting with the first row and inserting a value in a column if the first node links to it. The is repeated for the "i"th row and the "i" th node. The link matrix is sparse and values cluster on the diagonal, where links within a single network appear. Hence, sparse matrix solution techniques apply. Despite this, the data is stored compactly.

Cross-Link Computation

The 2nd SoSA challenge is determining values for links between layers or sides. These are the off-diagonal sub-matrices. Options range from operator-intensive situation awareness to automated computation.

Cross-link Options	Advantages	Disadvantages
Situation Awareness:		
- visual Indicator	- explicitly indicates a link	- places burden on planner
- qualitative Link	- collaboration shorthand	- difficult to manually quantify
- suggests possible link	- no software to maintain	- no audit trail
-planner decides importance	- reflects current practice	- no consistency check
Template-filling:		
- database interaction	- eases link quantification	- no decision context
- graphic user interface	- provides an audit trail	- difficult to quantify links
- requests a possible link	- allows some consistency	- no real computer smarts
- suggests a link value	- minimizes software	-no reach-back to similar links
Case-Based Reasoning		
- finds similar links	- leverages computer memory	- requires stored cases
- quantifies similarity	- allows knowledge tailoring	- difficult to specify context
- allows refinements	- easier to edit than create	- no support for new links
- learns link computations	- well-accepted practice	- needs knowledgeable user
Agent-based Simulation		
- interacts noes and links	- considers whole system	- need simulation expertise
- based on rule & constraints	- shows ensemble behavior	- link values defy explanation
- links based on equilibria	- allows what-if analysis	- not a primary technique
- allows sensitivity analysis	- war-gaming prelude	- not a "systems perspective"
Cross-link Functions		
- similar to transfer functions	- provides context & meaning	- multi-discipline expertise
- rely on underlying physics	- more intuitive explanation	- models are a challenge
- models the link	- metrics for quality enforced	- doesn't insure equilibrium
- adjustable parameters	- solid basis for link values	- difficult to maintain/extend

The options may be used individually, are not mutually exclusive, and in fact are mutually synergistic. The planner is the ultimate decision authority and accesses automated support to the extent preferred, subject to level of expertise, the granularity of the planning effort, and available time. As an example of how these planning options are combined during a collaborative planning mission, a political analyst suggests a link (Option 1) between a political leader and a social interaction in an Internet forum and partially completes a template (Option 2) with the information available. The planning staff of the supported commander uses an extensive database to retrieve cross-link computations previously used in similar missions (Option 3). The retrieved data has URLs to war-games (Option 4) and underlying cross-link functions (Option 5). The planner can then modify a retrieved case for the current situation by re-running the war-game with tailored parameters and/or reworking the cross-link function model.

Based on this example, actions to specify cross-links are manually initiated, and rely on a case-based approach, with template-filling, wargaming, and cross-link functions providing supporting functionality. Details on each of the options are now provided.

Situation Awareness

Although this function is typically satisfied with icons on a geographic map, for SoSA we augment the map view with a layered network view. As earlier indicated, many tools (NetViz, Pajek) help visualize networks and the SoSA task did not require building new ones. It is however useful to configure the existing tools to show multiple layers and multiple networks sides; for example, a social layer with nodes from a political layer seen from a combined friends and adversary perspective.

Template-Filling

This fairly minimal software capability provides a user convenience. A template allows a database query based on whatever keywords are entered. It contains lists of previously entered links and may suggest possible links. This facilitates link quantification based on experience, and provides an audit trail. It does not, by itself, provide a context for, or a computation of, the link strength and it does not relate it by a quantitative similarity metric to other links.

Case-Based Reasoning

A case is a link calculation, with its associated characteristics that provide context. This (Option 3) cross-link specification finds similar cases in the data base, along with the context within which link strengths are computed. Link similarities are quantified. Successful use of links retrieved from the database actuates a feedback mechanism; for example, Hebbian learning adds one point of "worth" for each successful use of a link case and deducts one point of worth for each unsuccessful use of a link case.

This option allows us to leverage the computer's memory to "remember" a similar link strength calculation and to quantify its degree of similarity to the cross-link of interest. For each case in the database, the similarity metric is computed as a weighted sum of similarities between the desired characteristics and the characteristics of the case in the database. For example, if the desired "from node" layer is the political layer and the case in the database matches this, the score for that characteristic is the highest score. The similarity metric is the same as the scoring in a trade study.

Case-based reasoning also provides a management technique. It registers manually-generated links (Option 1), finds similar template-filling (Option 2) results, obtains feedback from use of selected cases (Option 3), finds similar wargaming runs (Option 4), and identifies similar cross-link functions (Option 5). It brokers stored cases based on desired link characteristics and learns from usage.

Agent-based Simulation

This is the only tool that allows direct understanding of the interplay between friends, adversaries, and unaligned sides of a system-of-systems. The interaction is based on rules and constraints provided to the simulator. Links weights are determined directly from the simulation. We can also do "what-if analyses". Simulation is stated as a "helper technique". It doesn't postulate links, but instead allows links, constraints, and rules postulated by other techniques to be numerically computed and validated. Most importantly, simulation provides data for creating cross-link functions by curve-fitting.

Cross-Link Functions

This is the most difficult cross-link specification technique, and the one with which we have the least experience. Consequently, most of the work on SoSA is done here. A cross-link function is similar to a transfer function[151], a mathematical representation of the relation between the input and output of a (linear, time-invariant) system. The difference is that our cross-link functions may be non-linear and they may vary with time or other parameters.

Current Cross-Link Function Capability: Our belief networks and social network analysis tools (for example, Pajek) allow links that vary between [0,1]. We allow two scalar link values: they quantify the influence that belief and disbelief in a node have on another node. They remain constant for evidence fusion and propagation. Nodes may have multiple links of various (unconstrained) types. Our framework accommodates link specification by layer, side, value, and allows characteristics to be associated with links.

Desired Cross-Link Function Capability: We need cross-link functions that range from constants provided by engineering judgment to model-based functions based on relevant system variables. These are determined offline and stored in a database, but available for analyst tailoring and modification as required by the SoSA planner or analyst.

- **Interpolate Related Links**. In many cases, we will have intra-layer nodes with an established link value that are similar to nodes in another layer. In the example shown, we have social-to-social and a political-to-political link values and we want a social-to-political cross-link value. It is not unreasonable to think that the social-to-political cross-link value will be somewhere between the established social-to-social and political-to-political values.

- **Convert Nodes via Chain Rule.** To find a value for the cross-link between S1 and P2 in the example, we rely on a known or estimated sensitivity: the belief in a specific political node relative to belief in a related social node (Belief P1 / Belief S1). This allows us to (effectively) convert the social node to a political node, and provide a known political-to-political link. In practice, the sensitivity factor multiplies the intra-layer link factor. As indicated, we can also convert Belief P2 to Belief S2 and use a social-to-social link.

For example, suppose we have empirical data that relates reported social satisfaction as

---

[151] http://en.wikipedia.org/wiki/Transfer_function, accessed 07/12/2007.

a function of political funding. Based on a curve-fit to this data, we see that the sensitivity factor is (24% per $M). The curve fit [S = .24 * P] need not be linear.

- **Nomogram.** The advantage of this portrayal is that it simultaneously shows the relationships among all of the layers. The portrayal is extensible to show adversary and unaligned layers as well. As with the two earlier techniques, an assumption is that link values for similar intra-layer nodes are available or can be postulated. A straightforward process is to determine, on average, the strengths of connection between nodes in one layer versus another to provide the heuristics shown in the nomogram. Moving the slider on one layer rearranges the slider position on all others. Note that links are not symmetric: A link to B doesn't imply B link to A. The nomogram gives the analyst deeper insight into the relative strengths among layers.

- **Back-propagation.** This constraint-based approach is an existing belief network capability. This works extremely well in creating belief networks. Suppose the average belief is .57 in the initial network. After adding a node from another layer (with an associated degree of belief) and a link (with unknown value), the analyst feels that this should increase the belief in the top node to .88 and therefore "overrides" the top node by inserting this value. Based on the (averaging) combination rule, this produces a value (.69) for the new link. BNE currently allows multiple links and nodes to be added simultaneously and reconciled by adjusting link values using the back-propagation algorithm from artificial neural network theory.

Summary

This paper provides background on the SoSA problem, identifies a solution strategy, and presents preliminary ideas for separating the bigger objective into smaller sub-tasks. A detailed approach is presented, with emphasis on data schema, multi-network formulation as a two-dimensional problem, and potential schemes for cross-link computation and modification.

Chapter V-5
Defensive Space Control

We focus on two operational objectives. The first objective is coping with information overload. Mostly unstructured information flows into Command Centers from theater. Since unstructured content does not lend itself well to traditional types of automated processing, how does a commander wade through the data, digest it, and act on an operationally meaningful timeline? The second operational goal is Determining Adversary Intent. We collect information on adversary activities. We know when his ships pull into and out of port, we watch troop movements, and when our adversary turns on his radar systems, we know about it. But these are just the adversary's activities. What do all these activities mean? What is the adversary thinking? Specifically, what are the adversary's goals and objectives?

Our scenario (Figure V-5.1) is a Weapons of Mass Effect (WME) scenario: "What is the likelihood that our adversary is going to use a direct ascent ASAT carrying WME?" Our scenario follows the evidence trail from a seaport where an Iroma freighter is unloading what we believe are long range ballistic missiles for transport to an airfield, a known storage site for nuclear material.

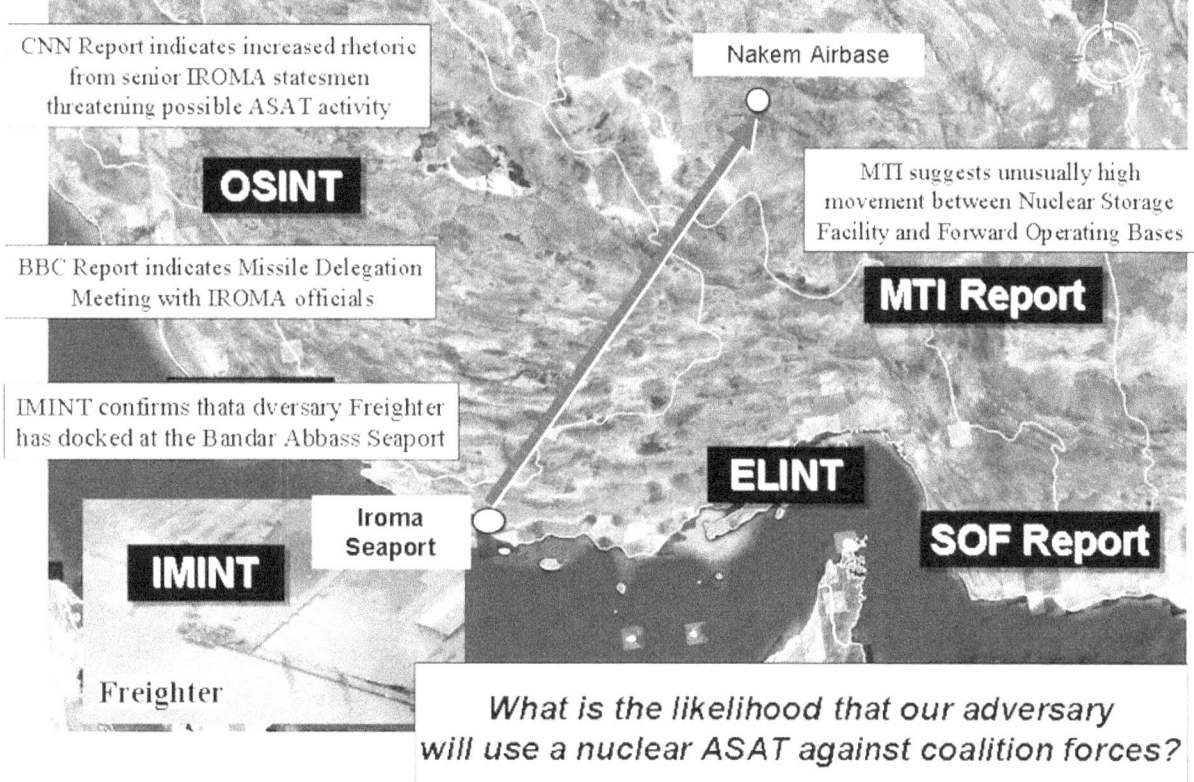

Figure V-5.1: Anti-Satellite Scenario

The next information is Imagery Intelligence. We have an image of the seaport which shows that a rogue nation freighter has docked there. We process the analyst report associated with that image. We also have a Special Operations Force report and an Electronic Intelligence report. Finally, we receive a Moving Target Indicator report stating that there is an unusual amount of movement between the airbase and the nearby Missile Operating Bases. We now believe there is a high likelihood that an ASAT missile event is imminent, and we put out a "high interest event" alert to all participants in the network.

Our processing thread is geared to Intelligence Preparation of the Battle Space. We do this 10 hour job in 10 minutes with our tools by combining what analysts do best with what computers do best. Inputs are unstructured data sources. Processing entails managing the mission, extracting information, and posting a high-interest event. Products include an adversary course of action, resulting in a belief network (Figure V-5.2) which is used to predict adversary intent and data mining to refine the belief network, thereby providing Level 4 data fusion[152].

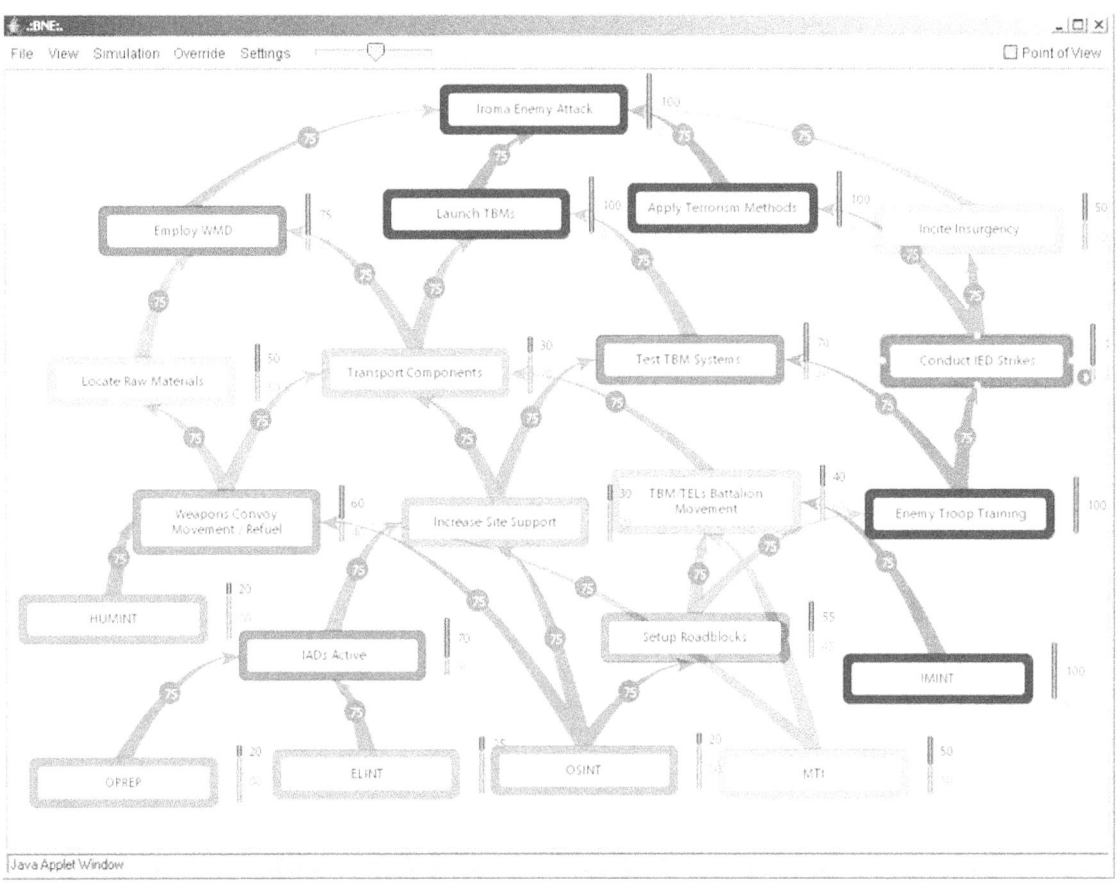

Figure V-5.2: Belief Network for Iroma Attack

Technically, a long processing thread begins with natural language processing on unstructured text. The extracted information is automatically posted to both belief networks and high-

---

[152]https://trac.v2.nl/.../Rethinking%20JDL%20Data%20Fusion%20Levels_B.., accessed 01/20/2015.

interest event maps. Data mining provides new hypothesis - here a LASER dazzler event - as a node to the belief network, This corresponds to JDL Level 4 Data Fusion, or Process Refinement. Integration is achieved using web services in a Service Oriented Architecture. We use the Sonic Enterprise Service Bus to orchestrate three types of C2 modules: desktop applications, applets, and thin clients.

Function	Description	Product/Application	Source
User Interface	Portal with editable content	Liferay	COTS
Mission Manager	Situation Awareness, Application Navigation	Mission Manager Configuration Builder	COTS w/AJAX
High Interest Event Map	Posts icons indicating events to maps	High Interest Event	COTS w/AJAX
Information Extraction	Parses and classifies unstructured text	General Architecture for Text Engineering	Open Source
Anti-Satellite database	Repository for structured data	MySQL	Open Source
Pattern Matcher	Extends rule induction to match patterns	Weka	Open Source, custom add-ons
Planning	Develop Courses-of-Action	Strategy Development Tool	GOTS
Correlation	Uses rule induction to discover patterns	Weka	Open source custom add-ons
Belief Network	Assess uncertainty in adversary	Belief Network Editor	Custom
Replan	Updates plans during conflict	Case-Based Repair	Custom

Chapter V-6
Sensor Fusion

Introduction

The data fusion concept that we describe provides high-confidence detection, localization, and identification of anomalies within background radiation measurements. The approach is novel, using data analytics[153] in combination with uncertain reasoning. Data Analytics leverages historical anomaly patterns and uncertain reasoning mathematically (Dempster-Shafer Theory of Evidential Reasoning) fuses observations and historical patterns to evaluate potential anomalies. Uncertainty in both the historical patterns and the sensor observations are represented as an evidential interval [belief, ignorance, disbelief] that allows explicit statement, and a geographic view, of what we don't know and what conflict exists in the data.

The data sources provide complementary and slightly overlapping data that are mathematically fused to produce high-confidence information on anomalies with low false-alarm rates. We first provide an overview of the technical approach. Multi-strategy fusion shows how evidence produces indicators of an anomaly and how indicators are combined with historical patterns of anomalies to confirm an anomaly. Each box in the fusion network is discussed. A detailed explanation of the underlying algorithms includes a detailed discussion of the fusion technique. A theoretical analysis with supporting data is given. Processing and implementation details are cited.

Technical Approach Overview

Moving platforms collect sensor data (Figure V-6.1). Software formats the data and sends it to a database where it is stored as a flat file. Periodically, historical rules are updated using a rule induction algorithm. Anomaly locations are computed using coordinate transformations. Discrimination is performed to determine what is causing the anomaly using a Bayesian classifier, augmenting rules, and analyst intervention when objects cannot be resolved. Evidence is queried by location, where locations are contained within predefined surveillance grids. Evidence is posted to a belief network for each location and time. Evidence is then fused and propagated to quantify belief that the anomaly exists at that location and time. Maps are updated with icons showing anomalies, and re-shaded to show when the grid is last visited.

---

[153] http://www.forbes.com/sites/huawei/2015/02/24/how-do-big-data-analytics-enhance-network-security/, accessed 02/26/2015.

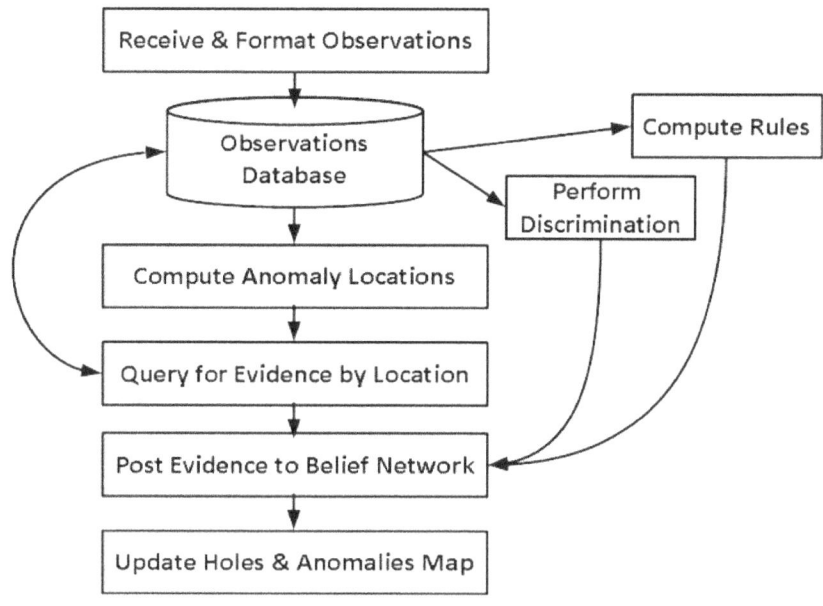

Figure V-6.1: Sensor Fusion Processing Flow

3. **Multi-Strategy Fusion:** Mathematical fusion of uncertain evidence it the primary thrust of the concept. The semantic network (Figure V-6.2), a network with meaning, ingests evidence at the source nodes (in gray), combines it with existing evidence (de-rated over time), and propagates it upward to the top (Confirmed Anomaly) node. Each of these nodes is discussed.

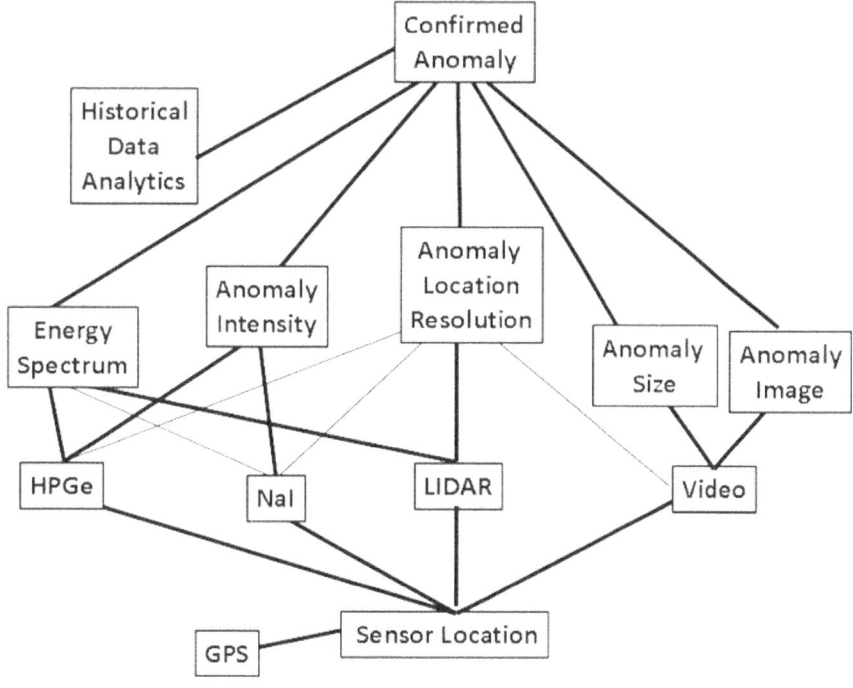

Figure V-6.2: Semantic Network for Data Fusion

3.1 GPS data provides position information (specifically, latitude, longitude, and altitude) and time for each moving sensor platform. GPS data also includes [belief, ignorance, disbelief] associated with GPS uncertainty (military differential GPS is more accurate than the civilian mode) that is supplied as a default value unless degradations have reduced confidence. In the latter case, an analyst adjusts the values.

3.2 Sensor Location: sensors need not be co-located on the same (air, space, land, or water) platform. Data records are formed for each time interval, nominally taken as one minute in duration, with each record containing information from a sensor that recorded data in that interval. A data record consists of the sensor identifier, observation time, and [belief, ignorance, and disbelief] in the accuracy of the sensor location. Given no degradation in GPS data, the evidential interval is nominally [100,0,0], indicating 100% belief in the accuracy of the sensor position with no ignorance or disbelief. This degrades over time on a moving platform if GPS updates are not received and processed.

3.3 HPGe: This high-quality radiation detector provides information on the azimuth, elevation, range[154] of each anomaly detected by the sensor at that time, sensor observables, and [belief, ignorance, and disbelief] associated with the observation. A simple similarity model computes [belief, ignorance, disbelief] as degradations from default performance based on the quality, volume, and resolution, and background signal and noise associated with the observations.

3.4 NaI: This medium-quality detector provides information on the azimuth, elevation, and range of each anomaly detected by the sensor at that time, sensor observables, and [belief, ignorance, and disbelief] associated with the observation. A simple similarity model computes [belief, ignorance, disbelief] as degradations from default performance based on the quality, volume, resolution, and background signal and noise associated with the observations. Anomaly intensity is expected to be nearly as good as that of HPGe, but the quality of spectrum data will be poorer and resolution may also be poorer. These will be reflected in the [belief, ignorance, disbelief] evidential intervals. The link between the NaI node and the Energy Spectrum node will also be weaker, indicating that NaI evidence will not be weighed as heavily. Full detail of data fusion and propagation are given in Section 4.

3.5 LIDAR: This sensor provides excellent information on the location of the anomaly because it is an active ranging sensor. It also provides excellent resolution for closely spaced objects because it has a narrow beam. It may also provide remote, long-range information on the spectrum of the anomaly[155]. A simple similarity LIDAR model provides the [belief, ignorance, disbelief] evidential interval for anomaly location confidence, which, barring performance degradation or spotty reporting, has a default of [99,1,0], indicating very high confidence in anomaly localization, very small ignorance, and no disbelief.

3.6 Video: High-resolution video can also provide anomaly position information, but is much more important as a way to get image and size information for the anomaly, perhaps from a drone. A similarity video model provides the [belief, ignorance, disbelief] evidential interval

---

[154] http://i-hls.com/2013/08/systems-to-detect-concealed-bombs-at-a-distance/ accessed 8/17/2013.

[155] http://www.sumobrain.com/patents/wipo/Remote-detection-radiation/WO2011017410.html, accessed 8/17/2013.

for anomaly location, size, and image confidence, which, barring performance degradation or spotty reporting, will have a default of [99,1,0], indicating very high confidence in anomaly localization, image, and size, very small ignorance, and no disbelief.

3.7 Energy Spectrum: this hypothesis, stated as "at this time, what is the confidence, expressed as degrees of belief, ignorance, and disbelief, that the energy spectrum of the anomaly is accurate". The [belief, ignorance, disbelief] is computed as the link-weighted contributions from (perhaps multiple observations in the same, or different, time interval) up to three sources: HPGe, NaI, and LIDAR. [belief, ignorance, disbelief] values from these sources are combined using the Dempster-Shafer Combination Rule to update the evidential interval for the energy spectrum hypothesis at each anomaly location at each time interval.

3.8 Anomaly Intensity: this hypothesis, stated as "at this time, what is the confidence, expressed as degrees of belief, ignorance, and disbelief, that the anomaly intensity (derived from "counts") of the anomaly is accurate". The [belief, ignorance, disbelief] is computed as the link-weighted contributions from (perhaps multiple observations in the time interval) two sources: HPGe and NaI. [belief, ignorance, disbelief] values from these two sources are combined using the Dempster-Shafer Combination Rule to update the evidential interval for the anomaly intensity hypothesis at each time interval.

3.9 Anomaly Location Resolution: this hypothesis, stated as "at this time, what is the confidence, expressed as degrees of belief, ignorance, and disbelief, that the specific anomaly location is accurately tagged with an object descriptor". The [belief, ignorance, disbelief] is computed as the link-weighted contributions from (perhaps multiple observations in the time interval) up to four sources: HPGe, NaI, LIDAR, and video. [belief, ignorance, disbelief] values from these sources are combined using the Dempster-Shafer Combination Rule to update the evidential interval for the anomaly intensity hypothesis at each time interval. Note that a bold line between LIDAR and Anomaly Location indicates that LIDAR is heavily weighted in the fusion process.

A Bayesian classifier is used as a "helper" algorithm when the association of closely-spaced objects in a video image is ambiguous. Details of the Bayesian classifier are discussed in Section 4. If the Bayesian Classifier does not provide a suitable identifier with probability above a pre-defined threshold level (say 75%), the anomaly is forwarded to an analyst for adjudication. Another option is to launch a drone to confirm the specific location.

3.10 Anomaly Size: this hypothesis, stated as "at this time, what is the confidence, expressed as degrees of belief, ignorance, and disbelief, that the anomaly size is accurate". The [belief, ignorance, disbelief] is computed as the link-weighted contributions from (perhaps multiple observations in the time interval) video. [belief, ignorance, disbelief] values from this source is combined using the Dempster-Shafer Combination Rule to update the evidential interval for the anomaly size hypothesis at each time interval.

3.11 Anomaly Image: this hypothesis, stated as "at this time, what is the confidence, expressed as degrees of belief, ignorance, and disbelief, that the anomaly image is accurate". The [belief, ignorance, disbelief] is computed as the link-weighted contributions from (perhaps multiple observations in the time interval) video. [belief, ignorance, disbelief] values from this source is combined using the Dempster-Shafer Combination Rule to update the evidential interval for the anomaly image hypothesis at each time interval.

3.12 Historical Data Analytics: first, an explanation of this box. Historical data analytics is a hypothesis that summarizes and leverages previous anomaly patterns. It retrieves a rule that is relevant to the current anomaly because of similar energy spectrum, intensity, location, size and image. For example, the database may have instances of previously verified anomalies with the same spectrum and image. These known anomalies provide evidence, based on their similarity, for evaluation of the current anomaly.

This hypothesis, stated as "at this time, what is the confidence, expressed as degrees of belief, ignorance, and disbelief, that historical data analytics is accurate". The [belief, ignorance, disbelief] is computed as the link-weighted contributions from (perhaps multiple observations in the time interval) five hypotheses: energy spectrum, anomaly intensity, anomaly size, and anomaly image. [belief, ignorance, disbelief] values are derived from the accuracy and support of the rule as discussed in Section 4.

3.13 Confirmed Anomaly: this hypothesis, stated as "at this time, what is the confidence, expressed as degrees of belief, ignorance, and disbelief, that there is an anomaly". The [belief, ignorance, disbelief] is computed as the link-weighted contributions from (perhaps multiple observations in the time interval) the six hypotheses shown. [belief, ignorance, disbelief] values from these indicators are combined using a multiplicative rule (since evidence of all hypotheses is required) to update the evidential interval for the confirmed anomaly hypothesis at each time interval.

In summary, an anomaly is confirmed based on historical patterns and indicators that characterize the current anomaly based on fused sensor data.

4. Algorithm Descriptions: Seven types of algorithms are orchestrated to provide the multi-sensor data fusion solution. These are location calculations, map shading, sensor similarity models, statistical transformations, observation association using a Bayesian classifier, data fusion using belief networks, and pattern matching using rule induction algorithms. Location of the anomaly is computed by vector addition of source location and observation vectors. Map shading shows the latency of anomaly detections surveys. Statistical transformations include conversion of historical data from accuracy and support metrics to [belief, ignorance, disbelief] evidential intervals. Belief networks apply algorithms to mathematically fuse data at nodes and to propagate evidence upward to produce indicators and an outcome. Pattern matching consists of performing rule induction on a historical data set and using pattern completion to compute the probability that a current anomaly is valid based on the strength of associated indicators. These are discussed in turn.

4.1 Location Calculations: Given a sensor position consisting of geodetic latitude $\{L_S\}$, longitude (λ_S), and altitude $\{h_S\}$, and the azimuth (A), elevation (E), and range (R) to the anomaly, the anomaly geodetic latitude (L_A), longitude (λ_A), and altitude (h_A) are computed (this is just a sketch of the equations needed) as:

Compute preliminary variables:

Geocentric Latitude: LSc = arctan[(1-e^2] tan LS]
Radius: r ~ hS + RE*[1-.(1/2)*k*(sin LSc)^2 + (3/8) *k^2* (sin LSc)^4]
where k=.0066 and RE = earth's radius, and the radius approximation is accurate to ~ 2 ft.

Convert to position components in a earth-centered, earth-fixed (ECEF) coordinates:
x = r * cos(λS) * cos (LS) , y = r*sin(λS) * cos(LS) , z = r*sin(LS)

Compute position components of the vector from the sensor to the anomaly:
XA = R*cos(λA)*cos(LA) , YA = R * sin(λ A) * cos(LA) , ZA = R* sin (LA)

Add vector components:
X = x + XA , Y = y + YA , Z = z + ZA , R = SQR(X^2 + Y^2 + Z^2)

Convert to map coordinates:
Anomaly Longitude: λA = arcsin(Z/R)
Intermediate calculation: Anomaly geocentric latitude: LC = atan(Y / X) ,
Anomaly geodetic longitude: LA = arctan[tan LC/(1-e^2)]
; Anomaly altitude: hA = R - RE*(1-e)/ SQR[1 + cos^2((1-e)^2)-1]

Map Shading: this novel idea is based on Carl Sagan's observation[156] that "Absence of evidence is not evidence of absence". This notion is especially critical for anomaly detection in background radiation data because these anomalies could indicate a threat. A map is gridded to identify, using colors gradations, the areas that have and have not been recently assessed. The color scheme is as follows:

Blue: 0 <= last survey < 1 day
Green: 1 <= last survey < 2 days
Yellow: 2 <= last survey < 1 week
Orange: 1 week < last survey < 1 month
Red: 1 month < last survey < 1 year
Violet: never surveyed

Note that this is a notional specification to add concreteness and that the times are meant to be easily adjustable. A discussion of "holes" and icons to represent these areas of uncertainty is described, based on previous work

Holes: represent geographic areas on uncertainty. Holes allow analysts to maintain a dynamic status of what they "don't know." For example, the area between the two towns is filled with icons that depict our intelligence holes. Since the holes represent what we don't know, the icons change color when we have performed reconnaissance or gathered intelligence in a particular area. The ground reconnaissance holes disappear as the sensors advance through their position. In addition, the user may eliminate SAR or UAV holes by performing the appropriate type of reconnaissance in an area where these holes exist. When holes are eliminated through reconnaissance anomalies which are present in the location of the hole are displayed.

---

[156] http://c2.com/cgi/wiki?AbsenceOfEvidenceIsNotEvidenceOfAbsence, accessed 8/18/2013

Following surveillance missions, the holes slowly re-appear over time. The color of a particular icon represents the time that has passed since intelligence was performed in a particular area and, therefore, our level of uncertainty, which increases over time.

Sensor Similarity Models

Case based reasoning[157] is used to compute sensor evidential intervals as a function of the characteristics of the sensor system, the environment, distance between sensor and anomaly, and the anomaly being detected. The result is the evidential interval of the pre-stored "case" that best matches the current sensor observation. For example, suppose LiDAR XYZ detects an anomaly at 2000 meters that produces a Signal-to-Noise Ratio of 15 decibels. A case on file (or computed from a performance model[158]) has similar parameters (and the highest similarity score, which is merely a weighted sum of matching characteristics) and is assessed as having an evidential interval for object resolution of [87,10,3]. LiDAR XYZ inherits this evidential interval for this observation based on similarity. HPFe, NaI and video sensor evidence is computed using this same similarity idea.

Observation Association

We propose a semi-automated software system that uses a Bayesian Classifier[159] (BC) augmented by rules and analyst intervention where necessary. The software conversationally elicits responses that define the scope and intent of the mission with follow-on machine learning processing[160] (a BC learns from adjudicated anomalies) with iterations that refine results.

A semi-automated software system with a conversational interface filters (Step 1) anomaly data and invokes a BC that operates on the data to learn to detect, discriminate, and tag an object that is close to other objects. Initially, an information technology professional provides a data base (Step 2). This consists of known anomalies, characteristics of anomalies, and an indication as to whether they are natural, threat objects or suspicious. If more expressive power is desired, objects can be classified with the specific spectral lines. The user inputs a file with the objects of interest into the Bayesian classifier to find anomalies (Step 3). Databases are optionally modified (Step 4). Results are refined through iteration (Step 5)

Step 1: Data Filtering. Create a data set that includes a user-specified set of objects with spectrum, counts, size, location, an image, and a classification (non-radiator, natural radiator, anomaly, unknown, potential threat,.... This information is elicited from the user using drop-down menus that have boxes to be checked, including an "all" box. Based on the log in credentials of the user and associated user data, an Email request, approved by the user, is sent to other partners whose data is required.

---

[157] http://web.media.mit.edu/~jorkin/generals/papers/Kolodner_case_based_reasoning.pdf, accessed 8/19/2013.
[158] http://archive.org/details/nasa_techdoc_19950005977, accessed 8/19/2013
[159] http://www.cs.waikato.ac.nz/~remco/weka.bn.pdf, accessed 8/19/2013.
[160] http://www.cs.huji.ac.il/~nir/Papers/FrGG1.pdf. , accessed 6/1/2013.

Step 2: Construct Data Base: this is a job for a computer science professional. Bayesian classifiers, both freeware and Commercial Off The Shelf, are readily available; for example, Weka provides nine different Bayesian Classifiers[161] (the naive Bayes classifier is recommended). Adjudicated anomalies data is input to the data base as a flat file.

Step 3: Bayesian and Rule-Based Inference. Process the data to find specified types of anomalies. A BC approach accomplishes the processing. "Naïve-Bayes classifiers[162] are simple yet powerful. Its efficiency has witnessed its widespread deployment in real-world applications including medical diagnosis, fraud detection, email filtering and web-page prefetching. One key contributing factor to NB's efficiency is its capability of incremental learning from qualitative data. To accommodate a new training instance, NB only needs to update relevant entries in its probability table. This often has a much lower cost than non-incremental approaches that have to rebuild a new classifier from scratch in order to include new training data." BC provides impressive expressive power, including identification of specific types of anomaly, and a probability of the classification being correct.

Rule-based Augmentation: A BC works by comparing each unclassified anomaly against a database of already classified objects. For closely spaced objects, but not individually anomalous, BC runs into a problem because it does not treat collections of anomalous objects. Fortunately, there is a simple solution: anomalies are preprocessed using simple rules to aggregate for object mixes (if a known non-radiator is in proximity to a radiator, then the anomaly is the radiator), Duplicate anomalies (by counting the number of duplicates), and other such resolution issues. Preprocessed results produce rule-based classifications that replace Bayesian Classification. Results are displayed and the anomalies are exempt from further processing.

Step 4: Data Base Modification (Optional). The experienced user or the Information Technologist modifies existing data sets to create new files. Many BCs use incremental learning, so only additions need to be processed, rather that an entire file. Updates are advertized to partners and included in new software releases.

Step 5: Iteration: Refine the data set under consideration, process using different or modified data, or perform BC using a subset of the variables. Iteration continues, perhaps with entry of missing or corrupted variable values, until desired results are obtained.

Advantages of this method are that it is has expressive power to identify nuances of an anomaly. BC is computationally convenient because the user tailors the software system by checking boxes. BC provides an explanation for suspicious anomalies by explicitly showing the type of anomaly. The BC is augmented with a rule-based reasoner to handle anomalies where a radiator are individually tagged, but the aggregate unresolved object is not. A rule-based reasoner can also tag an anomaly by spectral line detected; for example, plutonium.

Statistical Transformations

---

[161] http://www.cs.waikato.ac.nz/~remco/weka.bn.pdf, accessed 6/1/2013.

[162] http://sci2s.ugr.es/keel/pdf/specific/congreso/LNCS06IncrDiscr.pdf., accessed 6/1/2013.

The evidential interval [Belief, Ignorance, Disbelief] is computed for each piece of evidence from the five evidence sources and the Historical Data Analytics boxes in Figure 2. This evidential interval is obtained from rule induction[163] on historical data and is based on the accuracy and support metrics. Accuracy is the percent of instances that confirm the assertion, and support is the number of instances available. Accuracy provides the mean (m) and the number of samples provides (via a simple equation) the dispersion (d). For example, in a survey, a candidate may receive 35 % +- 3% margin of error of the endorsements: m=35, and d=3. The evidential interval is the set of fractional values for Belief (B), Ignorance (I), and Disbelief (D) such that B + I + D = 1. These are computed from the mean and dispersion as follows:

$$B = m - d, \quad I = 2 * d, \quad D = 1 - B - I.$$

Data Fusion

Fusion of evidence provides a focused solution in cases where a decision maker is faced with discrete bits of information that are inherently uncertain. The Dempster-Shafer Belief Network algorithm is the basis for my approach to data fusion. Advantages are that it: 1) provides intuitive results; 2) differentiates belief, ignorance, and disbelief; and 3) resolves conflicts and 4) allows data aging (decreases in belief and disbelief over time increase ignorance). Insights into the behavior of the belief network are obtained from an analytical and a Technology Readiness Level[164] 4 software implementation. Mathematical and empirical properties are explored and codified. Prototype software is the Belief Network Editor (BNE), as discussed in Chapter III-8.

Belief Network Display Interface: An important aspect of the test bed is providing a visual method (Figure V-6.2) for determining whether the prototype is behaving as predicted or desired. To this end, we developed graphical displays showing the results of the data fusion process.

Displays that present abstractions of belief networks are misleading. The assumptions inherent in the model are not explicitly shown. This is a dangerous over-simplification in belief networks because the user relies on the graphics to make potentially risky decisions. We designed the displays to show explicit belief and disbelief for all nodes. Links vary in thickness to show the degree to which they influence nodes. Automated explanations for nodes and links are also available.

The work described has broad applicability to decision making in circumstances where evidence is uncertain, incomplete, possibly conflicting, and arrives asynchronously over time. The Dempster-Shafer theory of evidence has been found to have very useful mathematical properties - in particular, an inverse has been discovered that allows fusion equations to be solved arithmetically. In addition, the derived properties of these belief networks collectively suggest intuitive application of the technique as a general-purpose fusion engine for uncertain

---

[163] Converts structured data to IF, AND, AND,... THEN rules, for example, www.cs.waikato.ac.nz/ml/**weka**/). The supervised variable, which provides the solution, is Anomaly[true, false].

[164] http://sourceforge.net/apps/mediawiki/gridlab-d/index.php?title=Technology_Readiness_Levels, accessed 8/19/2013.

reasoning. A novel feature of our implementation is the addition of a back-propagation algorithm that allows the user to override fused belief and disbelief in nodes or link values. The back-propagation algorithm adjusts precursor node and link values to reconcile the network. Thus, the network learns from user training data in the form of overrides.

Pattern Matching

Inductive reasoning produces general rules from a set of specific instances. It is one of the most powerful data mining techniques. Arguably, the rule induction algorithm is the "killer application" that got the field of data mining started. One of the earliest algorithms is Quinlan's C4.5 rule induction algorithms. we use a later version, dubbed J48, and available from the Weka data mining library. The algorithm automatically discovers rules in structured data sets. It is a supervised learning algorithm: the user specifies the variable that is the result of the rule.

Decision trees algorithms are based on divide-and-conquer approach to the classification problem. They work in a top-down manner, seeking at each stage an attribute to split on, that separates the classes best, and then recursively processing the partitions resulted from the split. The basic rule induction method, the Prism method, is described in the book by I.H. Witten and E. Frank[165]). We use the Weka rule induction algorithm cited in Witten and Frank and discussed in Chapter III-9.

Rule Induction generates only 'correct' rules, measured by the accuracy formula p / t. Any rule with accuracy less than 100 percent is incorrect, since it assigns examples to the target class in question that do really belong to the target class. The Prism method continues adding clauses to each rule until the rule is correct. The outer loop iterates over the classes, generating rules for each class in turn, re-initializing each time to the complete training set of examples. The algorithm always starts with an empty rule which covers all examples, and then restricts it by adding new conditions (attribute-value pairs), until it covers only examples of the desired target class. At each stage of adding the condition to the rule, the best attribute-value pair in terms of p / t (accuracy) is chosen. If there are more attribute-value pairs with the same value of p / t, then the one with greatest coverage is chosen.

Operational Analysis

A scenario describes a use case as a concrete example of how this concept works in practice. The setting is the city in Iroma. The year is 2016. Allied forces are sweeping the city, including the U.S. Embassy, to detect and characterize sources of radiation. Six HMMWV's (Hummers) drive along primary roads twice a week to detect anomalies; that is, unanticipated sources of radiation that could be nuclear devices.

The Multi-Strategy Data Fusion (MDF) system is undergoing field trials. Sensors are all colocated on the moving HMMWV's, along with client-side software that provides tasking updates and a means of transmitting observations. The tactical command center hosts the server, database, and the processing algorithms. Alternatively, observations can be preprocessed on the

---

[165] http://www.cs.waikato.ac.nz/~ml/weka/book.html, "Data Mining: Practical Machine Learning Tools and Techniques with Java Implementations", accessed 8/23/2013.

vehicles to reduce observation transmit bandwidth, with all but the historical data analytics done on the HMMWVs.

Raw observations are filtered on the HMMWV and those with background radiation below a certain pre-specified level are not reported. Each radiation observation consists of a time, a sensor type, sensor location (derived from GPS data), observation vector (range, azimuth, and elevation of the anomaly), counts versus background energy (for HPGe, NaI, and LiDAR (with modifications) versus observation time. LiDAR and video data contains a time, a sensor type, sensor grid location, observation vector, and image (video) or backscatter cross section (LiDAR) versus observation time. Evidential intervals are formed for all observations based on historical patterns of the underlying uncertainty, as discussed earlier.

The rule induction algorithm is executed periodically (for example, every 4 hours) to update the historical patterns of anomalies. This insures that the rules reflect current situations. An example rule is: IF sensor = HPGe AND grid#= 283 AND Image = building AND Counts = High, AND Spectral Line = Plutonium THEN Anomaly = True (532/540), which means 532 instances of the 540 instances meet the rule. The observations database has a fixed schema and is searchable based on pre-specified, but flexibly alterable, queries. The database also contains fused data on the confidence related to anomalies and historical patterns.

Anomaly locations for above-threshold radiation readings are computed and the database is queried for previous information, if any, on this potential anomaly. Evidence is provided to the belief network and fused with existing evidence. For example, suppose the historical database on NaI radiation detectors establishes an evidential interval of [belief = 92, Ignorance = 8, Disbelief = 0] for the current evidence (that an object at the specified location and time is valid). Further suppose that the location has not been visited lately and that the evidential interval is [belief =0, ignorance = 100, disbelief = 0]. This is trivially fused with the current evidence to yield [92,8,0].

Evidence is then propagated from the NaI node upward where it is multiplied by link weights and fused with evidence about Energy Spectrum, Anomaly Intensity, and Anomaly Location. The fusion and propagation continues until the top box (Confirmed Anomaly) is updated. A typical result of a successful anomaly evaluation is that the Confirmed Anomaly confidence is [96,2,2] which means we have a 96% belief that the anomaly is present, with 2% ignorance, and 2% disbelief.

Processing Time Estimate

So, how long will it take to process the observations from initial sample to final result? The process is fully automated, except for human intervention, to perform localization based on a video feed or discrimination based on a LiDAR signature. Each of these is expected to take about 30 seconds and this will dominate processing time. All other processing, for an anomaly at a single location, takes less than 30 seconds on a commodity personal computer. This includes communication with the server for mapping and historical data analysis. A processing time budget provides an estimate based on experience with the Automated Reasoning and Information Extraction Software for a homeland security remote cargo inspection mission.

Summary

A concept for solving the moving vehicle sensor fusion problem is presented. Novel aspects of the concept are:

- Evidential reasoning deals with sparse, uncertain, and sometimes conflicting data. Belief in evidence is de-rated over time as the evidence gets stale
- Map shows holes where surveillance is not current
- Radioactive sources are remotely identifiable from LiDAR
- Historical data is used to advantage: to double-check anomaly evidence with historical data analysis and to set evidential intervals based on similarity to previous measurements.

The concept (Figure V-6.3) combines GPS position data with sensor measurement vectors to localize the anomaly, combines HPGe, NaI, and LiDAR radiation data to characterize the spectrum and intensity of the anomaly, fuses LiDAR and video to resolve closely spaced anomalies, leverages video to get an image, and verifies an anomaly by combining spectrum, intensity, size, location, and image information with historical patterns.

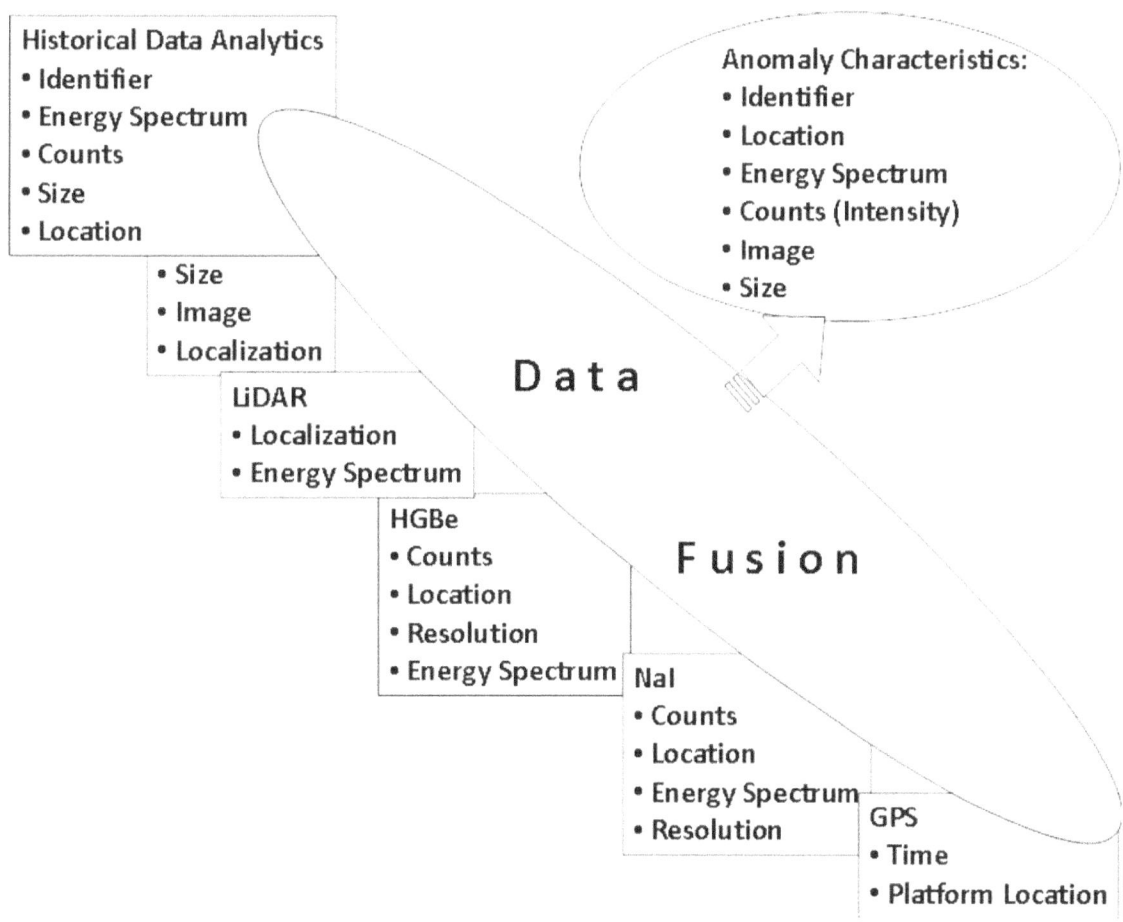

Figure V-6.3: Data Fusion Summary

Chapter V-7
Computer Network Defense

Our goal is to describe a decision loop consisting of automated decision support tools to close the Monitor, Assess, Plan, and Execute (MAPE) loop faster than our adversaries. We execute counter operations to avoid a distributed denial of service attack (DDOS) faster than the adversary can respond. Level 1 through Level 4 Data Fusion is also described. For example Level 4 Data Fusion combines data mining (Figure V-7.1) and data fusion to achieve Joint Director of Laboratories Level 4 Data Fusion[166] (Process Refinement) by discovering and adding new nodes and links to the Belief Network.

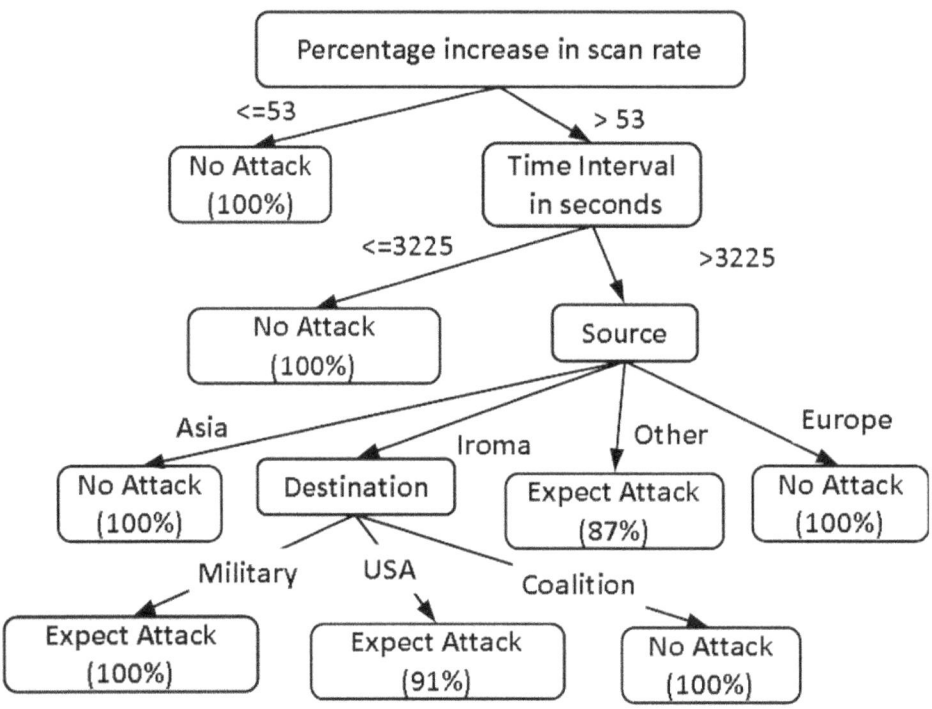

Figure V-7.1: Rule Induction Provides Attack Patterns

Introduction

The scenario is Computer Network Defense. We begin with routine operations and the MAPE loop. We monitor, assess, plan, and execute to get products in place so that we can respond quickly during crisis. The result of routine operations is to task for additional information whereas the result of quick reaction operations is to avoid a DDOS attack by executing counter measures.

We monitor the situation from a web portal. This portal shows high interest events we are occurring and allows us to launch applications. Now moving to situation assessment, we have a

---

[166] http://en.wikipedia.org/wiki/Data_fusion, accessed 02/26/2015.

very powerful Rule Induction data mining algorithm. This is the first major discovery algorithm to launch the field of data mining, approximately 25 years ago.

Rule Induction uses structured data from our data set. The Greedy algorithm discovers rules in the form of "if and then" rules. Now for a more intuitive view we discuss the rule tree. The highlighted rule states that if the Percentage Increase in Scan Rate is greater than 53% and the Time Interval is greater than 3225 seconds, and the Source is Iroma, and the Destination is Military, then Expect an Attack, based on 100% of the data records. Please note that in this tool and others, you'll see that we keep track of uncertainty.

Shifting gears, let's talk about planning. We have the Case Based Planning (CBP) tool that allows us to view a directive from higher authority. We then compare the words in this unstructured text with the dictionary words that are of interest. The results (Figure V-7.2) are the words shown in the six categories on the right: Region, Organization, Offensive, Defensive, Frequencies and Bandwidth. Also, the analyst has an option to add additional words. Suppose we find similar off the shelf plans and we see that the first case is 96.7% similar based on comparison of words in the directive and words in the on the shelf plans that are reflected in the dictionary. We liken CBP to an engineering trade study, the plans are the options and the matching words are the selection criteria. The highest score is a weighted sum. We select the suggested plan with the goal to edit it to the current circumstances rather than creating a new plan.

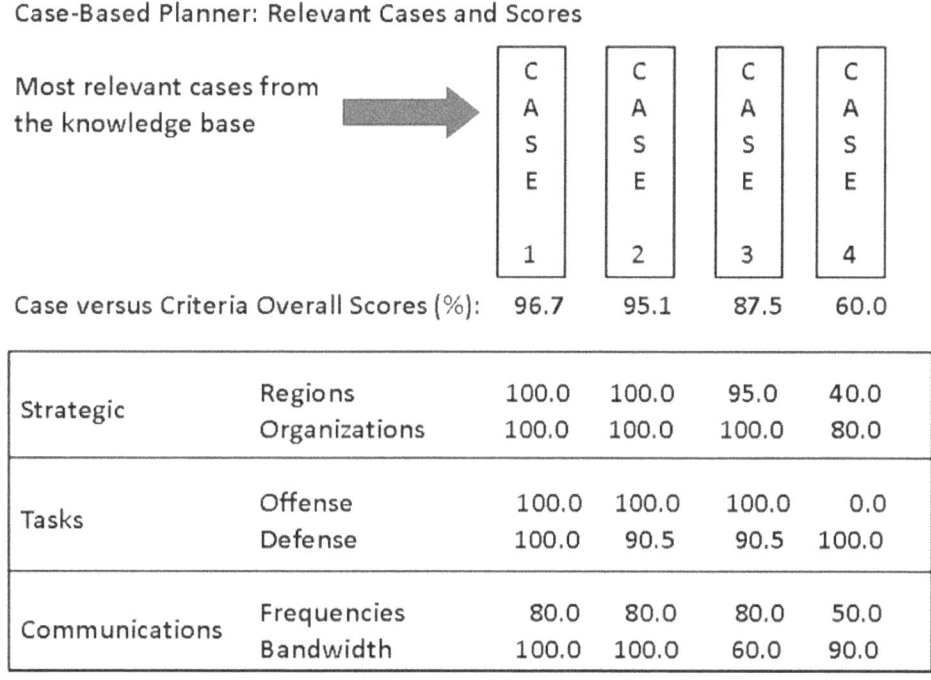

Figure V-7.2 Case-Based Planner

The plan that best matches the directive is built using the Strategy Development Tool, a Government-off-the-shelf product that is used at Information Warfare Planning Capability and throughout Theater commands. The mission is Computer Network Defense. Hierarchically, we have three courses of action. Each COA has objectives, affects to achieve, measures of effectiveness, and tasks. Templates provide slots for adding context, for example the description, the point of view and the command relationships.

As a convenience, we automatically port this plan to PowerPoint for approval. The PowerPoint briefing is customized to the decision makers preferences. We show the name of the plan, facts, and assumptions. We also show the analysis and Commander's intent and extract the first five COAs. This simple tool saves time, provides consistency, and minimizes transcription errors.

Courses of Action

- COA #1: Computer Network Defense
- COA #2: Restore Surveillance Network Readiness
- COA #3: Stop Distributed Denial of Service Attacks

We can also specify timing planned and actual assessments. These tasks will later be used for dynamic plan updates. From a timeline view we can see how the tasks are scheduled. This is important for de-conflicting or synchronizing cyber tasks to be sure they do not interfere with one another or interfere with mission operations.

The last module to discuss is the COG Articulator. This is produced by a Subject Matter Expert with the goal to predict adversary intent and to show in a nodes and links fashion how we intend to respond. An alternative view, in the form of a belief network (Figure V-7.3), is electronically generated from the COG Articulator.

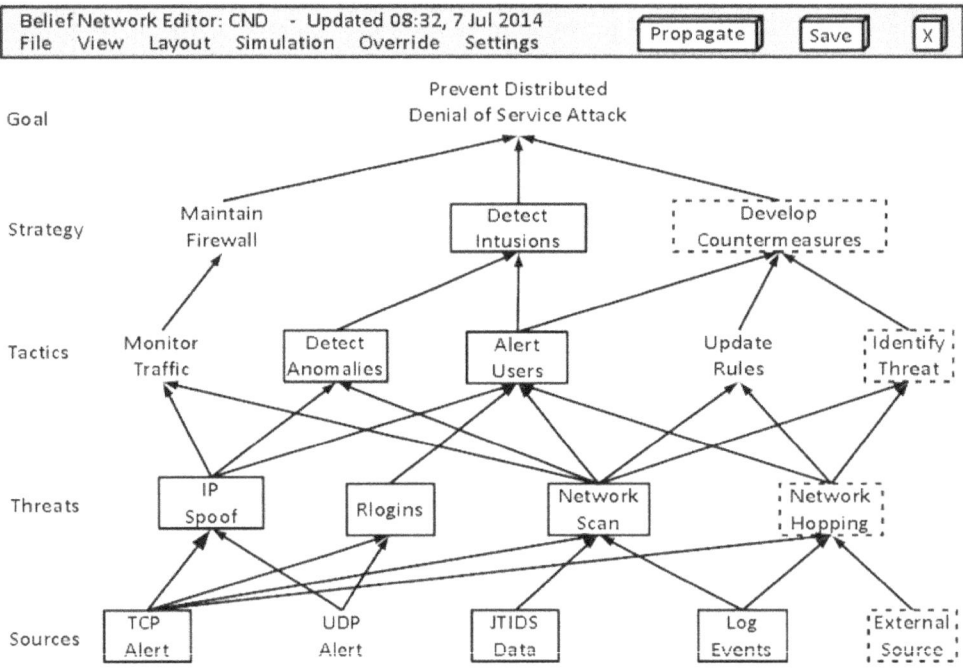

Figure V-7.3: Computer Network Defense Belief Network

The Belief Network view shows the same information. Along the bottom we have five sources of evidence. These nodes reflect our degree of belief, unknown, and disbelief. As evidence arrives it is fused with the bottom nodes. The evidence propagates to the top node, Prevent DDOS Attack

The evidence ages over time based on a persistence value more specifically both belief and disbelief give away to ignorance. For example, wisdom may last forever, but the weather may only hold for four hours. At the Identify Threat node we find a COA recommendation which results in additional information about the status of a network and recent network anomalies. By clicking on the node and choosing COA recommendation, we update the database and execute that plan update and the action is logged.

We have now traversed the routine operations outer loop (Web Portal to Monitor, Rule Induction to Assess, Case-Based Planner to Plan, and COAs to Execute) and now we go to the quick reaction inner loop. Let's begin with Information Extraction. This tool uses natural language processing, a general architecture for text engineering called GATE. We have put a modern user interface (AJAX[167]). We chose a cyber event called IP Spoof. We parse the who, what, when, where, why, and how; detailed reporter questions. Most importantly the hedge words, "unknown" and "probably". To get the overall tone of the message we average these to provide a default pedigree of the evidence, at least from the author's perspective.

Extracting the information really isn't enough to help the analyst so now we automatically post an icon to a map based on extracted information to Monitor the situation. We will also automatically update a Belief Network. First, the map has a High Interest Event tool to monitor quick reaction operations. We publish an event and apply the extracted characteristics to an outbound message. Accept the message in the High Interest viewer and pops up on the timeline. As the counter reaches that time the icon appears. This is a special icon that has many different characteristics of an event or activity. It has three dimensions of position, speed, time, time duration, description, label, color identifying class of activity (red = adversary), the size indicates importance, blink indicates urgency, and importantly the shape indicates uncertainty.

Let's now revisit our assessment. Here we are going to leverage the correlation; that is, the relationships we earlier found, and do pattern matching (earlier in Figure V-7.1) with incoming evidence. So in particular we've identified the source of these attacks as Iroma, the destination is a military network, and two important parameters: the percent increase in our adversary's scan rate and the associated time interval being over an hour. We see that the pattern match mode, which by the way we've added to this Rule Induction algorithm, indicates that based on past reports we should expect a DDOS attack with a 100% probability because in this simple example all previous cases met this rule.

Given this assessment that a DDOS attack is imminent we shift to our Belief Network. We post the Assess Threat evidence to the Belief Network. As we use a Bayesian classifier, also from the Weka tool set, we update the Assess Threat with a mix of belief and disbelief. Next, we execute an effective countermeasure and as time goes by our analyst receives evidence that the

---

[167] http://en.wikipedia.org/wiki/Ajax_(programming), accessed 02/26/2015.

countermeasure is successful and we then update the node to show the countermeasure is successful by changing the belief value. And now we have thwarted a DDOS attack.

Chapter V-8
Uncertainty Management –Army Scenario

This chapter details the scenario and software tools that demonstrate the effects of uncertainty in decision making. The objective of the Uncertainty Management work is to develop an interactive scenario to demonstrate the concept of "holes," geographic areas of ignorance or uncertainty, as they apply to decision making and uncertainty management. To demonstrate this concept the following objectives are also derived:

- A means of visualizing uncertainty or holes
- A means of controlling all scenario actions
- Creation of a scenario to be loaded for demonstration
- Implementation of user actions to obtain intelligence and alleviate uncertainty
- The integration of a belief network and the ability to dynamically adjust the belief and disbelief values of each node
- The integration of a Kiviat diagram to visually assessing dominant uncertainties
- The creation of a path replanning tool to control the movement of troops
- Creation of a schedule evaluator tool to calculate the probability of success based on scenario data and uncertainty
- Integration of information extraction tools and the creation of supporting messages

To meet these objectives, a scenario is developed and software tools are created and integrated to create an interactive platform for the demonstration. The following sections discuss the scenario itself, the software implementation and features, and the objectives which are achieved in the development of this demonstration.

Scenario

A scenario illustrates the concept of holes in uncertainty management. Events occur in the fictional country of Iroma. The scenario depicts operational maneuvers of a joint, interagency, multi-national force, which is charged with securing Seaport of Debarkation (SPOD), defeating enemy main force units, opening a route for relief supplies, conducting stability operations, and providing security for s relief convoy against enemy attacks.

The friendly or blue troops include four brigades and one command unit. The brigades include the following: 2nd MEB, 3/2 ID (SBCT), 1/25 ID (SBCT), and TF Merlin

The 3/2 ID deploys from the south and performs the main effort of the task. The 1/25 ID deploys from the north. These two brigades are charged with securing their routes and rendezvousing at a previously determined location. In route, the 3/2 ID encounters mines, an enemy ambush, and bridges destroyed by enemy troops. The goal of a rendezvous between the northbound 3/2 ID and the southbound 1/25 ID remains the same. In addition, troops encounter enemy troops, mines, and destroyed infrastructure along their routes. The primary difference is that this demonstration also depicts the effects of uncertainty on decision making. Uncertainty is represented as a "hole" or a geographic area of ignorance.

The southern brigade (3/2 ID) begins to move north, as the northern brigade (1/25 ID) begins its movement south on pre-specified routes.

Holes

Represent geographic areas of uncertainty as map overlays. Holes allow analysts to maintain a dynamic status of what they "don't know." In this demonstration we represent three different types of uncertainty, or holes. These are:

- Ground Reconnaissance Holes
- Synthetic Aperture Radar Holes (SAR)
- Unmanned Aerial Vehicle Holes (UAV)

The area between the two moving brigades is filled with a number of these semi-transparent icons that depict our intelligence holes. Since the holes represent what we don't know, the icons disappear when we have performed reconnaissance or gathered intelligence in a particular area. The ground reconnaissance holes disappear as the brigades advance through their position. In addition, the user may eliminate SAR or UAV holes by performing the appropriate type of reconnaissance in an area where these holes exist. When holes are eliminated through reconnaissance, enemy troops and infrastructure which are present in the location of the hole are displayed.

Following reconnaissance missions, the holes slowly re-appear over time. The opacity of a particular icon represents the time that has passed since intelligence was performed in a particular area and, therefore, our level of uncertainty, which increases over time. It shows an area where troops have passed through, and therefore where ground reconnaissance has been done. We can see the various opacities, and therefore uncertainties, of the holes as well as the enemy troops and mines revealed in the area.

Software Implementation

The scenario described above is implemented in software using a variety of tools. The main program, known as the Dashboard, displays a visualization of the scenario described above as well as tools for a number of different user actions.

Dashboard

Contains the main program functions (Figure V-8.1). The center of the screen contains the Situation Awareness geographic display which depicts a visualization of the scenario and intelligence holes. The Dashboard also contains a number of control and interactive user features including a schedule evaluator, a kiviat diagram, a belief network editor, status bars, and list of available user actions. These features are described individually in the sections that follow.

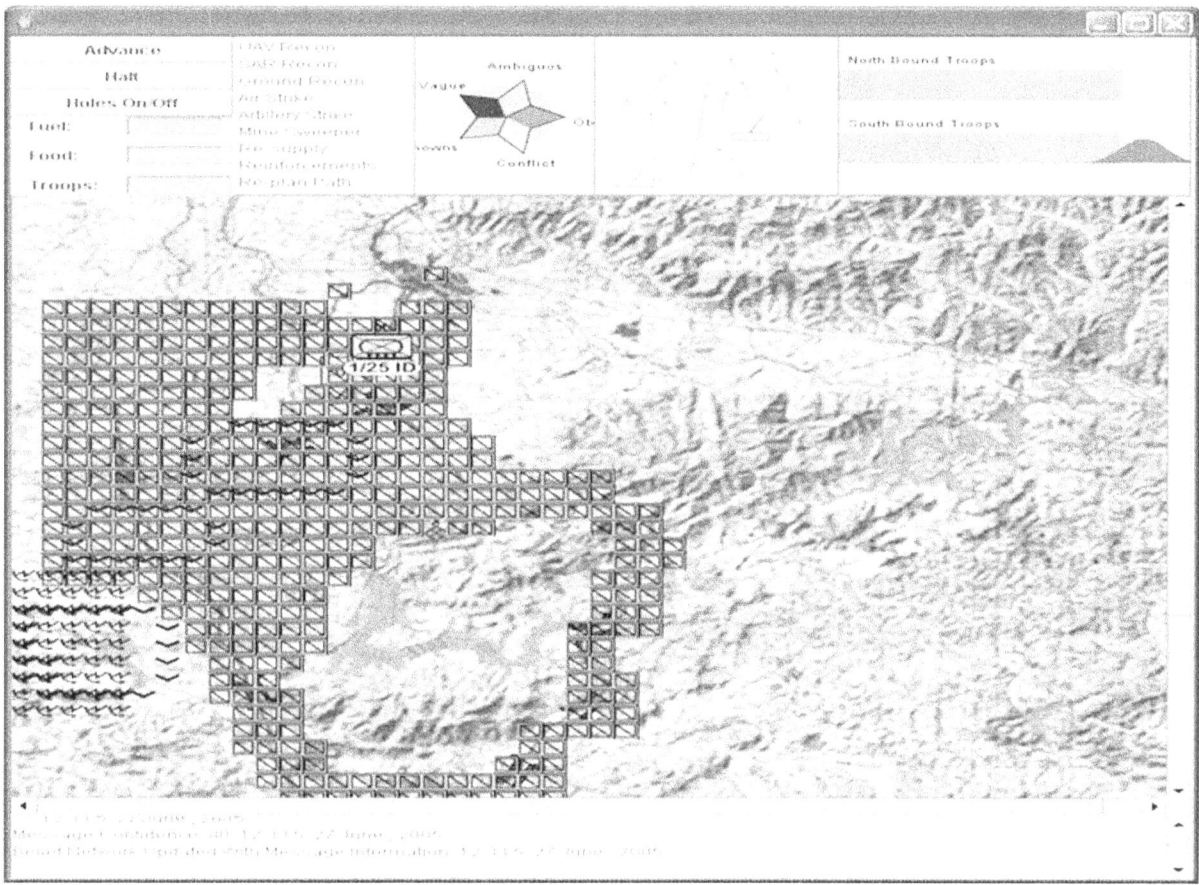

Figure V-8.1: Dashboard for Army Scenario

High Interest Event

This geographic display is seen in the center of the Dashboard screen. The display provides visualization of the holes. All elements of the scenario: blue troops, red troops, infrastructure, holes, etc., are displayed in the window. Each scenario element is represented as an icon defined in GIF files. The functionality of the display allows us to dynamically move these icons to a specified position throughout the scenario and to, in the case of holes, dynamically adjust their opacity.

In addition to the visible scenario objects, grids also provide elevation, water, and road information for the terrain being traveled by the brigades. This information is used in the calculation of the brigade's probability of arrival at the link-up point.

Scenario Controller

The actions and movements of the scenario components are managed by a scenario controller. The controller manages all scenario activities including troops movement, user controlled actions, and holes visualizations. A timer is created in the controller. At each time interval all of the grids representing visual objects are updated. Each grid contains its own update method specific to the types of objects contained in it. Any grid updates which result in icon movement or appearance changes are reflected in the High Interest Event.

Troop Movement

Only two brigades in the scenario, the 3/2 ID and the 1/25 ID, move throughout the scenario. These troops move along a pre-set path towards the link-up point, with the 1/25 ID advancing from the north and the 3/2 ID advancing from the south. The path of the 1/25 ID remains unchanged throughout the scenario. However, the path of the northbound 3/2 ID unit is dynamically changed by the user using the Path Replanner tool. By default, the troops move one step along their path every time step, as dictated by the Scenario Controller. However, their speed can be changed depending on landscape elements and enemy combatants they encounter.

User Controls

A number of user controls allow the user to dynamically control the actions of the northbound troops, the 3/2 ID brigade. The user accesses these action controls by clicking on the buttons and/or list located in the top left corner of the dashboard.

The following actions are available to the user:

- **Simulation Stop:** The keystroke CTRL-S pauses and re-starts the entire scenario.
- **Advance/Halt Buttons:** These buttons control the movement of the brigades. As the names imply the Halt button causes the movement of all blue troops to stop and the Advance button causes them to start up again. Even though the blue troops are halted, all other activity including holes updating, enemy position updates, and scheduled messages will continue as normal.
- **Holes On/Off:** This button turns the holes icons ON or OFF. The user may wish to turn the icons off to more clearly see the map and troop paths. However, even when the holes are turned off enemy troops and infrastructure that have not yet been uncovered by reconnaissance are not displayed.
- **Ground Recon:** Upon "halting" the brigade advancement a scouting unit can be sent out from the position of the 3/2 ID. Clicking on this option, and then clicking anywhere on the map results in a unit icon being displayed. The unit icon moves in a direct line across the map until it reaches the point selected by the mouse click. As this icon moves through areas that contain ground reconnaissance holes the holes disappear. Following reconnaissance, the holes reappear, or become more opaque, as time passes, indicating an increasing level of uncertainty.
- **Unmanned Aerial Vehicle Recon:** This works on the same concept as Ground Recon, the difference is that when this action is initiated, a mouse click on the area of UAV Holes is needed to setup the path of the recon. Clicking on a point to the upper left of any of the UAV Holes icon sets displays a UAV Path icon along with a UAV Drone icon. The Drone icon then follows the path and clears the UAV Holes in that area, thus uncovering any enemy presence within its path.
- **Synthetic Aperture Radar Reconnaissance:** As with the first two types of reconnaissance this button initiates the SARS Recon action. A mouse click on the area with SAR holes places the icon that simulates a satellite pass-over and clears the related

SAR holes in the area, with the same result as the first two reconnaissance actions.

- **Air Strike:** After initiating this action a mouse click on some portion of the map (preferably on a known enemy position) creates an air strike action. To indicate that this action is being performed, the air strike icon then moves from the command position in the south to its target. Any enemy located at the indicated attack position is destroyed.
- **Artillery:** After initiating this action a mouse click on some portion of the map (preferably on a known enemy position) creates an artillery strike action. To indicate that this action is being performed, the artillery strike icon appears at the target position. Any enemy located at the indicated attack position is destroyed.
- **Minesweeper:** After initiating this action a mouse click on some portion of the map creates a minesweeper action, which is originate at the 3/2 ID brigade and moves via a straight line to the destination position indicated by the mouse click. To indicate that this action is being performed, the minesweeper icon moves across the map to its destination position. Any mines located in the path of the minesweeper are destroyed.
- **Resupply/Reinforcements:** These actions are also meant to be used in conjunction with halted brigades. Initiating these actions results in supplies or reinforcements, respectively, being sent from the command post to the location of the 3/2 ID brigade. The actions are indicated by resupply and reinforcement icons which travel across the map from the command post to the location of the brigade. These actions affect the food, fuel, and combat power status of the brigade, as indicated by the progress bars in the top left of the Dashboard screen.
- **Replan Path:** This action works in conjunction with the brigades being halted. When this action is initiated the Path Replanner tool is displayed in a new window. After the path has been set in the Path Replanner tool, one must click on the Advance button to move the troops along the new route.

All of the actions described above create an event in the controller. The handling of these events is specific to the action performed, but all are handled in the Scenario Controller.

Holes Visualization

In addition to movements, the Scenario Controller also maintains the state of each of the holes. The holes are set by the controller to become more opaque over time, and to become invisible if reconnaissance is performed. The opaqueness of each hole icon is directly related to the amount of uncertainty associated with the geographic location the hole represents. The more opaque a hole icon is, the less we know about that particular area.

Schedule Evaluator

The top right area of the Dashboard contains a Schedule Evaluator. The Schedule Evaluator contains two progress bars, one for the southbound 1/25 ID brigade and one for the northbound 3/2 ID brigade. The distance that each of these troops are from the link-up point is indicated. In addition, the probability of the each brigade's arrival at the link-up point is indicated by the mean and variance of the Gaussian curve on the right side of each progress bar.

Methods are identified to dynamically change the mean and variance of the Gaussian curves at each time-step. These calculations take into account the geological features, enemy troops, and infrastructure in the brigade's path. They also account for uncertainty if holes exist in the brigade's path.

Path Replanner

This is a tool for specifying and dynamically changing the path of a brigade in a given grid structure. The Path Replanner tool is initiated by the Path Replan user action as described in the previous section. Once the Path Replan action is initiated, the Path Replanner tool appears in a new window. When initiated, the Path Replanner display appears and shows the current path of the 3/2 ID brigade fixed on a grid and overlaid across the map. The brigade's current path is indicated by shaded grid blocks. The current position of the brigade is indicated by the last shaded grid block in the series. The mouse is then used to click on grid spaces to indicate a new route. Clicking on the Accept button confirms that the brigade will follow the new course. Once the Accept button is pressed and the Path Replanner tool is closed, pressing the Advance button starts the brigade moving along the new path.

Scenario Development Tool

To develop the scenario, we need to create the grid layers specified above, as the data for each grid layer is read in from data files at the initialization of the scenario. To create these data files, the Path Replanner tool is augmented as a stand-alone program which captures the coordinates of the grid spaces highlighted by the user along with the properties associated with that particular location.

This data file development tool works in the following manner. The user clicks on the radio button associated with the grid he wishes to specify, that is, elevation, water, or road. For the elevation and water types, the user selects from the list indicating elevation change or water depth. Then, the grid space clicked on in the map area takes on the type and properties indicated by there choices. The Save button records this data to a file. In this manner, a user specifies the properties for each grid layer in a data file.

Belief Network Editor

A small belief network image also appears at the top of the Dashboard screen. Clicking on this icon creates a new window (Figure V-8.2) containing the belief network developed for the demonstration. The Belief Network Editor (BNE) is developed separately from this project and is integrated here in its fully functional form. So, the user can add and remove nodes, change belief, disbelief, and uncertainty values, change links, and run simulations as if the BNE is a stand-alone program.

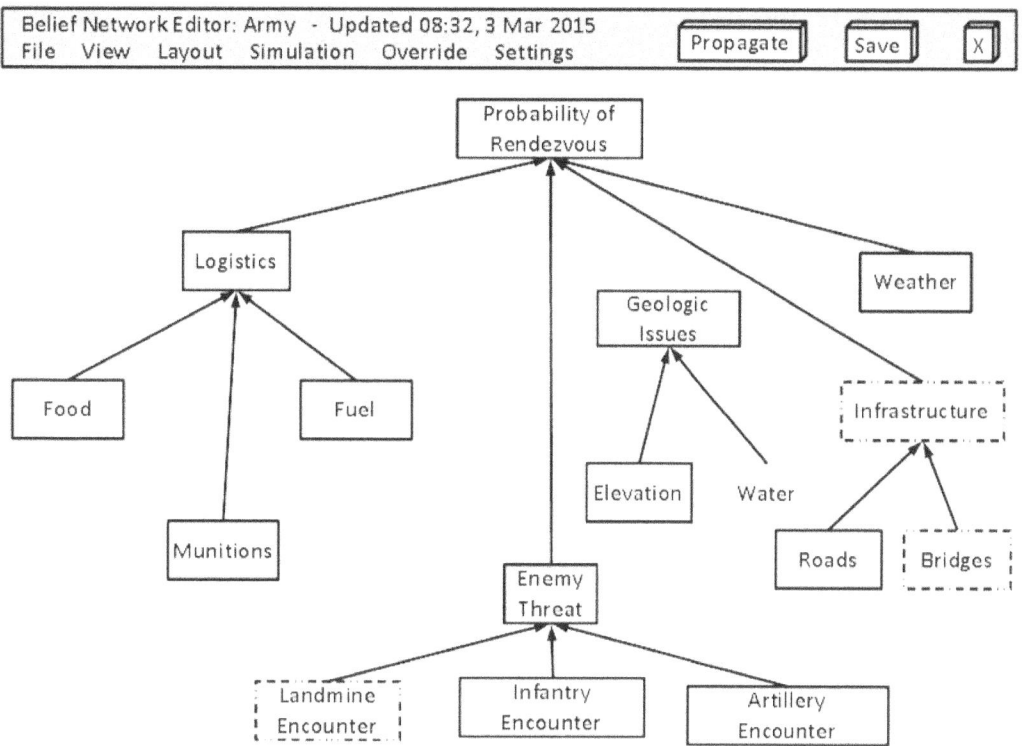

Figure V-8.2: Army Scenario Belief Network

In addition to updating the belief network in this window, we implement methods to dynamically change the belief, disbelief, and uncertainty values of each node based of the position of blue troops, the known data regarding their path, and the level of uncertainty, or holes, around them. When these values are changed, the belief and disbelief values of each node are propagated through the belief network to calculate the Probability of Rendezvous.

Kiviat Diagram

Also available in the Dashboard is a Kiviat diagram shown earlier. The kiviat diagram allows the user to visualize various types of uncertainty in the belief network nodes. Each type of uncertainty visible in the diagram is associated with one or more nodes in the belief network. Clicking on any side of the Kiviat diagram displays the BNE with the nodes associated with that type of uncertainty highlighted. Adjusting the uncertainty value of a belief network node dynamically changes the Kiviat diagram.

Results

We integrate a method for visually representing uncertainty in a command and control scenario. The uncertainty is integrated into our belief network and the schedule evaluator. This integration shows how levels of uncertainty effect our calculations of a brigade probability of arrival or mission completion, and how this information affects the decision making process.

The specific tasks are as follows:

- Build scenario
- Create of scenario file to generate timed events and messages
- Augment Path Replanner tool to be a stand-alone program for grid data specification
- Generate data files generated for elevation, roads, water, holes, red troop positions and paths, blue troops positions and paths
- Visualize holes and opacity update methods
- Design and implement update methods for all grid objects
- Design a Controller to update and maintain all grid objects position and visual representations
- Design Controller to interface with the High Interest Event display
- Create methods to read data files representing all grids at the initialization of the program
- Construct belief network
- Implement belief network dynamic control methods
- Implement Schedule Evaluator dynamic display
- Implement Path Replanner tool
- Implement Brigade supply (food, fuel, combat power) displays
- Create Kiviat diagram

Chapter V-9
Pattern-Preserving Extrapolation of the Space Catalog

Need

For future systems, such as air traffic control upgrades, high-speed trains, and interstate highway additions, increased traffic loads are envisioned – and the system designer must show that these increased loads can be accommodated. In space surveillance missions; for example, the Space Fence[168] system sizing studies require the use of an extended satellite catalog (SATCAT), hereafter called the "large catalog". As of today[169], there are 17,145 cataloged Resident Space Objects. We extrapolate from these objects, based on historical trends and other statistical characteristics of satellite orbits, to produce a suitably large catalog.

Objective

Four goal-oriented tasks guide this work to a readily justifiable large catalog:

1. Identify factors that are relevant in extrapolating to a large catalog. These factors include the specific characteristics of the Space Fence (for example, minimum detectable object size), historical trends, orbital regimes, popular orbits, launch trends, and debris-producing event trends.
2. Model and propagate the factors to a future year; for example, when the Space Fence is slated for operations.
3. Define "augmenting objects" by randomly choosing satellites from the current population and perturbing their orbits according to the models.
4. Produce output summaries that allow the user to understand the content of the extended catalog as compared to the current catalog.

Approach

- Task 1: **Factors** are identified, discussed qualitatively, and their relevance is explained
- Task 2: **Models** are produced for the year 2015, from a distribution from which randomized characteristics are drawn.
- Task 3: **Augmentation** is defined by Monte Carlo simulation of randomized orbit characteristics.
- Task 4: **Output** provides summary statistics of the large catalog, an electronic version of the large catalog, and a filtered catalog with objects <= 3,000 km in altitude.

Results

We construct a large catalog that has approximately the same proportions of objects in the Low Earth Orbit (LEO), Medium Earth Orbit (MEO) and High Earth Orbit (HEO), which includes

---

[168] http://www.wpafb.af.mil/news/story.asp?id=123331461, accessed 01/20/2015.
[169] http://www.celestrak.com/satcat/boxscore.asp, accessed 3/28/2015..

Geosynchronous Earth Orbit (GEO) altitude regimes. We account for the fact that the numbers of objects in these orbital regimes are growing at different rates. Results are cited by task.

We also study maximum and minimum observation rates per day, as seen from a ground-based sensor fence, and conclude that they are not widely dispersed from the mean observation rate. A small variance results because the largest component of the satellite population is small debris (5 -10 cm. object not currently cataloged) which is randomly distributed along orbital paths.

Task 1: Factors that are relevant to producing a large catalog are identified. These include the projected number of objects as a function of: Object type, Orbit type, Orbit altitude, Orbit inclination, Orbit right ascension, and Object Size.

Object types

Objects are created by two primary mechanisms: results of space launches (payloads, associated objects, debris, rocket bodies) and by on-orbit collision. Mirroring the nomenclature of the Satellite Boxscore (which provides the status of objects in the current space catalog), the three types of objects considered are: Payloads (PL), Rocket Bodies (RB) and Debris (DB). This information is encoded in the International Designator field: A = payload; B, C = Rocket Body; and D, E, F,.. = Debris. PL and RB populations will likely experience a slight upward trend. However, DB may increase and possibly cascade (increase exponentially) in the next 5 -10 years. Another source of "new" objects is small debris: always been there, but too small to have been cataloged.

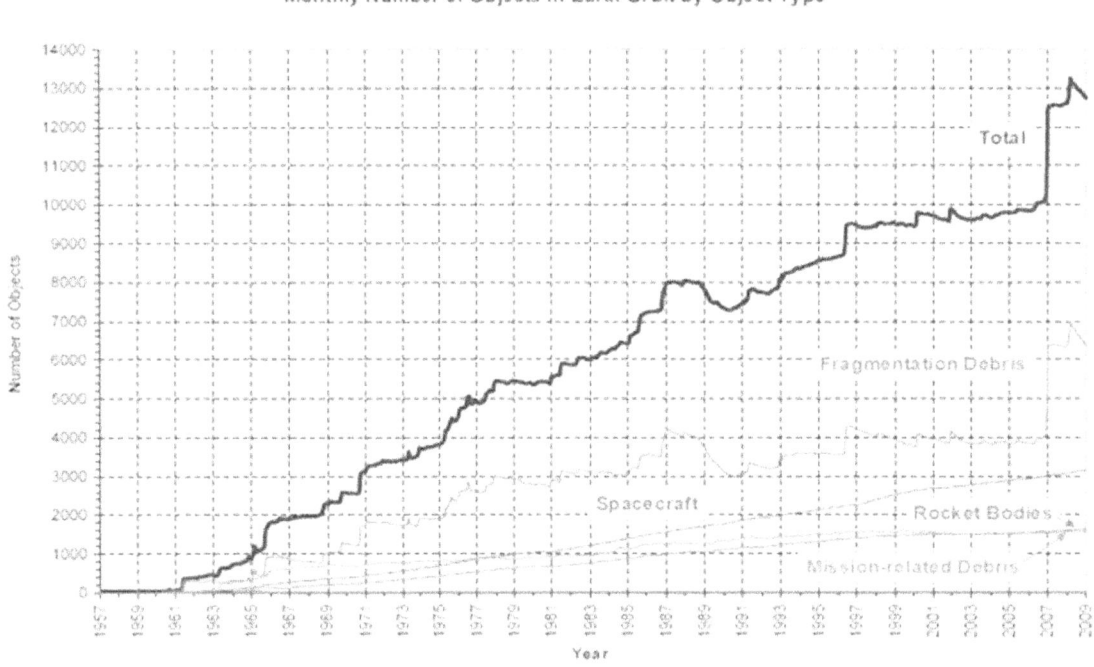

Figure V-9.1: Satellite Growth by Type

Thus, the large catalog should have a large amount of small sized debris. The growth of the catalog over time, with an update based on the current catalog, is shown by object types are (Figure V-9.1). Payloads, rocket body, and debris populations have grown roughly linearly with time.

Orbit types

Orbits are not uniformly distributed in Earth orbit. There are only a few popular orbit types. The most populated orbits are:

- Nearly-circular, low-inclination, low-altitude (below 200 km.),
- Highly-inclined, retrograde sun-synchronous,
- MEO constellations (above 2000 km., below 35786 km.)
- High-eccentricity semi-synchronous (12 hour), and
- HEO-synchronous (24 hour).

From Figure V-9.2, the GEO belt (about three earth diameters distant) is distinct. Also evident is the large number of LEO objects. We also see 12 hour MEO orbits that are inclined to the GEO belt.

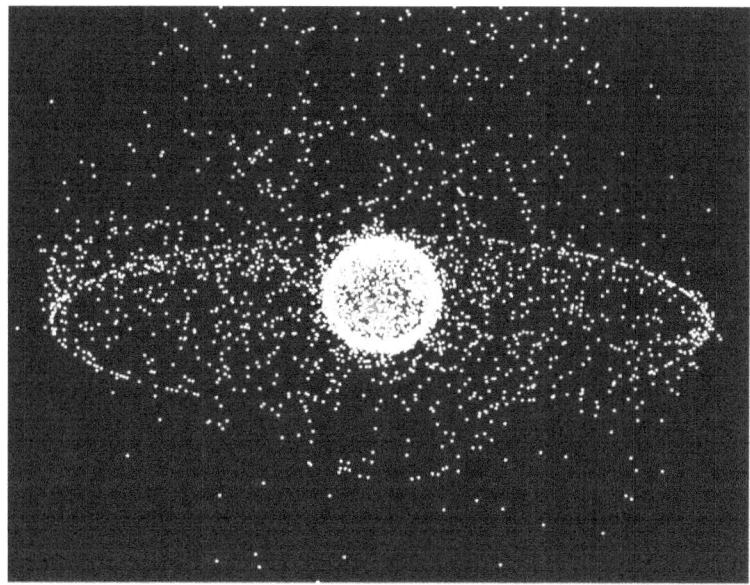

Figure V-9.2: Earth Orbiting Objects

Orbital altitude

Objects are not uniformly distributed in altitude. In fact, all categories of objects (payloads, rocket bodies, and debris) are mostly clustered in a few altitude bands: LEO, the sun-synchronous LEO belt, a few preferred MEO altitudes, and a HEO belt with most payloads and debris at geosynchronous altitude. The large catalog must mimic this altitude distribution. For the purpose of our analysis, we define three altitude intervals: LEO (0 – 2000 km.), MEO (2000-35000 km.), and HEO ($\geq$ 35000 km.). For the mean motion, the relevant orbital element, 2000 km. equals 11.3354 revs/day, and 35786 km. equals 1 rev/day.

The extrapolated number of objects in each of these orbit regimes shows quantitatively the presence of much debris in low earth orbit. A more detailed look at the distribution of LEO objects (PL, RB, but mostly DB) by altitude shows a spike at and above sun-synchronous altitudes (typically 600-800 km.).

Orbital inclination

Objects are not uniformly distributed in inclination. Spikes in inclination coincide with equatorial orbits (0° - 20° inclination), critically-inclined orbits[170] (63.4°) 12-hour orbits, polar orbits (~90°), and sun-synchronous orbits (~98°). For sun-synchronous satellites, inclination is a function of altitude. The perigee/apogee versus inclination plot (Figure V-9.3) shows inclination clusters. Inclination bands have the same distribution for the current catalog and the extrapolated catalog

Orbit Right Ascension

The times when objects pass over a location on earth are uniformly distributed for debris, rocket bodies, and most payloads. Figure V-9.2 shows, especially for LEO objects, a snapshot, or glimmer, of this uniform distribution. Active payloads in sun-synchronous, semi-synchronous, and geosynchronous orbits are exceptions. Augmenting objects that are active payloads of these types are phased within a constellation but should be phased randomly with regard to current objects. For the recent Chinese ASAT event, the fragmentation pattern shows the uniformly random distribution of debris.

The orbit phasing parameters are the mean anomaly and the right ascension of the ascending node. Here, we focus on the right ascension of the ascending node. It is expected to be non-random, especially for HEO (geosynchronous) payload and drifting defunct satellites. We did an impromptu analysis to show that this is the case..

Object Size

Object sizes are not uniformly distributed. The augmented catalog should have a large number of objects (~80,000) between 5 and 10 centimeters which were not previously detectable with enough regularity to catalog them. The distribution of objects by size closely follows Zipf's Law[171]. This discussion justifies the inclusion of a large number of small debris objects. Size is an optional parameter to be extrapolated because it is not a Two Line Element parameter. We calculate it and put it in the 2nd line (Line 1) of the TLE in place of the international designator.

Task 2: Models are devised for six factors identified in Task 1. The models provide an extrapolation of trends to a time frame of interest.

---

[170] http://www.scielo.br/scielo.php?script=sci_arttext&pid=S1678-58782005000400004, accessed 5/7/08
[171] http://en.wikipedia.org/wiki/Zipfs_Law, accessed 5/7/08.

Object type model

The scaling factors for payloads, rocket bodies, and debris are derived by extrapolating payloads and rocket bodies to a future year. The debris population is then computed as #DB = Desired SATCAT size - #RB - #PL. The rationale is that a large amount of small (5 – 10 cm.) debris is in orbit and will be observable by the Space Fence. For example, the #PL (for the year 2008) is given by:

$$\#PL/(Year - 1960) = 3257/(2008-1960) \Rightarrow \#PL = 67.854 * Year - 132994 \Rightarrow \#PL = 3596.$$

The #RB and # debris objects for the extrapolated catalog are computed using the same formula. As discussed earlier, a large number of objects (94%) are debris.

Orbit type model

The satellite catalog has one record with multiple fields for each object. The orbit type is specified by orbit altitude (h), which is related to the mean motion (n), the graitational constant (U), and the Earth's radius (R_E):

$$h = (U/n^2)^{1/3} - R_E$$

For the current SATCAT, we sort the three object types (PL, RB, DB) into three altitude regimes (< 2000 KM), (2000 to 35678 km) and (> 35678 km) as earlier defined. This produces a 3x3 matrix with 9 entries corresponding to the number satellites of each type (PL, RB, DB) in each altitude regime (LEO, MEO, HEO). The sum of the entries is the desired large catalog size.

Orbit altitude

We sort LEO objects into altitude sub-bands. MEO and HEO have many fewer objects and are augmented by perturbing orbits drawn from wider frequency bins.

Orbit inclination

For the LEO altitude sub-band, we sort the SATCAT for inclination bands for Payloads and Rocket bodies. Debris is assumed uniformly distributed, (we verified this assumption by analyzing the SATCAT for the debris distribution). The augmented catalog has the same histogram spikes as the current catalog. Note that the inclination bands may be different for LEO sub-bands, MEO, and HEO versus PL, RB, and DB. Payloads maintain orbit, whereas rocket bodies and debris drift in inclination.

Sun-Synchronous Orbits

For these LEO orbits, altitude determines inclination. Typical sun-synchronous orbits are about 600–800 km in altitude, with periods in the 96–100 minute range, and inclinations of around 98°. Inclinations are linked to altitudes by a rule: if the number of revolutions/day is (nearly) an integer between 7 and 16 then compute the sun-synchronous inclination corresponding to the altitude.

Orbit phasing

The orbit phasing parameters, mean anomaly and right ascension are modeled as random number draws (0 to 360 degrees). A quick-look analysis of the right ascension shows the distribution in right ascensions at HEO with stable and unstable longitude bands. The stable regions are called satellite graveyards[172] because defunct geostationary satellites drift to these regions and stay there.

Size

The current space catalog, which has a diameter versus size distribution that roughly follows Zipf's Law, is extrapolated (Figure V-9.3) to produce a large catalog. Diameter is estimated are more accurate than RCS which varies (by as much as three orders of magnitude) with object orientation. Here are the steps:

- Determine domain (.05 <= diameter <= 200) and range (1<= number <=90,000)
- Plot a log-log-scale, the relationship between diameter and number of objects is linear.
- Sampling: draw a uniformly-distributed random number between 1 and 90,000
- Convert it to its log10 equivalent (take the logarithm of it). This is the y-axis (N) value.
- Insert into the second equation to compute the x-axis (D) value.
- Convert this from a logarithm to a diameter in meters: Diameter (meters) = 10D
- Result: After performing a random draw, in the range [.05, 200], for each element in the catalog, converting that number to a N value, computing the corresponding D value, and converting the D value to a diameter, each catalog element will have a diameter.
- If an object has a known diameter (for example, the International Space Station), the known diameter replaces the randomly drawn diameter

Task 3

Augmentation to sub-catalogs defined above is accomplished using Monte Carlo sampling. We draw each of the required number of new objects from a uniformly distributed random distribution, perturb the altitude and inclination within the distribution defined (from the maximum and minimum values found) for each sub-catalog. We also uniformly perturb the mean anomaly and right ascension. The procedure relies on [min, max] sub-catalog values gleaned from the current catalog.

Initialize Application:
- Accept User input: Catalog Size (for example, 100,000), Catalog Date (for example, 2020)
- Read current catalog: accept default or pop up an browse window
    (for example, catalog_21_2008_02_29_am.txt)

---

[172] http://en.docsity.com/en-docs/ext/The_Geostationary_Orbit_Part_1-Wireless_and_Satallite_Communication-Lecture_Slides_, accessed 02/27/2015.

Partition current catalog into sub-catalog:
- determine object type (payload, rocket body, debris) from International Designator
- determine orbit type (Low, Medium, or High earth orbit) from number of revolutions/day
- define and populate inclination sub-catalogs for each object and orbit type
- define and populate inclination sub-catalogs for each altitude sub-catalog

Compute the number of additional element sets needed for each sub-catalog
- Compute fraction of the the extrapolated catalog for Low, Medium, High earth orbits
- Multiply by expansion factor formulas

For each sub-catalog:
- Draw an element set at random
- Perturb the altitude (via the mean motion) uniformly within [min , max] range
- Perturb the inclination (sun-synchronous rule is invoked here within [min , max]
- Perturb the mean anomaly and right ascension [0 l, 360 degrees]
- Perturb eccentricity and argument of perigee [min , max]
- Set other parameters; for example size, per earlier defined conventions
- Append perturbed elements to the catalog file

Task 4: Output is displayed to the user in the form of summary statistics. Ranges of values for the new objects included in the electronic version of the large catalog are also defined.

<u>**Target Year: 2020**</u>   <u>**Target Catalog Size: 100,000 objects**</u>

Processing Metrics:
- Catalog Load Time: 0.203 seconds
- Time to Extrapolate from 13,321 to 100,000 objects: 1.23 seconds
- Time to write extrapolated catalog to Two-Line Element file: 10.64 seconds

Count by Object type (Input/Extrapolated) catalogs:
- Total Objects: 13,321 / 100,000
- Payloads: 3,404 / 3,941
- Rocket bodies: 3,935 / 5,007
- Debris Pieces: 5,982 / 91,052

Value Added

We analyze an extrapolated NASA catalog. Existing code is modified to allow three different histograms (Figure V-9.3) per chart to be shown. Cursory analysis indicates that the NASA catalog preserves some patterns, but not others (altitude, altitude/inclination, right ascension). Size is not provided in the NASA catalog (although RCS at 10 cm. wavelength is), so we use our algorithm.

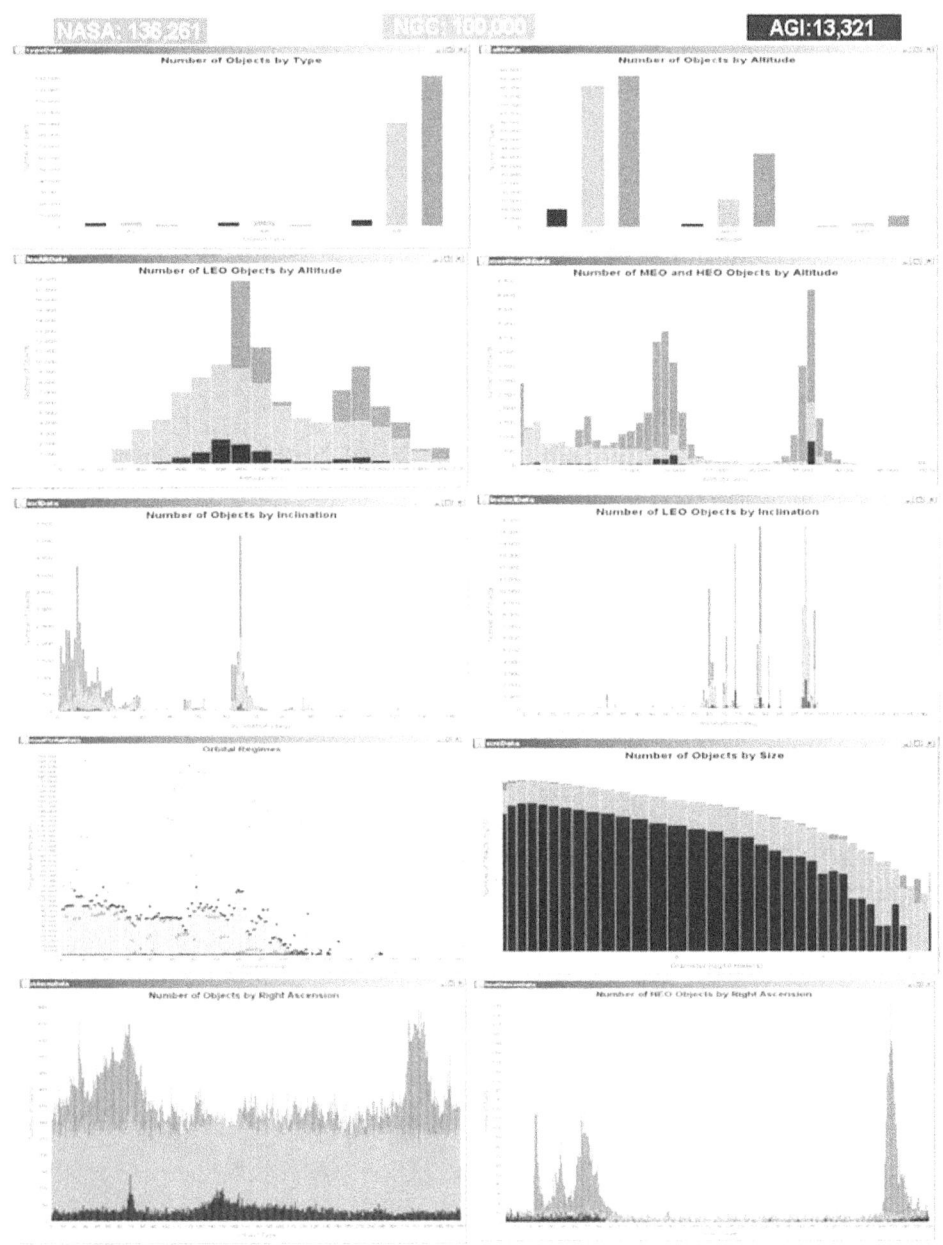

Figure V-9.3: Superposition of Results for Three Catalogs

Summary

A procedure is defined for extending the current satellite catalog to a large catalog. Inputs are the desired catalog size and date. Processing consists of segmenting the current catalog into sub-catalogs corresponding to object types and characteristics. The sub-catalogs are extrapolated to the size of the desired catalog. Monte Carlo sampling of sub-catalogs and perturbation of orbital parameters within bounds obtained from the current catalog produce new objects. These are appended to the current catalog. Output consists of a summary display and histograms of characteristics of the current and large catalogs. An electronic file of the large catalog, which has the same format (except that size is substituted for International Designator in Line 2) as the current catalog, is provided.

Future

User Interface: Software is developed and tailored to read the current catalog, parse it, find parameter ranges, perform pattern-preserving extrapolation, and write the resulting large catalog and a filtered file (alt <=3,000 km.) to a file. A user interface would make this task accessible to the non-programmer.

Detecting and Verifying patterns. These results allow visual inspection to verify that the extrapolated catalog preserves patterns. A rule induction algorithm could ingest the content of the current catalog and automatically produce a rule tree that show rules that are automatically induced from the data set. A straightforward conversion of the TLEs to ".arff" format is required. The extrapolated catalog could then be processed by the same rule induction algorithm to automatically assure the pattern is preserved.

Chapter V-10
Starship Cybernetics - Transition from Niche Decision Support to Pervasive Artificial Intelligence

Today, computers win at chess, interact superficially with users, perform diagnostics in narrow domains, and manage well-defined assets. While computing power is increasing exponentially, software lags. Starship computing requires smarter, more pervasive technology. Rather than a human orchestrating applications, a diffuse machine intelligence that monitors, assesses, plans, and executes a dynamic closed-loop system is needed. This cybernetics environment contrasts sharply with niche decision support applications currently in use.

A step toward a unifying framework for multi-strategy reasoning is instantiated as an executable, self-aware knowledge base. The ontology of this knowledge base is a semantic network - a hierarchical, characteristics-based graph with concept nodes and multiple link types that quantify the effects that one link has on another. Such a network is sufficiently powerful to represent rule induction, abduction, deduction, neural networks, genetic algorithms, search, probabilistic reasoning, planning, and simulation. Combinations of these reasoning strategies, driven by an underlying knowledge base that is executable and self-aware, are discussed and results are demonstrated.

The unifying framework for multi-strategy reasoning is demonstrated in a powerful context: automated discovery of unknown unknowns. This technology, pushed forward with real work now, transitions us from smart niche applications to pervasive closed-loop processing for dynamic systems. Although humans are particularly adept at absorbing new concepts and integrating them into an understanding, computers do this poorly. By representing known concepts in a belief network that serves as an executable knowledge base and abductively identifying new concepts and links from new evidence, the knowledge base adds previously unknown concepts.

Introduction

Imagine. The year is 2120. Starship One is about to begin an interstellar journey to Epsilon Erdani ~10.5 light years distant. The software comprising the Artificial Intelligence Module is undergoing final checkout. No surprises are anticipated - the software has been evolving in a virtual, simulation-driven laboratory for 11 years. Back in 2015, computers won at chess, interacted superficially with users, performed diagnostics in narrow domains, and managed well-defined assets. While computing power increase exponentially, software lagged. Starship computing requires smarter, more pervasive technology. Rather than a human orchestrating applications, a diffuse machine intelligence that monitors, assesses, plans, and executes a dynamic closed-loop system is prototyped, beginning in 2015. This cybernetics environment, the topic of this paper, contrasts sharply with niche decision support applications in use during the early 21st century.

The starship Artificial Intelligence Module (AIM), discussed here, contains a unifying framework for multi-strategy reasoning. The framework is instantiated as an executable, self-aware knowledge base. The ontology of this knowledge base is a semantic network - a hierarchical, characteristics-based graph with concept nodes and multiple link types that quantify the effects that one link has on another. Such a network is sufficiently powerful to represent rule induction, abduction, deduction, neural networks, genetic algorithms, search, probabilistic reasoning, planning, and simulation. Combinations of these reasoning strategies, driven by an underlying knowledge base that is executable and self-aware, are discussed and results are demonstrated for two relevant applications.

A key concept for the unifying framework for multi-strategy reasoning is demonstrated in a powerful context: automated discovery of unknown unknowns. This technology, pushed forward with real work now, transitions us from smart niche applications to pervasive closed-loop processing for dynamic systems.

Methodology

Concepts for a starship AIM are derived using a cohesive systems engineering approach. A top-down and bottom-up approach starts with mission requirements leading to a processing architecture, followed by a specification of the decisions the interstellar AIM must make. This drives interface displays between the AI and onboard or earth-bound humans. A knowledge structure that is sufficiently rich to support multi-strategy reasoning is the basis for integrating algorithmic software.

This systems engineering approach is grounded in current industry practice and has been shown on many military, intelligence, and commercial tasks to provide traceability from requirements to working prototype software that readily evolves as requirements mature.

Results

A systems engineering overview of the starship Artificial Intelligence Module is presented. Project Icarus mission requirements[173] and artificial intelligence considerations for Computing and Data Management are addressed. A cybernetic, closed-loop decision process is discussed. Display interfaces and algorithms are integrated with a self-aware, executable knowledge base.

Concepts are described, including an ironic look at human versus computer strengths (Figure I-2.1), the notion of pervasive cybernetics for AIM, the need for software evolution, the emergence of self-awareness in AIM, the role of multi-strategy reasoning to deal with uncertainty, and the new idea, automated discovery of unknown unknowns, to deal with unforeseen situations.

Two applications of AIM are explored: anomalous behavior and orchestration of algorithms. Anomolus behavior in the navigation and guidance system provides an opportunity to troubleshoot curved space astrodynamics. Algorithms are orchestrated to identify and assimilate information obtained through unexpected subsystem interactions.

Systems Engineering Overview

---

[173] 1 R.K,Obousy et. al., Project Icarus: Progress Report on Technical Developments and Design Considerations, JBIS Volume 64 No. 11/12, pages 358-371.

Systems engineering is the fundamental process from which the need for decision support tools is derived. Requirements are the starting point for the systems engineering process. Ideally, we receive these requirements from our customer. Practically, we often translate needs to requirements and identify solutions to problems. The goal of all engineering, we are taught, is to reduce risk. Here, risk is defined as a function of the probability of an issue occurring and the impact of the issue. To do good systems engineering, an understanding of probability and mission impact is essential.

Mission Requirements

Systems engineering is based on the definition, allocation, and satisfaction of requirements. Functional requirements specify what the system must do. Performance requirements define how well the system must perform. Interface requirements identify the other systems that the system of interest must interact with. Mission Requirements are a summary of Project Icarus Terms of Reference:

- **Mission:** unmanned probe, delivering scientific data about a target star, associated planetary bodies, and the interstellar medium
- **Technology:** current or near-term, with spacecraft lauch as soon as credibly possible
- **Mission Duration:** as fast as possible, < 100 years and ideally much sooner
- **Destinations:** spacecraft must be designed to allow for a variety of target stars
- **Propulsion:** must be mainly fusion-based; for example, Project Daedalus[174]
- **Encounter Time:** must allow deceleration for increased encounter time at the destination.

Computing Architecture

Referencing AIM specifically a knowledge base receiving preprocessed information that is tagged, or classified, with contextual information. Post-processing algorithms, a few categories of which are shown, process the knowledge content to find patterns of change, fuse information, and update plans as the basis for action.

Decisions

Operational processing requirements result is a set of tasks, or decisions, allocated to AIM. These decisions are central to the process of determining what decision support tools are required, their degree of automation, and the functionality they provide. Fortunately, across many mission domains, including interstellar travel, the same decisions must be made. At a high level, these decisions are: what's going on? what to do?, do it?, and how well did we do? This forms a Monitor/Assess/Plan/Execute repeating decision loop that is usefully segmented into a routine loop and a quick-reaction loop.

The design of a mission thread is critically dependent on the decisions required Because of

---

[174] http://www.bis-space.com/what-we-do/projects/project-daedalus, , accessed 3/29/2015.

this, it is worthwhile to define a decision. What does it mean for a computer or a human to decide? What is the tangible or practical result of a decision? Can a decision and its impact be measured? The answer to these questions is closely connected to the concept, discussed earlier in Chapter I-3, of a decision as a transition from a current state to a desired state, according to a plan.

Displays

A story metaphor graphically represents knowledge. The story begins with situation monitoring. The crew chief is briefed on the status of the spacecraft and the crew monitors health and status data. As changes occur, the operations center receives tasking directives from higher authority, input from the Artificial Intelligence Module, and expert input from engineers. The crew shifts focus to mission planning and plan optimization. Authority to update a plan is then received, the action is taken aboard the spacecraft, and assessment evidence is received at the operations center. The crew fuses this evidence to perform mission assessment.

	Space Task	Space Rationale	Ground Task	Ground Rationale
M O N I T O R	Raw state-of-health data monitoring.	On-board sensor measurements are readily available.	Process state-of-health anomaly data monitoring.	Human expertise for anomaly, trends, and recurring problems.
	Stored data monitoring seconds => days	Required for context, but on-board storage and processing are limited	Stored data monitoring months => years	Requires human visual, analytic, and commonsense
A S S E S S	Immediate, real-time situation assessment	Timeliness, availability are critical. Belief network treats uncertainty and data aging.	Evaluate processed data by exception.	Non-real-time human expertise leveraged to assess anomalies
	Choose stored dynamic plan updates	Countermeasure and mission safing timelines to prevent failures are short.	Review telemetry pass plans & mission strategy	Need human-guided trade studies with multiple "soft" criteria
P L A N	Learn the immediate worth of plan updates.	Simple to modify the worth of a (case-based repair) fix based on belief network state	Originate and revise dynamic plan updates.	Requires human common sense and wisdom.
E X E C U T E	Implement commands up-links	Standard operating procedure	Issue non-real-time commands	Needed to change spacecraft parameters
	Exercise on-board preplans	Constrained by ground-managed rules of engagement	Act on telemetry / Upload pre-plans	Act on mission and state-of-health data / Expertise and timing are key

Algorithms

This is what makes AIM useful. These applications provide what computers do best; for example, math, memory, and search, and humans don't. Having said this, the use of algorithms is highly constrained: their only role is to provide information to make decisions. An algorithm is an effective method for solving a problem. Most practical decision-making problems are not amenable to "brute force" approaches because the underlying problem is known to be "non-polynomial hard" (NP-Hard). This means that the problem cannot be solved in polynomial (quadratic, cubic, quartic, etc.) time exponents of the parameters of the problem. This fact - that most practical decision-making problems are NP-Hard - makes algorithms the critical component of AIM.

Algorithms accept data and register this content in a knowledge base. In fact, algorithms operate on the knowledge base to make it executable. Once data is registered in the knowledge base, multi- strategy reasoning algorithms perform automated manipulation, distillation, and discovery on the data and post the results to displays. A sample set of decision algorithms consists of background and foreground tasks. Background tasks do the "under the hood" computations that "set the stage" for foreground tasks that are more highly user interactive. Knowledge management provides content in context, fog-of-war perturbs already uncertain data, kinematics propagate spacecraft and sensor geometry through the scenario timeline, and decision tool controllers accept operational contexts and scenario control parameters in advance of user interactions. Foreground tasks are arranged by decision type and connected to form a mission thread.

Knowledge Structure

The semantic network concept, literally a network with meaning, provides a unified framework for interfacing all algorithms through the knowledge base. Semantic networks are like to stories. The knowledge base is structured as a characteristics-based hierarchical ontology. Belief networks are already represented as semantic networks; data mining algorithms are a mix of tables, headers, and graphs; and the plan update algorithm consists of tabular content.

Knowledge is captured in semantic networks consisting of nodes and links, perhaps representing an spacecraft system's status over time. A story fragment is a portion of a semantic network that contains nodes connected by links. Nodes have evidence records that are stored as flat files (more recently they are stored as data services). Evidence records are linked to the textual context or structured data file from which content is extracted. This provides a drill-back from high level mission impact inferences to the raw sensor evidence. Stories have nodes that state hypotheses and these have degrees of [belief, unknown, disbelief]. Hypothesis characteristics are linked to nodes by the node name. Stories have links that define the impact one node has on another. Links have node values, which may be either supporting or detracting, for both beliefs and disbeliefs.

Concept

Since this paper explores AIM ideas for a future starship, results include underlying concepts that may usefully drive future prototyping of the cybernetics system. These include a discussion of what humans do well versus what computers do well, a look at the state-of-the are in spacecraft autonomy, the notion of self-awareness in a knowledge base, how algorithms need to be orchestrated to solve difficult problems, and a relatively new idea - automated discovery of unknown unknowns.

Pervasive Cybernetics

Spacecraft autonomy is increasing gradually. Many unmanned probes, including the recent Curiosity[175] landing on Mars, provide evidence that robots can collect and transmit useful scientific data. However, a much more significant use of artificial intelligence, in the form of a probabilistic control system, allows driver-less cars to successfully complete a difficult route through the desert[176]. This idea is adapted to spacecraft and provides a concept for pervasive cybernetics - defined as artificial intelligence provided to a closed loop control system.

Multi-Strategy Reasoning

The unified framework for multi-strategy reasoning didn't initially include intelligent software agents, introspection, and simulation. We did not, at the time, attempt to incorporate intelligent software agents or simulation into the semantic network framework. An impediment to combining plans and graphs, with virtual simulations, which are typically used in an off-line or background mode, is that the underlying applications are used by different communities of interest. Decision support tools that automate real-world command and control missions are embedded in the natural world; however, simulations do not get nature "for free". Simulations must model all relevant aspects of nature. Agents and simulations are now incorporated – see Chapter IV-2.

Earlier, we introduced the "Guess and Check" idea. A planner, based on a directive from higher authority, "guesses" anomalous behaviors and a planning tool produces graphical views for situation monitoring and mission assessment. Another planner constructs a simulation with Monte Carlo numerical experiments simulating confidence intervals, and executes it to provide an independent "check" on the anomalous behaviors and planned reactions and can also perform what-if analysis to identify emergent behavior, perhaps in the form of unintended effects. We want to extend the "Guess and Check" idea to assure that all of our tools are explicitly modeling the same scenario.

Applications

Two applications are described that combine the concepts discussed in the previous section. Anomalous behavior of a spacecraft is addressed by leveraging the executable knowledge base and automated discovery of Unknown unknowns. Algorithm orchestration is based on a routine and quick reaction sequence of the monitor, assess, plan, and execute decision loop.

Orchestrated Algorithms

---

[175] http://www.nasa.gov/mission_pages/msl/, accessed 02/27/2015.
[176] http://www.darpa.mil/newsevents/releases/2014/03/13.aspx, accessed 03/29/2015.

The scenario is space weather. Data from the Space Weather Forecast Center is input to a rule induction algorithm (Figure V-10-1). The resulting rule tree automatically provides evidence to the belief network to assess the problem as space weather, and specifically as scintillation.

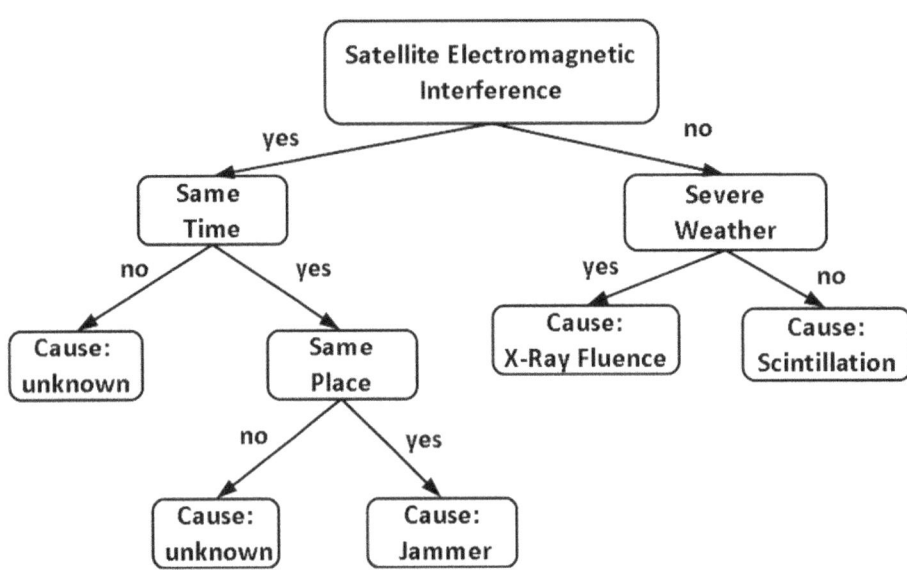

Figure V-10-1: Rule Induction Applied to Space Weather

Anomalous Behavior

The scenario is intermittent failure of a spacecraft communications system. A belief network diagnoses system problems. As new evidence is added to both the belief network and the data mining tool, a new node (System Overload) and new links are discovered. This proof-of-principle of automated discovery of unknown unknowns resolves the anomalous behavior by identifying a communications system overload as the cause.

Summary

A systems engineering approach, described as a decision-centered methodology is proposed for the Artificial Intelligence Module. A computer architecture that features a self-aware, executable knowledge base that fosters multi-strategy reasoning is proposed. Algorithms are orchestrated using automated discovery of unknown unknowns. Finally, two space-related applications, anomalous behavior and algorithm orchestration, are presented.

Discussion

The concepts discussed in this paper are, in most cases, prototyped in space missions, including collision avoidance, space weather effects on communications, and spacecraft command and control. A common measure of maturity, the Technology Readiness Level[177], is used to rate the concepts. For example, Level 3 is prototype software.

Concept	TRL
Decision-Centered Methodology	7
Computing Architecture	7
Multi-Strategy Reasoning Algorithms	3 – 5
Executable Knowledge Base	5
Algorithm Orchestration	4
Automated Discovery of Unknown Unknowns	3
Self-Aware Knowledge Base	1

Time and distance are the most constraining aspects of interstellar travel. Even the closest stars are many light-years away. We are experimenting with a time-distance nomogram (Figure V-10.2) that shows in a very simple way how long it takes to get there. The user clicks on the starship mass, engine thrust, and, the fraction of time the starship is coasting. Fixing these, the constant acceleration and total required velocity increase (Delta-V) is computed. The user enters either destination distance (and gets the travel time) or the travel time (and gets the destination distance). Alternatively, the user can input, Thrust, Coast Fraction Destination Distance and Travel Time and get the allowable starship mass. This is called a fisheye nomogram because the input and output variables are lined up across the center of the dislpay and magnified for easy reading.

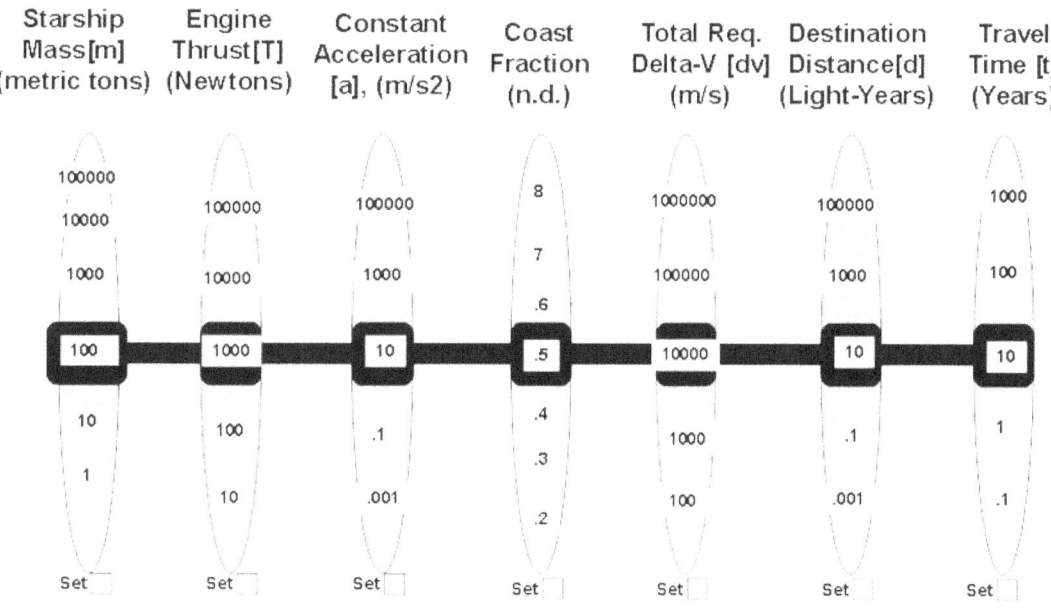

Figure V-10.2: Fisheye Nomogram for Interstellar Time – Distance Calculations

---

[177] http://www.dtic.mil/ndia/2003systems/nolte2.pdf, accessed 02/27/2015.

Next steps include attention the self-aware knowledge base. This has received the least attention and consequently has the lowest technology readiness level. A self-aware knowledge base, that "knows what it knows", can introspectively determine what additional content it needs to make decisions and autonomously evolve by adding new concept nodes and links to the knowledge base through automated discovery of unknown unknowns.

Another task for the near term is to configure a knowledge base for the starship domain. This will begin with a semantic network that relates sensor, health & status, and mission data to projected probability of mission success. Intermediate nodes will hierarchically represent subsystems; for example, communications viability.

Chapter V-11
Longevity Predictor Based on Deep Belief Networks

Idea

User (Questionnaire) => Filtering (Case Based Reasoning) => Inference (Deep Belief Net)

Objectives

Define a novel approach for predicting individual longevity. Provide sufficient detail for the reader to understand the innovative aspects of the idea and appreciate its feasibility.

Terminology

This paragraph defines specific terms used in this paper as they are tailored to longevity prediction.

- **Case Base Reasoning (CBR)** obtains a structured data set with records that are most similar to the characteristics of an individual as obtained through a questionnaire. A longevity estimate for each record is assumed to be included in each record. The database from which cases are filtered may be the result of joining multiple databases, expert opinion, or the results computed by multiple models.
- **Deep Belief Network (DBN):** a hierarchical semantic network (a network with meaning) that consists of nodes stating hypotheses and links indicating which nodes influence which other nodes. The network is trained layer-by-layer using backpropagation and evidence is propagated using fusion algorithms.

Approach

Four major tasks are:

Case Base: use CBR to construct and populate a case base with characteristics pertinent to each individual. Relevant cases are based on an individual's responses to a questionnaire[178] consisting of about 40 questions (emphasis on adding questions to discover health ailments and other negative conditions) a case base is interrogated to find the most similar cases and a similarity score is computed. The score for each relevant case is based on the matching characteristics between the individual and a case.

Deep Belief Network: design, construct and train a DBN. Rather than jumping from a large set of individual's characteristics to a life expectancy score from the case base, the characteristics are decoupled into hierarchical layers (Figure V-11-1) to form a deep belief network. Each characteristic is assigned to a hypothesis contained in the DPN, based on rules or, equivalently, a table look-up.

---

[178] http://gosset.wharton.upenn.edu/mortality, accessed 07/10/2014.

Train: compute DBN link values by back-propagating cases from the case base. The back-propagation algorithm from Artificial Neural Network theory is used to train the DPN based on relevant cases. Essentially, values for hypotheses in Layer 1, the Characteristics layer, and Layer 2, the Activities layer above it, produce link values for those links between Layer 1 and Layer 2. This is repeated for Layers 2 and 3 and the process continues until the links to the top layer are computed. We have had success with simultaneous specification of selected hypothesis values across multiple layers, and single or multiple inter-layer link computation. Values of Characteristics are never modified by the back-propagation process.

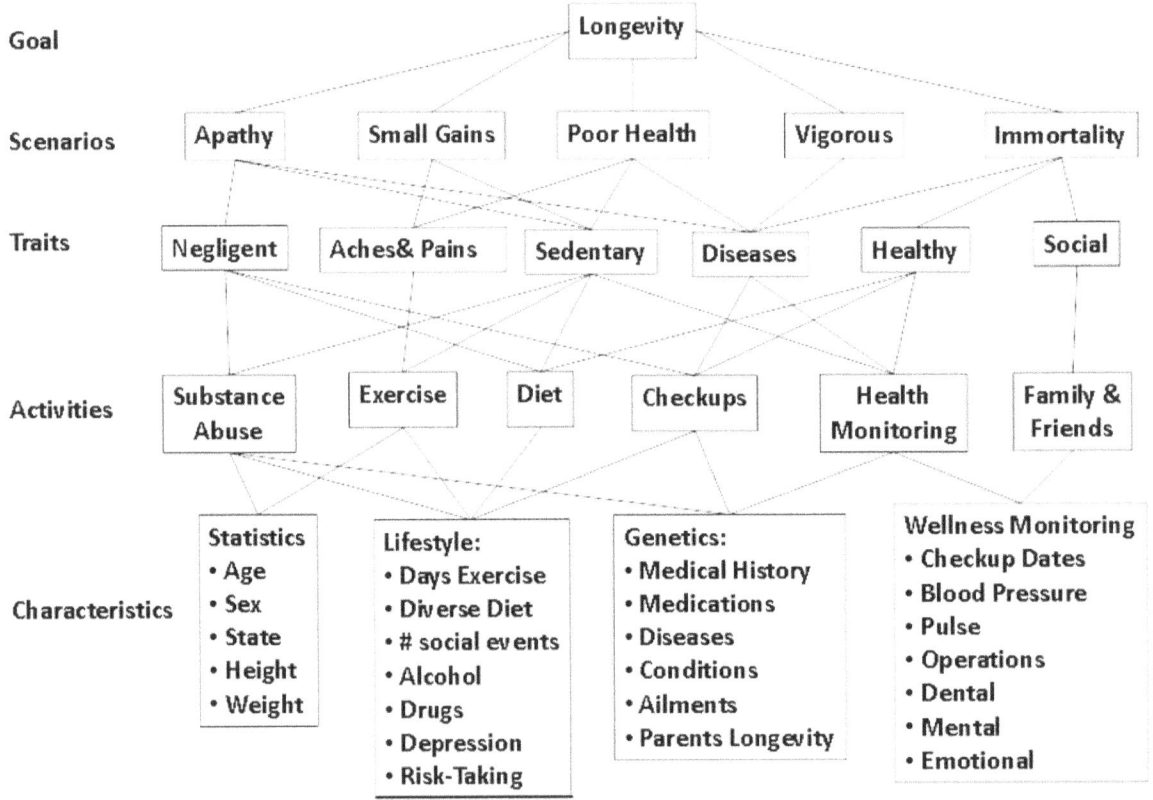

Figure V-11-1: Deep Belief Network Sketch

Compute: Execute the DBN with an individual's characteristics to calculate a life expectancy and related statistics. The user can modify nodes to do "what-if" analysis. The individuals responses to the questionnaire drive the DPN as hypotheses are initialized and data is fused in nodes and propagated through links to determine the value of life expectancy, which is the top node.

Longevity Data Sets: The search for life expectancy data focused on data that predicts life expectancy and related statistics for people in the current population.
- The Center for Disease Control (CDC) offers many data sets, including the "faststats"[179] database of the U.S. As of 2010. Faststats states the top 15 causes of death.

---

[179] http://www.cdc.gov/nchs/fastats/life-expectancy.htm, accessed 07/11/2014.

- The United Nations Department of Economic and Social Affairs Population Division, Population Estimates and Projections Section offers World Population Prospects[180]. From the "Mortality Data" tab, here is a sample Excel spreadsheet summary:

Summary indices | Life Expectancy at birth ((e0) | The average number of years of life expected by a hypothetical cohort of individuals who would be subject during all their lives to the mortality rates of a given period. It is expressed as years.

The World Life Expectancy site[181] provides deaths per 100,000 for all countries, based on causes: alcohol, Alzheimer/dementia, breast cancer, coronary heart disease, diabetes mellitus, and 27 other causes. It also provides U.S. Causes of death by age and gender for the top 50 causes. Given that accidents are the 5th highest cause of death, we found the medical mistakes database[182,183] to be potentially useful as a source of cases. Finally, the site provides research and feature articles. The database from which relevant cases are drawn is a union of existing databases and data computed from existing longevity prediction models. Database design, expert opinion, and model choice are performed and reviewed by longevity prediction experts.

Rationale for Approach

Specificity, explanation, practicality, "what-if" analysis, and algorithm suitability are the concepts that drive the rationale for this approach.
- **Specificity:** Because the objective of the challenge is to predict longevity for an individual, characteristics for the individual are solicited via a typical questionnaire as the basis for producing a particularized case base for each individual. The rationale for filtering the databases for instances that are particularized to the individual is to provide fine-scale granularity for predictions, without over-fitting.
- **Explanation:** Case Based Reasoning is chosen as the inference method for filtering and organizing data because it offers as explanation: an individual may have many specific characteristics in common with a case and some that are different. Intermediate hypotheses serve to classify an individual by characteristics, activities, traits, scenarios, and ultimately predicted longevity. By tagging an individual with these longevity hypotheses an intermediate level of granularity provides an evidence trail from characteristics to longevity predictions, thus providing a deeper sense of explanation. For example, a 65 year old male, who runs daily (activity=fitness), monitors his health (trait= health conscious), can anticipate a long healthy life (scenario = vigorous) and live to be 98 (Longevity = 98).
- **Practicality:** An additional advantage of profiling an individual based on an interpretation of characteristics as intermediate descriptors is that it allows expert advice to be solicited for a relatively small set of "canonical specimens" rather than individual characteristics which represent a very large set (40 questions with 5 possible answers per case produces $5^{40} = 9.1 \times 10^{27}$ possibilities!).
- **"What-if" Analysis:** Another reason for intermediate hypotheses is that it provides the

---

[180] http://esa.un.org/wpp/Excel-Data/mortality.htm, accessed 7/12/2014.
[181] http://www.worldlifeexpectancy.com/, accessed 7/12/2014.
[182] http://www.ncbi.nlm.nih.gov/pmc/articles/PMC419427/, accessed 07/11/2014.
[183] http://www.ncbi.nlm.nih.gov/books/NBK43635/Figure/advances-mokkarala_103.f2/?report=objectonly, accessed 07/11/2014.

individual with the ability to tweak the model by modifying an intermediate hypothesis. For example, if the model shows a belief that the individual is a "fitness buff" is 30%, resulting in a longevity of 72 years of age, but the individual believes this is too low (even though he may have answered the questionnaire correctly), the individual can modify the entry (with help from a wizard that shows typical fitness buff levels versus time and type of activity) the fitness buff entry to 45%, which may result in a new longevity prediction of 76 years of age.
- **Algorithm Suitability:** Deep Belief Networks[184] (DBNs) are very powerful tools in machine learning. The assumption is that observed data is generated by the interactions of many different characteristics on multiple levels, corresponding to different levels of hierarchical abstraction. The DBN is trained layer by layer using the standard greedy back-propagation algorithm. Alternatively, a set of hypotheses can be set to specific values and a back-propagation done on one or more layers of the belief network, which results in link values being adjusted, non-specified nodes being modified. In no case are characteristics modified.

The network is executed by running the DBN application. DBN accepts inputs from the individual characteristics file produced by answers from the questionnaire. The Dempster-Shafer Combination rule fuses evidence, expressed as degrees of [B U , D] corresponding to [belief, unknown, disbelief] at each node. In cases where evidence is expert opinions based on a common set of inputs, simple averaging of belief and disbelief [B, D], with unknown computed as (1 − belief − disbelief) is preferred. Evidence is propagated up the hierarchical network from Characteristics => Activities => Traits => Scenarios => Longevity Prediction.

Results from an Example Scenario

In preparation for software implementation, Questions from a longevity predictor[185] are numbered and associated with nodes in the DBN, implemented using Belief Network Editor (BNE) software. The questions are summarized (Figure V-11- 2, Column 1). A sample set of user responses is entered (Figure V-11- 2, Column 2) and these are converted to [B , D] based on expert opinion and derived rules (Figure V-11- 2, Columns 3 and 4). Although the nodes have a short-hand label, they represent hypotheses for which degrees of [belief, unknown, disbelief] are computed. For example, the "Exercise" node means: what is the degree of [belief, unknown, disbelief] in the premise that exercise promotes longevity?

A representative case base with 9 cases, including longevity predictions, is created by randomly changing some entries (Figure V-11- 2, Columns 5 - 13), and averaged percentages of [~B, ~D] (Figure V-11- 2, last 2 columns) are computed.

The number of differences is shown (Figure V-11- 2, Row 42. A similarity metric (% matching) is computed (Figure V-11- 3, Row 43) for each row (number of matches between user response and case response divided by the number of cases). The similarity metric adjusts the [B, D] for each question in each case. The [`B, `D] values (Figure V-11- 2, last 2 columns and lower right) are used to train the DBN .

---

[184] http://www.cs.toronto.edu/~hinton/nipstutorial/nipstut3.pdf, accessed 07/11/2014.
[185] https://liveto100.com/ or http://gosset.wharton.upenn.edu/mortality, accessed 7/14/2014.

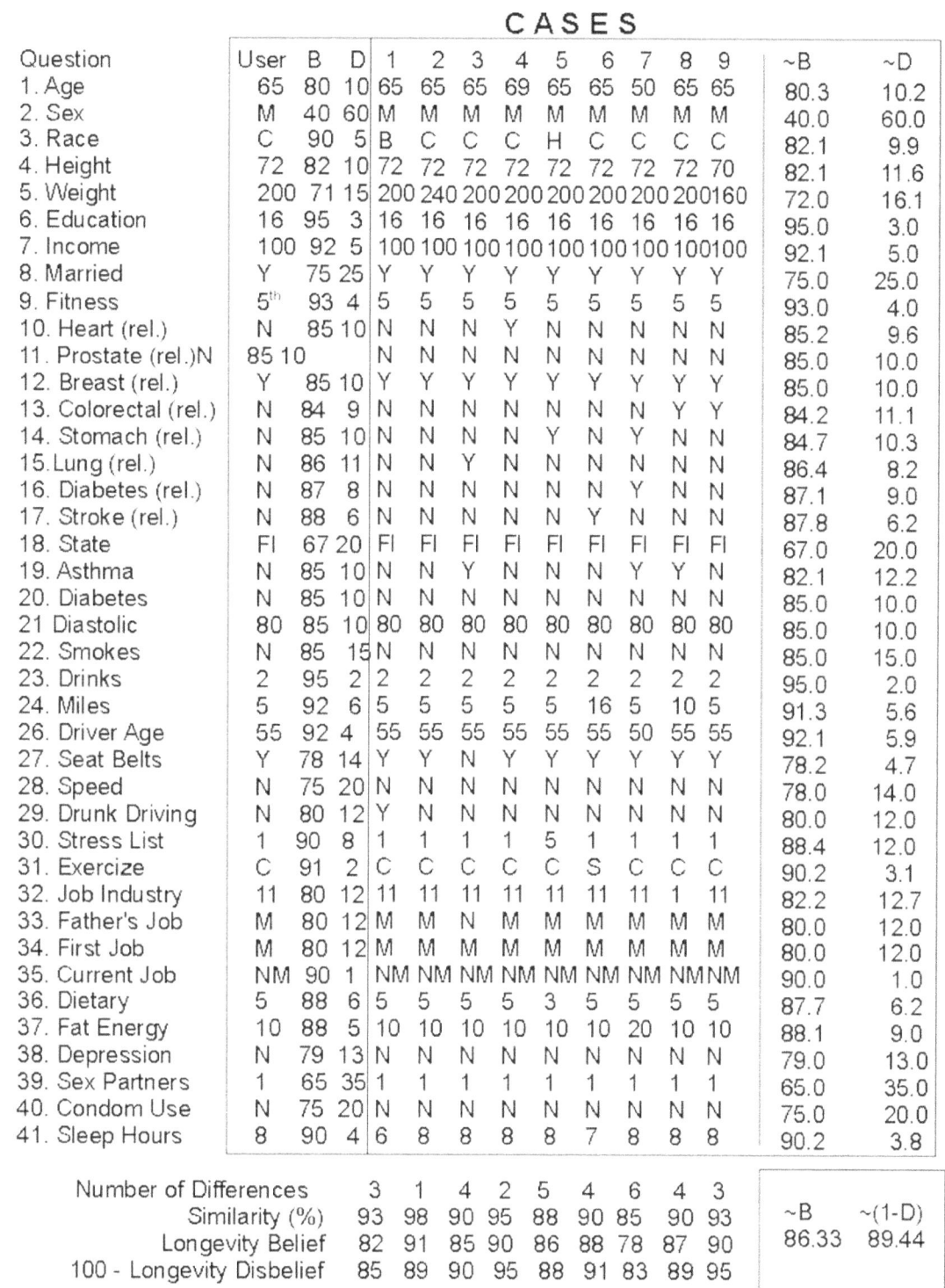

Figure V-11-2: Sample Data for Worked Scenario

The DBN is then used to forward propagate the users responses and this resulted in [B, D] for the longevity node. Based on this semi-automated proof-of-principle, automated input to a knowledge base (Protege) and the MyCBR[186] case based reasoner is attempted.

Example of a Deep Belief Network for Longevity Prediction

Screen capture of a belief in longevity of 90% and 6% disbelief is computed using input values shown for the bottom nodes. The back-propagation algorithm computes all link and intermediate node values. Note that detracting links connect Lifestyle to diet and Genetics to Substance Abuse, creating strong disbelief for those nodes and nodes they affect. Questionnaire content is mapped to DBN nodes. Protege Knowledge Base: the case base shown in Figure V-11-2 is entered into MySQL[187], read into a Protege plug-in, and ported to Protege.

Case Based Reasoning

A plug-in for Protege, MyCBR, provides the ability to filter a large number of cases and find those most similar to the user input to the questionnaire. An intermediary Protégé plugin, called DataMaster, reads the MySQL records and imports them as instances into Protégé. Slots (Characteristics) are used to compute the similarities, along with their weights. We don't define Similarity Measure Functions (SMF) for all the slots. We randomly pick several. The similarities are from a comparison of instance 11 versus the other 10 instances.

A nuance that we notice is that MyCBR provides the flexibility to define a different similarity metric for each characteristics. All similarities (S), however, are given in fractions ($0 <= S <= 1$). The overall similarity for a case, as compared to user questionnaire input, is simply the average of the similarities for the characteristics.

A synergy between the DBN and the CBR arises: The DBN can differentiate the belief network to determine how important each characteristic (lowest nodes in the DBN) is to longevity. Highest rate of change in longevity per node perturbation can be equated to highest weight of that characteristic in the overall similarity calculation, where the overall similarity is computed as the weighted average of similarities of characteristics.

Advantages

- **User Orientation:** The individual seeking longevity prediction data is the user. A user-oriented case base is derived from the answers the user provides in a questionnaire. Thus, the case base is specific to the user. Results are shown in a DBN (Figure V-11- 1) that allows to user to tweak the boxes to better reflect the user's view of their characteristics.
- **Novelty:** Both the user-specific case base and the DBN are innovative. The latter features backpropagation to automatically and flexibly train and rerun a network based on user tweaks to intermediate nodes.
- **Computational Scalability:** both the CBR and the DBN scale linearly with the number of cases and the number of characteristics. In fact, the DBN scales linearly with the number of cases, nodes, and links for back-propagation training.

---

[186] http://mycbr-project.net/, accessed 04/09/2015.
[187] www.MySQL.com, accessed 7/31/2014.

- **Reliability:** extremely accurate longevity predictions (U < 6%) are anticipated, based on the user-specific case base, use of expert input, and the powerful DBN algorithm. A significant advantage of DBN is that it reasons with uncertain, incomplete, and possibly conflicting information, with provision to reflect obsolescent data.
- **Use of Expert Knowledge:** The idea relies on experts to identify and combine data sources, rework the DBN (Figure V-11-1 is only an example to illustrate the possibilities), define "canonical cases", and potentially offer specific feedback to an individual on a fee for service basis.
- **Software availability:** Questionnaires[188], CBR applications[189], and the DBN[190] are readily available as freeware. The DBN described in this paper is available from the author. The code is written in Java and has been used for about the past 14 years in applications ranging from climate change prediction to anti-terrorism threat analysis to time-critical targeting for Air Force Command and Control. This latter software is rated as Technology Readiness Level 4 by the Johns Hopkins applied Research Laboratory.

Insights

- **Case Based Reasoning:** to demystify CBR, liken it to a trade study. CBR characteristics (also called indices) are trade study factors. CBR cases are trade study options. The CBR similarity metric is a trade study score. A CBR critic is a tweak in a trade study option. Given these insights, a CBR is easily produced in an Excel spreadsheet!
- **Differentiation:** The DBR has an option that automatically computes the sensitivity of each node to all other nodes. It computes the rate of change of one to another. This provides the change in predicted longevity versus the change in other hypothesis nodes, indicating which are most important for increased longevity! This can provide automated advice to the user; for example, to live 2 years longer, exercise at least twice versus once a week.
- **Non-stationary:** Statistically, the longevity prediction problem is non-stationary because the underlying probability distribution, both mean and variance, change over time as a complex function of known and unknown parameters. This corrupts classical (frequentist) statistical approaches and favors subjective probability approaches; hence, the Dempster-Shafer theory of evidential reasoning is chosen.
- **Input:** A user's answers to a questionnaire are converted to [B, U, D] for input to the DBN. This is accomplished by converting the input to an average, or expectancy value. For example, in 2015, if the user enters current age at 65, his birth year is 1950. From a table look-up[191], the probability of living to 100 (assumed maximum age) is 7.99% (males), 12.80% (females), and 10.40% combined. With a margin of error of +- 2.00%, the [B,U,D] interval for both sexes combined is [8.40%, 4.00%, 91.2%].

    Another example: the response to the question, number of stress factors, with a drop-down menu showing seven factors, can be 0,1,2,3,4,5,6,7,? The latter indicates "don't know". Delving into the logic behind the menu, with assistance from an expert, may

---

[188] https://liveto100.com/ or http://gosset.wharton.upenn.edu/mortality, accessed 7/14/2014.
[189] www.mycbr-project.net/, accessed 7/112/2014.
[190] https://github.com/rasmusbergpalm/DeepLearnToolbox, accessed 7/14/2014.
[191] http://discovertheodds.com/what-are-the-odds-of-living-to-100/, accessed 7/31/2014.

indicate that subtractions from maximum longevity (assume 100) are: 0, 2.0, 3.5, 5.2, 8.1, 12.8, 16.9, ? , respectively. Assuming a + - 3.0% margin of error, the [B,U,D] intervals are: [97,3,0], [95,5,0], [91.8,6,2,2], [88.9, 6, 5.1], etc. For all questions, a response of "don't know" results in an evidential interval of [0,100,0] corresponding, naturally, to 100% unknown!

- **Output:** The DBN output is the [B, U, D] at the longevity node, where B = belief, U = unknown, and D = Disbelief. Degrees of belief are scaled so that B + I + D = 1. These are scaled to life expectancy, in years, by multiplying by maximum life expectancy, say 120. For example, [.80, .15, .05] => [76 , 18 , 6] which states a belief that the individual will live to at least 76 (minimum), but may live 18 years longer. This is readily converted to a mean life expectancy (M) and margin of error (E): M = B + I / 2 , E = I.
- **How Many Cases:** When computing longevity predictions, too few or too many cases will result in large uncertainty. A golden section search will efficiently provide the optimum number of cases: the number that minimizes uncertainty (U).

Shortcomings

- **Domain Integration:** A questionnaire, CBR, and DBN have not been integrated for specific application to longevity prediction. However, we've managed a team of software engineers that integrated CBR and DBN applications with a template-filling tool as plug-ins to Protege[192].
- **Unstructured Text:** The questionnaire, CBR, and DBN deal with structured data. More information is available in the form of unstructured text and this is not addressed. However, we've prototype software that extracts information from unstructured text and fills templates (akin to questionnaires), and assigns hypotheses (Bayesian Classifier) to extracted templates. This provides input to the DBN from unstructured text. We also have text-based CBR software that finds similarities between texts. This has proven successful in finding on-the-shelf plans that meet planning directives.
- **Frequentist Statistics:** Although the current approach provides upper and lower longevity probabilities (B, 1 – D), a mean, and a margin of error, it does not provide quartiles or standard deviations. If desired, a modified methodology provides them: after training the DBN with similar cases, run the DBN with each case, weighted by degree of similarity to the questionnaire. The resulting outputs form a bell-shaped curve (confirmed empirically) from which quartiles and standard deviations can be computed.

---

[192] http://Protege.stanford.edu, accessed 0/14/2014.

Chapter V-12
Fusing Data to Estimate Total Uncertainty

Motivation: Too often, we assume that information seen on a computer screen is accurate, complete, current, and without conflict. Compounding this human tendency is automated processing of information that strips away all hints of underlying uncertainty.

Idea Description: We propose a general purpose fusion engine[193] (GPFE) that scales linearly with data size to combine 12 types of uncertainty, thus allowing a single quantitative measure of overall uncertainty. A taxonomy (shown earlier in Figure IV-1.1) shows uncertainties in data and process. Data may be uncertain because it is incomplete or inaccurate, while process uncertainties may be characterized as conceptual or dynamic. Each of these categories has specific uncertainties.

Use Case: The idea, though broadly applicable to all information processing, is illustrated using climate change data: In this domain, confidence is typically expressed qualitatively. Quantified measures of uncertainty in a finding are expressed probabilistically (from statistical analysis of observations, models, or expert judgment)." Hence uncertainties are not currently combined!

Step 1: construct a knowledge base using MySQL. In addition to the hierarchy, characteristics of each node in the hierarchy are also supplied to form a frame. A robust way to identify characteristics is to use a detailed version (55 to 60 fields) of the Reporter's Questions: who, what, where, when, why, how, how much, and how certain (this field is the most important one for this idea). So, for example, a frame contains who = plant name or responsible organization, what = emission molecules, when = year, why = analysis goal, how = analysis methodology, how much = tons of pollutants, and how certain = [belief, unknown, disbelief] for evidence related to each node. Note that the evidential interval [belief, unknown, disbelief] is the form into which all uncertainties are converted for evidence fusion.

Step 2: get data from both structured and unstructured sources. Both are readily imported into MySQL; using the Autonomy IDOL[194]. Unstructured data processing is more challenging: data triage consists of information retrieval, text segmentation, prioritization, and reports. Filtered texts are then "tagged" with a concept in the ontology using the nApps, augmented with (for example) the Weka Bayesian Classifier[195] and passed to the GATE information extraction application, improved with a Detailed Reporter's Questions template-filling tool. Another improvement in the template-filling tool is a table look-up to convert hedge words (such as maybe, probably, likely not) to [belief, unknown, disbelief].

Step 3: Fuse and propagate data. The ontology is sufficiently rich[196] to support orchestrated

---

[193] http://www.wikipatents.com/US-Patent-7424466/general-purpose-fusion-engine, accessed 5/23/2014.
[194] http://www.autonomy.com/products/idol, accessed 3/29/2015.
[195] http://weka.pentaho.com, accessed 5/23/2014.
[196] Talbot, P., Semantic Networks – A Unifying Framework for Multi-Strategy Reasoning, http://www.is.northropgrumman.com/about/ngtr_journal/assets/TRJ-2003/SS/03SS_TOC.pdf, accessed 5/23/2014

processing, based on imports/exports with the knowledge base. Processing chains include applications for reasoning with the 12 types of uncertainty. Data fusion consists of mathematically combining current and new evidence. The Dempster-Shafer Combination rule is the most powerful, but seven additional fusion algorithms (for example, averaging) are available.

Uncertain Reasoning: seven of the 12 types of uncertainty are highlighted:

Random uncertainties in climate data are due to fluctuations in the estimates of emissions: means and standard deviations are converted to [belief, unknown, disbelief] for GPFE fusion.
Measurement uncertainties are also derivable from climate data which represents a finite sample size and is therefore subject to a confidence interval about the mean. Modeling errors, a form of measurement uncertainty, are computed for climate change simulations. A third measurement uncertainty is in rule-based patterns computed from data mining tools. The evidential interval [belief, unknown, disbelief] is readily computed from the support and accuracy metrics.
Conflicting statements in Intergovernmental Panel on Climate Change[197] (IPCC) text and hedge word parsing produce [belief, unknown, disbelief]. The value of disbelief indicates the degree of conflict.
Missing Known information is easily identifiable as gaps in Reporter's Questions data. Belief networks and data mining algorithms tolerate missing data.
Missing Unknowns: missing nodes and links in the belief network are identified by data mining tools for inclusion (upper right of Figure 2).
Obsolete information currency is tracked from climate and IPCC data time tags. A degradation factor for both belief and disbelief is constructed by a regression on multiple years of data. The degradation factor is applied versus time when fusing the data.
Chaotic uncertainty, or sensitive dependence to initial conditions, is quantified using Monte Carlo simulation (numerical experiments) with climate change models. [belief, unknown, disbelief] is formed from Monte Carlo runs. A warning flag indicates circumstances when chaos dominates.

Total Uncertainty: GPFE mathematically fuses (for example, Dempster Shafer Combination Rule) and propagates the [belief, unknown, disbelief] for all evidence with all types of uncertainty. Propagating the evidence up the ontology produces the total uncertainty at the top node and provides the [belief, unknown, disbelief] evidential interval for the "anthropomorphic climate change?" hypothesis.

---

[197] http://www.ipcc.ch/, accessed 03/30/2015.

Chapter V-13
Monte Carlo Sampling for Benchmarking in the Cloud

Introduction

The challenge is to provide an implementable idea for running a benchmark in the cloud. Execution time is not guaranteed in the cloud due to varying hardware, unknown failure points and other uncontrollable uncertainties. This chapter focuses on a scoring system that relies on numerical experiments that is fair for all competitors and accounts for cloud constraints.

As required, the solution is sufficiently general to account for two outcomes of the benchmark match. Some customers will want to take the solution from a Map Reduce[198] type architecture, and maintain it on the cloud, running in much the same way as it ran during competition. Other customers will have standard single or multi-core processors that will execute the solutions off the cloud, even though the competition is run in the cloud.

The idea presented here is to solve the cloud computing problem outlined above by employing statistical techniques that are tailored to uncertainties in the cloud computing environment. Specifically, the number of Monte Carlo trials is minimized using a **competitive filtering strategy**. For each set of Monte Carlo runs, samples are obtained for each viable entry, mean execution time is computed, this expected value is compared with those of competing entries, and those that are "sufficiently close" to winning times are processed further. The winning entry is the one that scores the **fastest probable run time** a specified percentage of the time. An **elegant procedure** for computing this metric is to sample run times for entries a large number of times to get the percentage of times that each algorithm fares best. Finally, implementation insights, based on available architectures such as Map Reduce and Parallel Virtual Machine[199] and vendor tools for managing cloud computing environments are briefly discussed.

Approach

This idea is based on identifying and quantifying the uncertainties associated with a cloud computing environment, employing a competitive filtering strategy to minimize computing resources, and computing a scoring metric. The steps that constitute the technical approach are:

1. State **assumptions and ground-rules** for the cloud computing environment
2. Assess a **taxonomy of uncertainties** to identify those that are relevant
3. Define a tailored set of **statistical tests** for outliers (for example, failed runs), and run time closeness of competing entries, including the mean, variance, margin of error, and optionally the largest Lyapunov exponent. The latter is a measure of chaos – sensitive dependence to initial conditions.
4. Employ a competitive filtering strategy
5. Define the **scoring metric** for the benchmark that will produce the winning algorithm in cloud computing environments.
6. Design an **algorithm** to compute the winning entry. Key components are statistical tests,

---
[198] http://www-01.ibm.com/software/data/infosphere/hadoop/mapreduce/, accessed 03/30/2015.
[199] http://www.csm.ornl.gov/pvm/, accessed 03/30/2015.

a competitive filtering strategy to minimize Monte Carlo trials, and Monte Carlo sampling of distributions to compute a scoring metric.
7. Provide a **test case** to explain the idea
8. Implementation insights

Results

1. Assumptions and Ground-rules

Based on the stated challenges are:

- Neither the candidate code being benchmarked nor the computing environment can be restructured to guarantee that no uncertainties are present. The goal is to manage uncertainty, not remove it!
- Environment: Candidates for benchmarking are expected to be presented to the cloud during a time interval for which the computing environment is reasonably static in configuration. Stationarity is, however, not assumed, but lack of sufficient process stability may trigger "undecidable" indicators leading to mitigation actions.
- Recalculation: When the computing environment configuration changes, benchmarks will be rerun.
- Parameters in statistical tests, Monte Carlo sample sizes, and presentation of results are configurable.

2. Taxonomy of Uncertainties

A reasonably complete and structured set of uncertainties (shown earlier in Figure IV-1.1) is defined and assessed. The set of uncertainties is predicated on hypotheses and evidence. Generally, a hypothesis is an assertion about a concept – in this context, the hypothesis is whether the entry being benchmarked is, in fact, the best. Evidence is information, here it is run time data, that supports a hypothesis. Hypotheses and evidence are subject to many different kinds of uncertainty. These sources of uncertainty are defined and examples are provided.

Types of Uncertainty Defined and Assessed

- **Random.** A hypothesis or evidence may be dependent on random variables; that is, measurable quantities that randomly vary. A defining characteristic of a random variable, in classical statistics, is that it has a well-defined mean and standard deviation. An example of a random variable is the result of a coin toss. Evidence, in the form of run times, is subject to random uncertainty.
- **Measurement.** This is also known as systematic error. Examples are small samples, bias error, and uncertainties (confidence intervals) due to lack of a sufficient number of measurements. An example is polling error, often expressed as a confidence interval or margin of error: Candidate has a 78% acceptance, +- 3% margin of error (based on an assumed normal distribution, a sample size of 1000). Because there will be a finite number of Monte Carlo samples, evidence is subject to measurement uncertainty.

- **False.** This category of uncertainty captures the idea that we may not be sure that the hypothesis or the evidence is valid. Here, we assume neither the hypothesis nor the evidence is false.
- **Obsolete**. This uncertainty stems from the evidence not being up to date. Evidence ages over time. This induces temporal uncertainty that must be accounted for when fusing evidence arriving at different times. For example, last year's weather forecast is obsolete, and likely has no value. Here, obsolete data is discarded and cases are rerun when cloud computing configurations change.
- **Missing Known**. Hypotheses have characteristics, including quantified degrees of belief or disbelief, and answers to questions such as who, what, where, when, why, and how. A missing known is an empty slot indicating missing evidence about characteristic. Here, we assume required data is collected and available.
- **Vague**. This uncertainty arises from spoken language. It denotes a lack of crispness, or fuzziness in interpretation. Words like probably, tall, and soon are vague. Not applicable.
- **Conflicting**. This uncertainty is defined for evidence that contradicts a hypothesis. The lack of conflicting evidence is called plausibility (Pl). In an evidential interval [0, 1] the degree of conflict or disbelief is 1 – Pl. Here, we assume that run time data collected is non-conflicting.
- **Ambiguous**. This type of uncertainty reflects the fact that either evidence or a hypothesis may be understood or interpreted in more than one way. For example, "All is well" does not have sufficient context (with who, with what, where, when,...) to be interpreted unambiguously. Here, run time evidence is not ambiguous.
- **Missing Unknown**. This category of uncertainty is defined as missing hypotheses or links between hypotheses; that is, we don't know what we don't know. Here, we assume a single hypothesis (does a particular algorithm score best) and there are no unknown unknowns
- **Understanding**. Ideally, hypotheses and evidence should be intelligible. Uncertainty in the human understanding of a hypothesis or evidence may result in a fundamental cognitive problem. For example, "I can't understand what you're saying?". Because the hypothesis and evidence will be clearly defined, this is not a source of uncertainty.
- **Chaotic.** The indicator that this uncertainty is present is sensitive dependence to initial conditions. One measure of chaotic behavior is the Lyapunov exponent. Examples are the "butterfly effect": a butterfly flapping its wings in Aspen produces a snowstorm in Denver. Many nonlinear systems exhibit this behavior. This uncertainty will be tested for in cloud computing environments.
- **Undecidable.** This uncertainty applies to evidence, hypotheses, and formulations of problems that are ill-posed. The test for undecidability is whether an answer can be generated in a finite number of steps (a generalization of Godel's work). For example, if we are attempting to achieve a set of effects, but the environment is not rigorously specified, data may not reflect the process being executed. Here, the problem is well-posed, yet the cloud may not yield a sufficiently "tight" particle distribution, so this uncertainty is important.

These types of uncertainty are organized as uncertainties occurring in data or in processes. Uncertainties in data are distinguished according to whether they represent inaccurate or incomplete data. Uncertainties in processes are grouped according to whether they are conceptual or dynamic. There are data inaccuracies and dynamic process uncertainties. The solution presented addressed these four uncertainties: random, measurement, chaotic, and undecidable.

3. **Statistical Tests:** These are tied to the four uncertainties of interest as identified above. In all cases, statistical tests are based on well-known parameters. No "special statistics" are invented.

Random Uncertainties: The primary data set is a vector of size N of run times, in seconds, for each entry in a benchmarking competition (an entry being benchmarked is hereafter called "code"). The code has auxiliary information, such as an identifier, the cloud environment identifier, and the time that the run is submitted. Indicators of a failed or aborted runs are also collected. The primary data set is input to a web service in the cloud that computes the mean and variance according to well established formulas. The simplest test is to rank the performance of the codes by computing their mean run time. This is used later in the competitive filtering strategy.

- **Mean Test:** The lowest run time, as computed by the mean, or average, is the winner. However, if the number of samples is small or variability of the data is large, the computed mean may not give a true measure of performance.

Data is not assumed to be normally distributed; however, a data point is considered an outlier if it exceeds a specified number (Vmax) of variance.

- **OutlierTest** : if $V(i) > Vmax$ then discard the data point and
    recalculate variance for remaining data points

Initial screening is accomplished by ranking codes from "best" to "worst". The best code may be the one that executes fastest, the code that consumes the least amount of memory, or the code that handles the greatest variety of data sets. A weighted average of these and other attributes may be defined as the score (S). The filter test is levied on the ranked codes to determine which scores are within a specified (Sfilter) number of standard deviations (square roots of variances). Only those are kept for follow-on processing:

- **Filter Test** : if $V(i) > V(1) + Sfilter * S$ then filter out

Note: Variances and standard deviations is to remove outliers from with a data set and to remove entire data sets that have too large a standard deviation..

Measurement Uncertainties

The fact that we are executing a code a finite number of times (Monte Carlo sampling) induces uncertainty. The samples obtained may not represent the true statistics of the overall population. The Monte Carlo sampling error, for random sampling from a large population at 95% confidence, is given as error = $.98 / SQR(N)$, where the constant factor assumes a normal distribution.

- **Sampling Test**: if $N(i) > N(1) + .98 / SQR(N)$ then filter   (95% Confidence) **Note:** this sampling test is the basis for the competitive filtering strategy, discussed in Section

Chaotic Processes (optional): The degree of chaos in a system is measured by the Lyapunov exponent (L). If the largest exponent is positive, the system is said to be chaotic. Methods are available[200] for computing this exponent for small time series data sets. For our application, a code is executed sequentially over a time interval and the (start time, execution time) data provide the time series.

- **Chaos Test:** if $L(i) > 0$, for the "i"th data set, then the system is chaotic – and results are invalid. Note that this test can be executed as a background task during benchmarking trials or interspersed with benchmarking to assure that sensitive dependence on initial conditions is not invalidating results.

Undecidable: This uncertainty arises when the results produced from benchmarking trials do not admit of a "suitable" result. The benchmarking of a code is deemed undecidable if all runs end in failure or are aborted, if the variability is too large, if the margin of error is too large, or if the environment is chaotic.

- **Undecidable test:** if for all "i", $V(i)$ = failed, $V(i) > V_{max}$, $V(i) > V(1) + S_{filter} * S$, or $L > 0$, then V = undecidable. (Note that this decidability test serves to summarize results from all other tests. A data set must be decidable to proceed.)

<u>Notes:</u> Any or all of the uncertainties discussed can be addressed or omitted. If random uncertainties dominate, the code can be executed multiple times in a single trial. If measurement uncertainties are unacceptably large, more Monte Carlo trial can be performed. If the cloud is chaotic, this can be reported to the service provider.

4. <u>Competitive Filtering Strategy</u>: To minimize the number of Monte Carlo trials, a competitive filtering strategy is employed. This assures that only potentially viable entries are considered. During the first trial, all entries are executed, run times are computed, and ranked from fastest to slowest. The percentage above best entry (Figure V-13.1) is computed for all entries and tabulated. Since the margin of error for one trial is 100%, entries with run times greater than 100% of the fastest are filtered out: they have run times larger than the percentage that can result in a win, based on sampling error. The 2nd set of trials, with entries within 100% of the fastest, are then conducted, run times are tabulated, percentage above fastest are computed, and mean percentage above fastest is determined. Approximately 70% (.98 / SQR{2} = .69) of the remaining entries are retained. This competitive filtering process continues until the number of viable entries is sufficiently small. A useful stopping criteria is that the number of viable entries = the number of prizes.

A numerical example (Figure V-13-1) shows how this would work with 60 entries. To keep it simple, the entries are assumed to be uniformly distributed with percentages between 0% and 200% of the best entry (fastest run time). In practice, this will not be the case. The first set of trials allows half of the 60 entries to be filtered out – because they are > 100% of the fastest. After the 4th set of trials, only half of the remaining entries are viable (INT[.98/SQR{4}* 30] = 15). Finally, after the 9th set of trials, only 10 remain (INT[.98/SQR{9}*30] = 10. Since this equals the number of prizes, no further iteration is required. The 10 entries proceed to the scoring metric phase. Rather than computing percentage above best entry for all entries (60

---

[200] http://physionet.ics.forth.gr/physiotools/lyapunov/RosensteinM93.pdf, accessed 02/27/2015.

entries * 9 trials = 540 runs), this technique reduces the number of runs to 60+30+21+18+15+13+12+12+11+10= 202 runs, a savings of ~63% ({1 – 202/540} * 100 = 62.59%) for this example. This technique exploits the steep drop in the f(n) = .98/SQR(n) curve for small n to significantly reduce the required number of Monte Carlo trials.

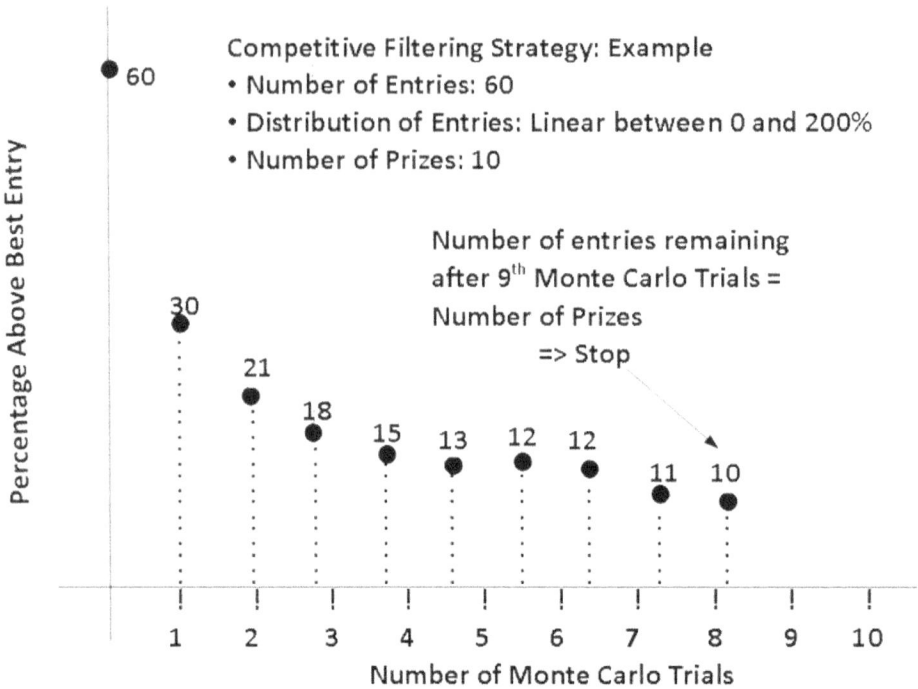

Figure V-13-1: Competitive Filtering Strategy Minimizes Monte Carlo Runs

5. Scoring Metric: Although, as indicated earlier, the scoring metric can be a weighted sum of multiple decision criteria. For simplicity, the metric developed is based on probability of fastest run time P[Rmin]. This is computed probabilistically, again using Monte Carlo sampling. Data points obtained for each code, which represent the actual run time values, versus a curve-fit to some assumed underlying distribution, are employed. A "Do Loop" selects, at random, a value from each data set. The code with the shortest run time is given a "point" (p). The process continues for some configurable number of trials (T). The probability of shortest run time is computed as:

$$P[Rmin] = p / T, \text{ where } p = \# \text{ points and } T = \# \text{ trials}$$

Notes: each code should ideally have the same number of valid run time trials. If not, three options are feasible: if "m" is the lowest number of valid run time trials, competitors with more than "m" trials can either choose their lowest "m" run times, or "m" trials can be drawn at random from each data set. The first technique favors robust codes, the second favors situations where the cloud environment, and not the code is at fault. The third technique is to keep the data sets uneven and draw a random trial from each – this can favor smaller data sets if larger data sets include a few extremely large values.

6. Algorithm Design: the goal of this section is to collect the tests and scoring metric and define an algorithm that computes the best score for benchmarking in a cloud computing environment.

Monte Carlo Sampling Algorithm

- For each cloud environment
    - Choose all codes to be benchmarked
    - Assign a random integer to each code, which provides priority for sequential processing in the cloud, if any.
    - Present all codes simultaneously to the cloud, along with priorities
- For each trial
- For each code
    - Execute each code times, recording run time (t)
    - Perform Mean, Outlier, Filter, Sampling, and/or Chaos tests for each code
    - Employ competitive filtering strategy
    - Store run time data for filtered codes
    - Optionally, filter data so all codes have the same number of trials
- Next code
- Next trial
- Compute Probability of Shortest Run time, rank codes, and assign prizes
- Next Cloud Environment
- Display results

7. Test Case: Suppose that there a six codes to benchmark. Assume that there is a single, stable cloud environment. Codes are assigned a priority integer, at random, to establish precedence for sequential processing in the cloud, and they are labeled C1 – C6.

- C1 is executed
- Codes C2 – C6 are interspersed and executed during this same time frame. No outliers are observed. Largest Lyapunov exponent is negative, so no chaos is apparent in the cloud.
- Means are (3.556, 3.678, 4.587, 2.777, 3.092, 3.523) for C1-C6, respectively.
- The number of entries (6) is approximately equal to the number of prizes (let's assume there are 5 prizes): no data sets are filtered – the competitive filtering strategy (Section 4) need not be employed.
- A default number of of Monte Carlo trials (10) is employed, resulting in a data set with 60 values.
- The first trial for each code produces, via random selection of a run time from each entry,(3.678, 3.872, 3.677, 2.930, 2.951, 3.421), resulting in a point for C4. After 1000 trials, the number of points for each code are (3, 5, 7, 963, 21, 1). The result is that C4 wins with a probability of shortest run time of 96.3%, which surpasses a predefined award threshold. Note that the award threshold could simply be the code scoring the most points.

8. Implementation Insights: although this idea focuses on a scoring system for benchmarking in cloud computing environments, parallel computing has developed over many decades and many architectures and tool are available to develop, monitor, and optimize performance. Examples of architectures are the Parallel Virtual Machine (PVM) and Map Reduce. Examples of tools are Solar Winds Virtualization Manager[201], Prober[202], and information on performance tuning[203]. The filtering and scoring ideas presented here are candidates for integration, as a post processor, into existing tools and architectures

Summary

An idea for benchmarking algorithm entries in a cloud computing environment is presented

- Relevant uncertainties and associated tests are identified.
- A Monte Carlo sampling approach is adopted.
- A competitive filtering strategy minimizes the number of Monte Carlo trials required
- A Monte Carlo scoring technique is presented
- A test case illustrates the methodology
- Implementation insights are briefly discussed

---

[201] http://www.solarwinds.com/products/virtualization-manager/vmware-management.aspx, accessed 01/20/2015.
[202] http://www.ppgee.pucminas.br/lsdc/artigos/goes_sbac02.pdf, accessed 01/20/2015.
[203] http://www.osti.gov/bridge/purl.cover.jsp?purl=/672021-0Qpl2z/webviewable/672021.pdf, accessed 01/20/2015.

Chapter V-14
Co-Orbital Anti-Satellite (ASAT) Scenario

Themes

We emphasizes three technical ideas in this application:

1. **Observe, Orient, Decide, Act (OODA) Loop**: decision support tools help close the OODA loop *faster* than our adversaries: we maneuver to avoid a collision faster than adversary counter-maneuver.
2. **Level 4 Data Fusion:** we combine data mining and data fusion to achieve Joint Director of Laboratories (JDL) Level 4 Data Fusion – Process Refinement – by adding new content to the Belief Network.
3. **Maneuver Slide Rule:** a nomogram allows loss of mission lifetime to be computed based on time-to-impact and maneuver geometry via sliders.

Introduction

The sun-synchronous belt, where the COSMOS - Iridium collision occurred[204], is getting increasingly crowded. This scenario postulates a covert ASAT that maneuvers stealthily to intercept our National Polar-orbiting Operational Environmental Satellite System (NPOESS) satellite. We'll orchestrate the scenario by traversing the OODA loop twice: once for routine operations (special tasking) and once for crisis action (NPOESS maneuver).

Directive from Higher Authority

>From: JFCC-SPACE
>To: Joint Space Operations Center
>
>C L A S S I F I C A T I O N : Unclassified
>
>NARR/ This is a Situation Report – Request JSPOC Commander's Estimate with Alternative Courses of Action for JFCC-Space consideration within 2 hours of receipt./
>
>/Situation/ The situation in Iroma is tense. Iroma launched an ASAT in 2015 and are threatening to repeat the event. Possible Iroma threats to U.S. Forces are direct ascent ASAT, Co-orbital ASAT, and high-power chemical Laser./

As we move through the scenario, the Mission Manager highlights situation changes and the belief network fuses uncertain evidence related to the situation. Evidence propagates upward to show mission impact. We use a data mining algorithm – rule induction – to identify new hypothesis nodes and post them to the belief network for process refinement (Level 4 Data Fusion). We begin with a situation report requesting that Courses of Action be generated based

---

[204] http://celestrak.com/events/collision.asp, accessed 01/21/2015.

on a space-related concerns in Iroma. Our Mission Manager (as shown previously) provides a high-level view, with stoplight indicators that currently all is well.

The belief network is retrieved by the Intelligence Analyst (J2). Initially, no evidence is posted to this Intelligence Preparation of the Battle Space tool. Evidence nodes along the bottom provide indicators of adversary activity which are then propagated upwards to produce indicators of adversary intent. We first see (Figure V-14.1) that Telemetry is OK. Next we see an issue with Space Surveillance Network (SSN) observations and Space Object Identification (SOI) data in sun-synchronous orbit. These appear to be related to the recent collision in this orbit regime.

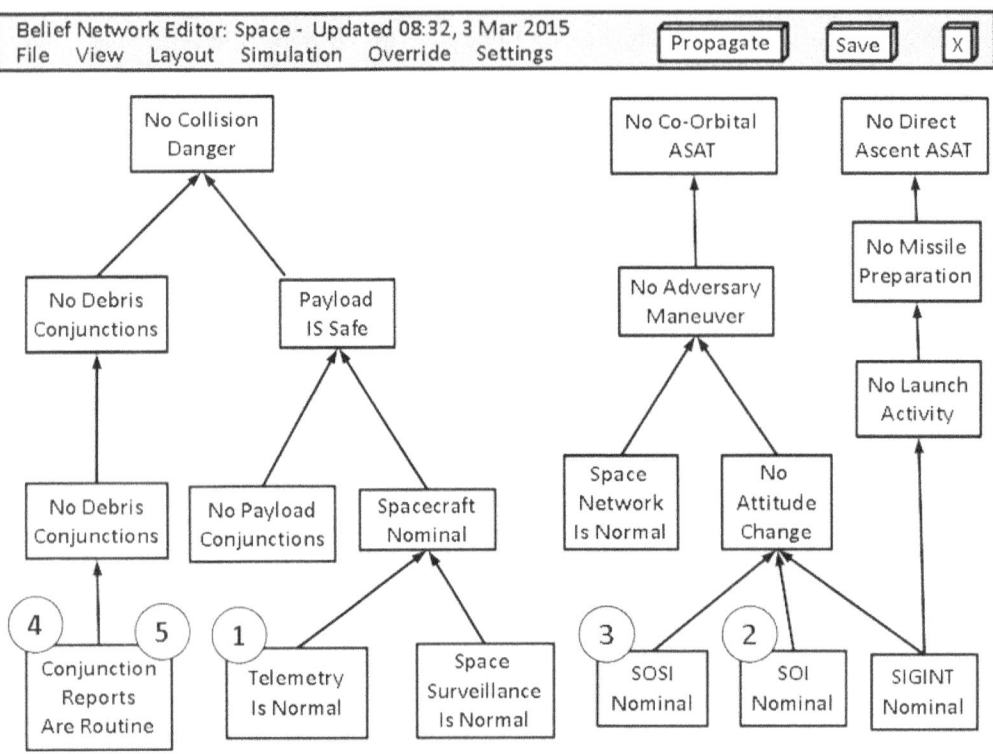

Figure V-14.1: Belief Network for Space Scenario

Transition

Based on receipt of a Maneuver Message, the J2 attempts to correlate the possible impact on our space assets. The J2 now uses a powerful data mining tool (Figure V-14.2) to understand patterns of adversary activity in the sun synchronous belt. Based on previous data, and the results of training exercises, this rule tree shows a correlation between a maneuver message and previous activity. It suggests a potential ASAT and indicates we keep an eye on weather satellites. Also shown is a correlation between sun-synchronous and hostile activity.

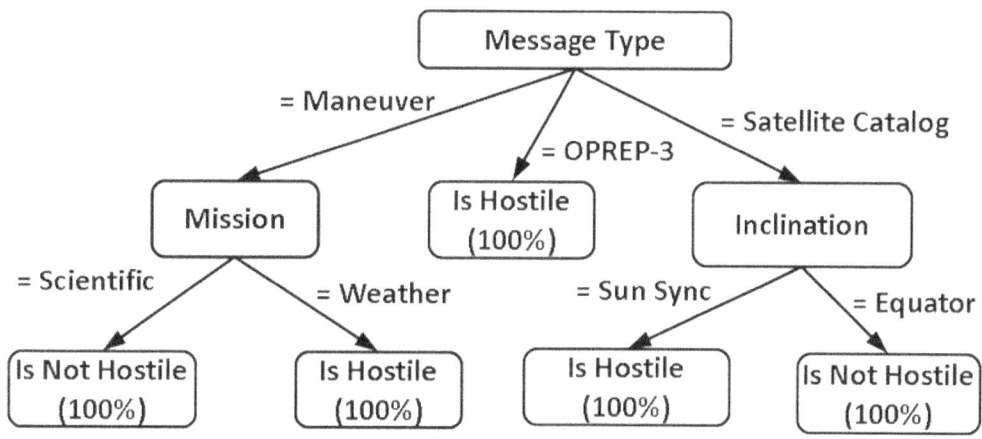

Figure V-14.2: Data Mining Produces Patterns in Maneuver Data

Meanwhile (Figure V-14.2) the belief network is updated to reflect receipt of the maneuver message and activity from the Iroma SOSI network.

The J2 brings up the Strategy Development Tool. This is an operational system of record from the Air Force Research Laboratory that we use to perform Space Intelligence Preparation of the Battle Space. A subject matter expert builds the plan and associated adversary Center-of-Gravity Articulator This automatically produces the belief network that we have been using.

We also review the special tasking schedule provided by our Auction Based Scheduler. The schedule shows sensor activity slots versus time.

The J2 makes refinements in the COG Articulator, updates the belief network and right clicks on boxes indicating lack of information. Specifically, we click on Conjunction Reports Are Routine and review dynamic plan updates, which also come from the COA that we built using SDT. The Command Director requests a special perturbation run on the rocket body that is presumed inactive to determine whether it is on a collision path with the NPOESS satellite.

We use automated information extraction to quickly ingest unstructured content. It is a Situation Report warning of satellites at risk of conjunction. This is posted to the belief network and our box score application. The maneuver message that produces the conjunction is now posted to the pattern matcher, and we see that, based on historical information and the results of training exercises, the maneuver, against the weather satellite is likely hostile.

We also review Space Object Identification data to see the attitude change in the suspected co-orbital ASAT. The supposed rocket body is shows stable motion solution rather than the typical tumbling motion. The stable attitude allows separation of an ASAT. We fuse the evidence of ASAT separation in the belief network. Next we again right click on "No Collision Danger" which we firmly disbelieve, drill down to a detailed description of this COA, and review the dynamics (Figure-14.3) of the decision with the maneuver planning slide rule.

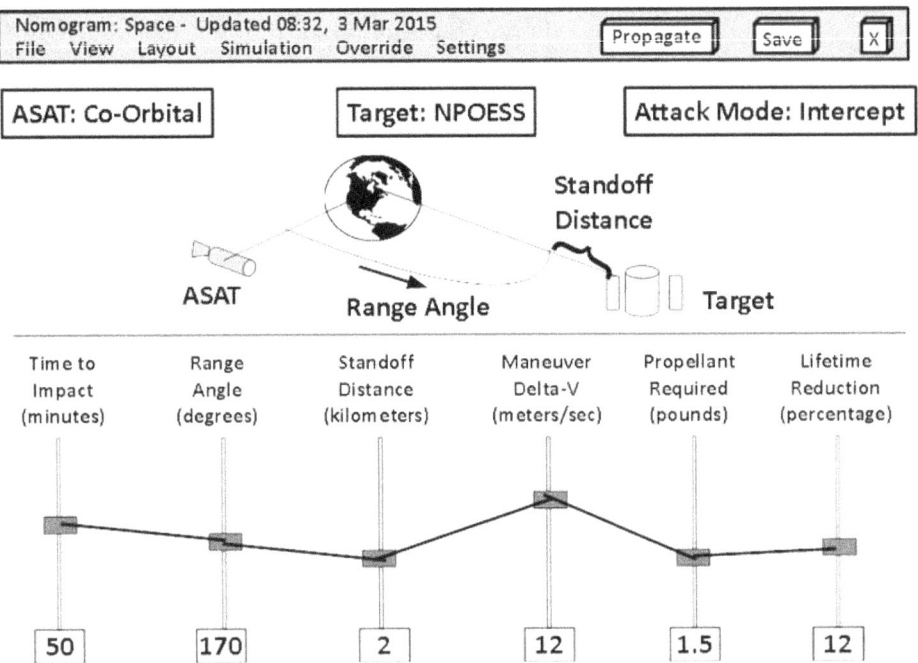

Figure V-14.3: Nomogram to Determine Maneuver Time versus Lifetime Reduction

The NPOESS maneuver is approved by the Owner Operator, the impulse is made, and a collision is avoided. At the belief network, the evidence is posted to a new node (Maneuver successful) that joins the two networks. As additional evidence of the avoided collision arrives, the belief network indicates the situation is normal.

Chapter V-15
Retail Problem Solver

Idea Description

Ever think that you are solving the same problems over and over again? Based on query of the existing data set to identify a problem, the retail problem solver retrieves problem-specific information, consisting of problem characteristics coupled with a solution, from a database. For example, a data set query on top-spending clients who have not made a purchase recently may be considered a problem. The associated solution may be for the salesman to call the client. The problem-solution pairs that best match problems of interest are collected. These are summarized in easily understood rules. The worth of each rule is then computed and converted to an intuitive display.

The goal of this idea is to use the specific data already available to identify problems. An automated (not automatic) system – a retail problem solver (RPS) - accepts user input consisting of characteristics of a retail problem, retrieves relevant solution information from a knowledge base, filters the information based on similarity of the retrieved solution to the user problem, induces rules to summarize the solutions, and ranks the solutions with a worth metric. Implementation (Figure V-15.1) is envisioned in a cloud-computing architecture with a knowledge base accessible by a mobile device.

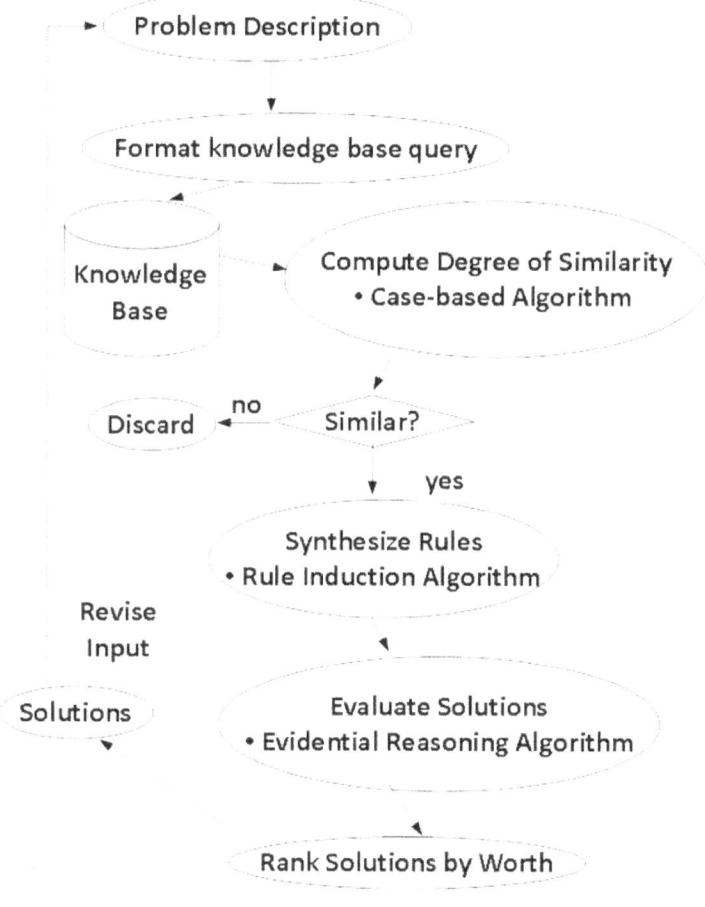

Figure V-15-1 Automated Problem Solver Flow Diagram

Three innovative features of the automated problem solver are: 1) the existing data set is mapped to a hierarchical, characteristics-based ontology that is augmented with a free text description of the problem - solution context. This provides the user with an explanation. 2) a case-based reasoning algorithm, likened to an engineering trade study, automatically filters information. 3) machine learning metrics, specifically accuracy and support, are converted to evidential reasoning output which is more intuitively understood by the user.

The retail problem solver idea is feasible from both a cost and technical perspective. The knowledge base, case-based reasoning, rule induction, and the data fusion algorithms have been individually prototyped in Java for a real world domain.

How it Works

Details of the conceptual design of an RPS are presented. The functional flow (Figure 1) is elaborated. The intent is to provide sufficient detail for understanding how the existing data set is leveraged, the novelty of the idea, it's usefulness, its technical feasibility, and its implementation cost and complexity.

Input

The user clicks on a "Retail Solutions" icon on the desktop of a laptop or mobile device. The input menu (Figure V-15.2) solicits answers to a highly tailored for of the Reporter's Questions (who, what, where, when, why, how, and how certain), that fully leverages the existing data set. After viewing solutions, the user can edit the characteristics to further focus the solutions provided.

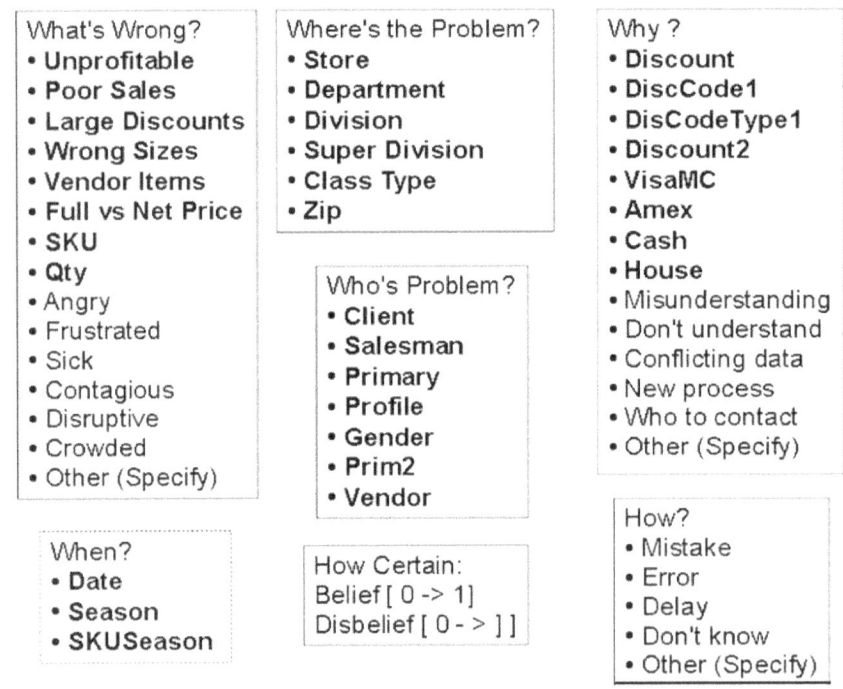

Figure V-15-2: Automated Problem Solver Questions

This idea uses the furnished customer data directly: it is driven by patterns in that data and results of the analysis of the customer data. For example, customer data may identify the following problems: low jewelry sales in July, low sales revenue in store 06880 in Summer, low profit in a department, etc.

An automated RPS "server-side" application resides "in the cloud". It is updated asynchronously, and does not impact users. The functional flow (Figure V-15.1) of the algorithm is straightforward, consisting of an integration of a knowledge base with analogical, induction, and probabilistic fusion algorithms. These are discussed below.

Knowledge Base: The existing data set is mapped to a knowledge structure that is a hierarchical, characteristics based ontology[205] consisting of problem – solution pairs and a link to a textual description of these pairs. Knowledge of recurring problems and associated solutions is solicited during the software development phase, perhaps by crowd-sourcing. A good example of crowd-sourcing commonsense knowledge is described[206] by MIT. The user interface guides user query (posing of problems) to the knowledge base so that a fixed set of characteristics are defined. The same interface allows those providing information to the knowledge base to add characteristics to problem-solution text as meta-data.

Reasoning by Analogy: This method of inference is also called case-based reasoning. In engineering circles, it is best explained as a trade study. The user enters the problem characteristics as a set of criteria. The knowledge base provides solution options as problem – solution pairs. The options are scored based on an optionally weighted sum of user criteria matching characteristics of each option. The score is a similarity comparison between user-specified characteristics of the problem and characteristics stored in the knowledge base for problem-solution pairs. The characteristics of problem-solution pairs are structured as frames in the knowledge base and made available to the rule induction engine.

Rule Induction: An algorithm that converts structured data to a set of IF, AND, AND,… THEN rules[207] operates on the frames of data produced by the case-based reasoning algorithm. The supervised variable, which provides the solution, is defined by a meta-data label for problem-solution frames and linked to the text narrative described earlier. The rule induction algorithm also provides two metrics for each rule: accuracy and support. Accuracy is the number of instances of the rule divided by the total number of rules having the same criteria. The support is the total number of rules having the same set of criteria. Accuracy and support metrics help determine the worth of a solution.

Probabilistic Fusion: this algorithm evaluates each rule based on four criteria: similarity to user problem characteristics, probability of success, accuracy, and support. The first step is to form an evidential interval from the rule induction metrics (accuracy and support). Accuracy provides the mean (m) and the number of samples provides (via a simple equation) the dispersion (d). For example, in a survey, a candidate may receive 35 % +- 3% margin of error of the endorsements: m=35, and d=3. The evidential interval [0,1] is the set of fractional values for Belief (B), Ignorance (I), and Disbelief (D) such that B + I + D = 1. These are computed from the mean and dispersion as follows:

$$B = m - d, \qquad I = 2 * d, \qquad D = 1 - B - I.$$

---

[205] http://protege.stanford.edu/, accessed 02/27/2015.
[206] http://conceptnet5.media.mit.edu/, accessed 01/21/2015
[207] www.cs.waikato.ac.nz/ml/weka/ , accessed 01/21/2015.

Once the evidential interval [B, I ,D] is initialized, the data fusion algorithm must handle the probability of success of the solution. Recall that this is entered during the software design phase to answer the "how certain" question in a problem – solution pair.

Probability of success entries may be incomplete, sparse, and possibly conflicting. Based on these requirements, the Dempster-Shafer combination rule for evidential reasoning, rather than classical or Bayesian statistics, is employed. Answers to the "how certain" parameter value in the problem – solution pairs in the knowledge base are converted to Beliefs and Disbeliefs. Values for "how certain?"can be entered as linguistic variables (for example, likely, probably, certainly, maybe not) and converted to belief and disbelief via a table look-up, or entered directly as belief between {0,1} or disbelief between [0,-1]. The Dempster-Shafer Combination Rule is used to fuse this evidence with the initialized values.

The similarity score of the solution is computed as the ratio of the number of correct word matches (user input versus problem-solution pair) and the total number of user inputs. This is also fused using the Dempster-Shafer Combination Rule.

Belief, ignorance, and disbelief are presented to the user as the worth of the solution. Belief provides the degree of certainty, ignorance indicates "don't know", and disbelief indicates conflict in the solution. If the user finds no compelling solution, input characteristics are modified.

THE END.

P.S. Track us online at http://64.93.121.68/index.html.

www.ingramcontent.com/pod-product-compliance
Lightning Source LLC
Chambersburg PA
CBHW080236180526
45167CB00006B/2305